Science and Engineering of Casting Solidification
Second Edition

T0203034

Doru Michael Stefanescu

Science and Engineering of Casting Solidification

Second Edition

 Springer

Doru Michael Stefanescu
Department of Materials Science and Engineering
The Ohio State University
Columbus, OH
USA

ISBN 978-1-4419-4509-9   e-ISBN 978-0-387-74612-8

Printed on acid-free paper.

9 8 7 6 5 4 3 2 1

springer.com

To my teachers:

my mother	my first and dearest
my father	my first metallurgy teacher
Prof. Laurentie Sofroni	my cast iron teacher
Prof. Suzana Gâdea	my doctorate advisor
Prof. Carl Loper	my American teacher
Max, my grandson	for recasting my understanding of this world

"The book of nature is written in mathematical language" - Galileo

PREFACE

We come to know about the world in two distinctive ways: by direct perception and by application of rational reasoning which, in its highest form, is mathematical thinking. The belief that the underlying order of the world can be expressed in mathematical form lies at the very heart of science. In other words, we only know what we can describe through mathematical models.

Casting of metals has evolved first as witchcraft, to gradually become an art, then a technology, and only recently a science. Many of the processes used in metal casting are still empirical in nature, but many others are deep-rooted in mathematics. In whatever form, casting of metals is an activity fundamental to the very existence of our world, as we know it today.

Foundry reports indicate that solidification modeling is not only a cost-effective investment but also a major technical asset. It helps foundries move into markets with more complex and technically demanding work. The ability to predict internal soundness allows foundries to improve quality and deliveries, and provides the information required to make key manufacturing decisions based on accurate cost estimates before pattern construction even begins. The acceptance of computational modeling of solidification by the industry is a direct result of the gigantic strides made by solidification science in the last two decades.

Yet, solidification science is of paramount importance not only in understanding macro- and microscopic changes during the solidification of castings, but is also the basis of many new processes and materials such as semi-solid casting, laser melting, powder atomization, metal matrix composites, bulk metallic glasses. This book is the second attempt by the author to synthesize the information that can be used for engineering calculations pertinent to computational modeling of casting solidification. It includes additional material on the fundamentals of rapid solidification and bulk metallic glasses.

This book is based on the author's more than forty years experience of teaching, research and industrial practice of solidification science as applied to casting processes. It is an attempt to describe solidification theory through the complex mathematical apparatus that includes partial differential equations and numerical analysis, required for a fundamental treatment of the problem. The mathematics is however restricted to the elements essential to attain a working knowledge in the field. This is in line with the main goal of the book, which is to educate the reader in the fast moving area of computational modeling of solidification processes. While the book is not intended to be a monograph, for the sake of completion, a special effort has been made to introduce the reader to the latest developments in solidification theory, even when they have no engineering applications at this time. In this respect this is a unique attempt to integrate the newest information in a text book format.

The text is designed to be self-contained. The author's teaching experience demonstrates that some of the students interested in solidification science are not fully proficient in partial differential equations (PDE) and/or numerical analysis. Accordingly, elements of PDE and numerical analysis required to obtain a working knowledge of computational solidification modeling have been introduced in the text, while attempting to avoid the interruption of the fluency of the subject. Numerous modeling and calculation examples using the Excel spreadsheet as an engineering tool are provided. The book is addressed to graduate students and seniors interested in solidification science, as well as to industrial researchers that work in the field of solidification in general and casting modeling in particular.

The book is divided in 15 major chapters. After introducing the length scale of solidification analysis in the first chapter, the reader is exposed to the basic concepts of driving force for solidification, undercooling, local equilibrium, and interface non-equilibrium from the thermodynamic perspective (Chapter 2). The following three chapters present a detailed analysis of the governing transport equations and their application at the macro-scale level to predict such features of interest in casting solidification as segregation, shrinkage cavity, solidification time and velocity, and temperature gradients. Numerical approximation methods with an emphasis on finite difference approximations are presented in Chapter 6 together with numerous examples of solidification modeling through analytical and numerical methods solved on the Excel spreadsheet. In this chapter, the reader is also introduced to the applications of macro modeling of solidification in today's casting technology.

Chapters 7 through 11 extend the transport equations to the study of microscale phenomena and the formation of casting microstructure. Nucleation is discussed from the engineering standpoint that is emphasizing possible methodologies for quantification in solidification analysis of castings. A detailed analysis of existing models for dendritic, eutectic, peritectic and monotectic growth is provided. Again, the emphasis is on the use of this knowledge to build computational solidification models. To achieve this goal, each section of this chapter includes a comprehensive discussion of the applicability and limitations of transferring the information available from steady state analysis to continuous cooling solidification.

Chapter 12 extends the concepts introduced earlier to the evolution of microstructure during rapid solidification. Rapidly solidified crystalline alloys and metallic glasses are briefly discussed. The solidification behavior in the presence of a third phase (gaseous or solid impurities) is covered in the 13th chapter.

Chapter 14 is dedicated to the fast moving field of numerical modeling of solidification at the micro-scale. Deterministic and cellular automaton models are covered in detail, while phase field modeling is briefly summarized.

The analysis of nucleation and growth at the atomic scale level, required for a complete understanding of solidification and the associated phenomena is presented in chapter 15. Since the current level of understanding does not permit the use of this information directly in computational modeling of solidification, the emphasis is on the physics rather than on engineering.

CONTENTS

NOMENCLATURE

$C,$ C_o	alloy composition	p	probability
C_S^*	interface composition in the solid	q	diffusion flux
C_L^*	interface composition in the liquid	r	radius (m)
D	species diffusivity (m^2·s^{-1})	t	time (s)
E	internal energy (J·mole^{-1} or J·m^{-3})	v	volume (m^3)
F	Helmholtz free energy (J·mole^{-1} or J·m^{-3})	v_a	atomic volume (m^3·atom^{-1})
G	Gibbs free energy (J·mole^{-1} or J·m^{-3})	v_m	molar volume (m^3·mole^{-1})
	gradient	ΔC_o	concentration difference between liquid and solid at the solidus temperature
H	enthalpy (J·mole^{-1}, J·m^{-3}, J·kg^{-1})	ΔG_v	change in volumetric free energy (J·m^{-3})
I	intensity of nucleation (m^{-3})	ΔH	change in volumetric enthalpy (J·m^{-3})
J	mass flux	ΔH_f	latent heat of fusion (J·mol^{-1}, J·kg^{-1}, J·m^{-3})
K	curvature (m^{-1})	ΔS_f	entropy of fusion (J·mol^{-1}·K^{-1} or J·m^{-3}·K^{-1})
	permeability of porous medium (m^2)	ΔT	undercooling (K)
	equilibrium constant (Sievert's law)	ΔT_c	constitutional undercooling (K)
P	pressure (Pa)	ΔT_k	kinetic undercooling (K)
	Péclet number	ΔT_o	liquidus-solidus interval (K)

Q	volumetric flow rate ($m^3 \cdot s$)	ΔT_r	curvature undercooling (K)
R	gas constant ($J \cdot mol^{-1} \, K^{-1}$)	Γ	general diffusion coefficient
T	temperature (K or °C)		Gibbs-Thomson coefficient ($m \cdot K$)
T_L	liquidus temperature (K)	Φ	phase quantity
T_S	solidus temperature (K)	α	thermal diffusivity ($m^2 \cdot s^{-1}$)
S	entropy ($J \cdot mol^{-1} \cdot K^{-1}$ or $J \cdot m^{-3} \cdot K^{-1}$)		dimensionless back-diffusion coefficient
V	velocity ($m \cdot s^{-1}$)	β_T	thermal expansion coefficient (K^{-1})
V_o	speed of sound ($m \cdot s^{-1}$)	β_c	solutal expansion coefficient ($wt\%^{-1}$)
c	specific heat ($J \cdot m^{-3} \cdot K^{-1}$)	γ	surface energy ($J \cdot m^{-2}$)
f	mass fraction of phase	δ	boundary layer, disregistry
g	volume fraction of phase	v	kinematic viscosity($m^2 \cdot s$)
$g, \boldsymbol{g}$	gravitational acceleration ($m \cdot s^{-2}$)		vibration frequency
h	heat transfer coefficient ($J \cdot m^{-2} \cdot K^{-1} \cdot s^{-1}$)	ρ	density ($kg \cdot m^{-3}$)
k	solute partition coefficient	λ	interphase spacing (m)
	thermal conductivity ($W \cdot m^{-1} \cdot K^{-1}$)	μ	growth constant
k_B	Boltzman constant		chemical potential ($J \cdot mole^{-1}$)
l	length (m)		dynamic viscosity ($N \cdot m^{-2} \cdot s$)
m	slope of the liquidus line ($K \cdot wt\%^{-1}$)	θ	contact angle
	mass (kg)	τ	momentum flux
n	number of atoms (moles)		

	superscripts		subscripts
het	heterogeneous	*cr*	critical
hom	homogeneous	*e*	equilibrium
m	molar	*eut*	eutectic
r	property related to radius of curvature	*f*	fusion
*	interface	*g*	glass
		het	heterogeneous
		hom	homogeneous

	subscripts		subscripts
E	equivalent, eutectic	i	component, interface
G	gas	k	kinetic
L	liquid	met	metastable
P	particle, pressure	n	atoms per unit volume
S	solid	r	property related to radius of curvature
T	thermal	s	surface, stability
c	constitutional, solutal	st	stable
		v	property related to volume

LENGTH-SCALE IN SOLIDIFICATION ANALYSIS

Cast copper is about 7000 years old. Cast iron, the first man-made composite, is at least 2500 years old. However, as the human species gradually moves from the "iron age" to the age of "engineered materials," of all metal forming processes, the casting process remains the most direct and shortest route from component design to finished product. This makes casting one of the major manufacturing processes, while casting alloys are some of the most widely used materials. The main reasons for the longevity of the casting process are the wide range of mechanical and physical properties covered by casting alloys, the near-net shape capability of the casting process, the versatility of the process (weight from grams to hundred of tons, casting of any metal that can be melted, intricate shapes that cannot be produced by other manufacturing methods), and the competitive delivery price of the manufactured goods. While castings are "invisible" in many of their applications, since they may be part of complex equipment, they are used in 90% of all manufactured goods.

Solidification is an inherent part of the casting and welding processes. During solidification, the as-cast structure of the casting is generated. Since many castings are used in the as-cast state (that is, without further thermal or mechanical processing), it follows that the structure which results from solidification, the as-cast structure, is often also the final structure of the casting. It also follows that the mechanical properties of the casting, which are a direct consequence of the microstructure, are controlled through the solidification process.

About one billion tons of metal is solidified worldwide annually. The applications of solidified materials include shaped castings, but also a large variety of semi-finished products such as atomized powders, continuously cast wires and sheets, continuously cast strands of various cross sections. It is obvious that the modern metal caster must be well educated in the science of solidification.

While solidification science evolved from the need to better understand and further develop casting and welding processes, today, solidification science is at the base of many new developments that fall out of the realm of traditional metal casting.

Castings are made with dimensions of a few millimeters up to tens of meters in length. It is natural, therefore, to assume that the important dimensions to use in describing castings are of that magnitude. However, as the microstructure of the casting (the structure which can be seen using an optical microscope) determines the properties of the casting, it, too, is important. Moreover, because solidification is the process of moving individual atoms from the liquid to a more stable position in the solid alloy lattice, the distances over which atoms must move during solidification are also important. For these reasons, the changes in solid-liquid (S/L) interface during solidification must be discussed at three different length scales, macro-, micro-, and nano-scale. However, solidification models often analyze solidification events at an intermediate scale, the meso-scale. These scales are described graphically in Figure 1.1.

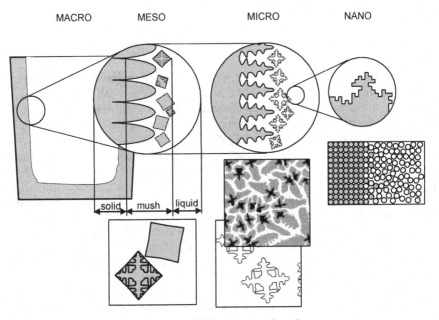

Figure 1.1. Solidification length scale.

The macro-scale (macrostructure): this scale is of the order of 10^0 to 10^{-3}m. Elements of the macro-scale include shrinkage cavity, cold shuts, misruns, macrosegregation, cracks, surface roughness (finish), and casting dimensions. These macrostructure features may sometimes dramatically influence casting and welding properties and consequently castings acceptance by the customer.

At this scale only two phases are assumed to exist, the solid and the liquid, separated by a sharp solid/liquid (S/L) interface. The computational models that describe solidification at the macroscale are based on the solution of conservation equations for mass, energy, species and momentum. The model output includes the temperature and composition (macrosegregation) fields when the energy and species diffusion equations are solved, and may include shrinkage prediction when the mass and momentum equations are also solved.

The meso-scale: this scale allows description of the microstructure features at grain level, without resolving the grain boundary. Generally, it can be considered that the mesoscale is of the order of 10^{-4}m. It is seen on Figure 1.1 that the S/L interface that appears as a line at the macro-scale is more complex when examined at the mesoscale. There is no clear demarcation between the liquid and the solid. In fact, three regions can be observed: liquid, mushy (containing both liquid and solid), and solid.

The computational models that describe solidification at the mesoscale are typically based on the Cellular Automaton (CA) technique. The computer is transformed into a dynamic microscope as the evolution of grain morphology can be outputted during the run of the computer model. However, standard transport models of the type developed for the macroscale can be combined with transformation kinetics model to predict microstructure evolution. Yet, while this type of models can calculate volumetric grain density, they rely on average properties and cannot typically output grain morphology.

The micro-scale (microstructure): this scale is of the order 10^{-6} to 10^{-5}m. The micro-scale describes the complex morphology of the solidification grain. In a sound casting or weld, mechanical properties depend on the solidification structure at the micro-scale level. To evaluate the influence of solidification on the properties of the castings it is necessary to know the as-cast grain morphology (*i.e.*, size and type, columnar or equiaxed), the length scale of the microstructure (*e.g.*, interdendritic arm spacing), and the type and concentration of chemical microsegregation. The CA technique or the phase field methods that are used for modeling microstructure evolution at this scale calculate all this information.

The nano-scale (atomic scale): this scale is of the order of 10^{-9}m (nanometers) and describes the atomic morphology of the S/L interface. At this scale solidification is discussed in terms of nucleation and growth kinetics, which proceed by transfer of individual atoms from the liquid to the solid. Currently there is no database correlating elements of the nano-scale with the properties of castings. However, an accurate description of the S/L interface dynamics requires atomistic calculations. The present knowledge and hardware development does not allow utilization of the atomic scale in applied casting engineering. Nevertheless, accurate solidification modeling may require at least partial use of this scale during computation.

During the last decade, solidification modeling has exhibited a sustained development effort, supported by academic as well as industrial research. The driving force behind this undertaking was the promise of predictive capabilities that will allow process and material improvement, as well as shorter lead times. The most significant recent progress has been incorporation of transformation kinetics, for both the solid/liquid and the solid/solid transformation, in the macro-transport models. The results of these efforts have materialized in a proliferation of publications and commercial software, some of which have penetrated the industry. Numerous claims are made regarding modeling methods accuracy and capabilities. They include prediction of casting defects, of microstructure length scale and composition, and even of mechanical properties.

With the advent of faster computer paralleled by the rapid development of numerical methods, the metallurgical aspects of microstructure evolution have finally become a quantitative engineering science. Indeed, in the broadest definition of engineering science we know only what we can predict through mathematical models. As late as 1975, the solution of the complete macro transport-transformation kinetics problem required for microstructure prediction was considered a "formidable problem" (Maxwell and Hellawell, 1975). Microstructure prediction was strictly an empirical exercise, where elements of the microstructure were correlated with processing and material variables. However, today, as proven by the numerous papers that tackle various phenomena occurring during solidification through mathematical/numerical modeling, the task has lost its reputation for inaccessibility. Nevertheless, the problem is far from a final solution. The complete casting process, from initial mold filling to the final stressed component, including defect prediction, has not been modeled yet quantitatively. The main reason is the tremendous computational requirements. Indeed, full casting simulation requires the solution of highly non-linear discretized equations that may involve a large number of continuum variables (10-20), a complex unstructured mesh (10^5-10^6 nodes), and substantial temporal resolution (10^3-10^4 time steps) (Cross *et al.*, 1998). Certainly, this task can only be addressed using high performance parallel computers.

A more general discussion of the time-space scales in computational materials science and the available simulation methods can be found in Raabe (1988). A short summary is introduced in Table 1.1.

Table 1.1. Space scales and methods in materials simulation (adapted after Raabe, 1998)

Scale, m	Simulation method	Applications
10^{-10} - 10^{-6}	Monte Carlo molecular dynamics	thermodynamics, diffusion, ordering structure and dynamics of lattice defects
10^{-10} - 10^{0}	cellular automata	recrystallization, phase transformation, grain solidification and growth, fluid dynamics
10^{-5} - 10^{0}	large-scale finite element, finite difference, linear iteration, boundary element methods	averaged solution of differential equations at the macroscopic scale (composition, temperature and electromagnetic fields, hydrodynamics)
10^{-6} - 10^{0}	finite elements or finite difference with constitutive laws considering microstructure	solidification, microstructure evolution of alloys, fracture mechanics

References

Cross M., Bailey C., Pericleous K.A., Bounds S.M, Moran G.J., Taylor G.A., McManus K, 1998, in: Thomas BG, Beckermann C eds., *Modeling of Casting Welding and Advanced Solidification Processes VIII*, The Minerals, Metals and Materials Soc., Warrendale, Pennsylvania, p.787

Maxwell I, Hellawell A, 1975, *Acta Metall.* **23**:229

Raabe D., 1998, *Computational materials science*, Wiley-VCH, Weinham, p.5-12

EQUILIBRIUM AND NON-EQUILIBRIUM DURING SOLIDIFICATION

Thermodynamics is a useful tool for the analysis of solidification. It is used to evaluate alloy phase constitution, the solidification path, basic alloy properties such as partition coefficients, slopes of liquidus and solidus phase boundaries.

2.1 Equilibrium

The free energy of any phase is a function of pressure, temperature, and composition. Equilibrium is attained when the Gibbs free energy is at a minimum (equivalent to mechanical systems for which equilibrium exists when the potential energy is at a minimum). Thus the condition is:

$$dG(P,T,n_i...) = \left(\frac{\partial G}{\partial T}\right)_{P,n_i...} dT + \left(\frac{\partial G}{\partial P}\right)_{T,n_i...} dP + \left(\frac{\partial G}{\partial n_i}\right)_{T,P,n_j...} dn_i + ... = 0 \qquad (2.1)$$

where n_i is the number of moles (or atoms) of component i. The partial derivatives of the free energy are called partial molar free energies, or *chemical potentials*:

$$\mu_i = \left(\frac{\partial G}{\partial n_i}\right)_{T,P,n_j...} \qquad (2.2)$$

At equilibrium, and assuming T,P = constant,

$$dG = \mu_i dn_i + \mu_j dn_j + ... = 0 \qquad (2.3)$$

For a multiphase system, a condition for equilibrium is that the chemical potential of each component must be the same in all phases (for derivation see inset):

$$\mu_i^\alpha = \mu_i^\beta \tag{2.4}$$

where the superscripts α and β stand for the two phases.

Derivation of the equilibrium criterion. Consider two phases, α and β, within a system at equilibrium. If an amount dn of component A is transferred from phase α to phase β at $T, P =$ ct., the change in free energy associated with each phase is $dG^\alpha = \mu_A^\alpha \, dn$ and $dG^\beta = -\mu_A^\beta \, dn$. The total change in free energy is:

$$dG = dG^\alpha + dG^\beta = \left(\mu_A^\alpha - \mu_A^\beta\right)dn$$

Since at equilibrium $dG = 0$, it follows that $\mu_A^\alpha - \mu_A^\beta = 0$.

Although equilibrium conditions do not actually exist in real systems, under the assumption of *local thermodynamic equilibrium*, the liquid and solid composition of metallic alloys can be determined using *equilibrium phase diagrams*. Local equilibrium implies that reaction rates at the solid/liquid interface are rapid when compared to the rate of interface advance. This concept has been shown experimentally to be true up to solidification velocities of 5 m/s.

Equilibrium phase diagrams describe the structure of a system as a function of composition and temperature, assuming transformation rate is extremely slow, or species diffusion rate is very fast. Two-component phase equilibrium in a binary system occurs when the chemical potentials of the two species are equal.

2.2   The undercooling requirement

The driving force of any phase transformation including solidification, which is a liquid-to-solid phase transformation, is the change in free energy. The Helmholtz free energy per mole (molar free energy) or per unit volume (volumetric free energy) of a substance can be expressed as:

$$F = E + P \cdot v - T \cdot S \tag{2.5}$$

E is the internal energy, *i.e.*, the amount of work required to separate the atoms of the phase to infinity, P is the pressure, v is the volume, T is the temperature and S is entropy. Thermodynamics stipulates that in a system without outside intervention, the free energy can only decrease.

The entropy is a measure of the amount of disorder in the arrangement of atoms in a phase. In the solid phase, the disorder results from the thermal vibrations of the atoms around their equilibrium position at lattice points. In the liquid phase, additional disorder comes from structural disorder, since the atoms do not occupy all the positions in the lattice as they do in solids. Indeed, the greater thermal energy at higher temperatures introduces not only greater thermal vibrations, but also vacancies. Immediately below its melting point a metal may contain 0.1% vacancies in its lattice. When the vacancies approach 1% in a closed-packed structure, the regular 12-fold coordination is destroyed and the long-range order of the crystal structure disappears. The number of nearest neighbors decreases from 12 to 11 or even 10 (the coordination number, CN, becomes smaller than 12, as shown in Figure 2.1). The pattern becomes irregular and the space per atom and the average interatomic distance are increased. Short-range order is instated. In other words, the liquid possesses a larger degree of disorder than the solid. Thus, the entropy of the liquid is higher than the entropy of the solid. The disorder resulting from melting increases the volume of most materials.

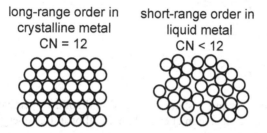

Figure 2.1. Schematic representation of long - and short - range order regions (solid and liquid metals, respectively).

A certain amount of heat, the heat of fusion, is required to melt a specific material. Since the heat of fusion is the energy required to disorganize a mole of atoms, and the melting temperature is a measure of the atomic bond strength, there is a direct correlation between the two.

Let us start our analysis of solidification by introducing a number of simplifying assumptions:

a) pure metal
b) constant pressure
c) flat solid/liquid interface, *i.e.*, the radius of curvature of the interface is $r = \infty$
d) no thermal gradient in the liquid

For constant pressure, Eq. (2.5) becomes the Gibbs free energy equation:

$$G = H - TS \tag{2.6}$$

where $H = E + P \cdot v$ is the enthalpy.

Eq. (2.6) is plotted in Figure 2.2. Since the slope of the line corresponding to the liquid free energy is higher (*i.e.*, $S_L > S_S$), the two lines must intersect at a temperature T_{ea}. This is the equilibrium temperature, at which no transformation (melting on heating or solidification on cooling) can occur. Under normal nucleation conditions, when the temperature decreases under T_{ea}, α stable solid will form. If nucleation of α is suppressed, β metastable solid will form at a lower temperature, under $T_{e\beta}$. If nucleation of both α and β are suppressed, metastable glass forms. The metastable γ solid can only be produced by vapor deposition.

The equilibrium condition Eq. (2.4) can be written for the case of solidification as:

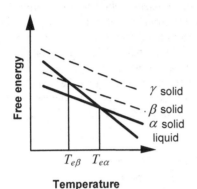

Figure 2.2. Variation of the free energy of the liquid and solid with temperature.

$$\mu_L - \mu_S = 0 \quad \text{or} \quad G_L - G_S = 0 \tag{2.7}$$

where the subscripts L and S stand for liquid and solid respectively. This means that at equilibrium the change in chemical potential or in free energy is zero. At the equilibrium temperature, if the two phases coexist:

$$\Delta G_v = G_L - G_S = (H_L - H_S) - T_e(S_L - S_S) = 0$$

Thus, one can further write:

$$\Delta H_f = T_e \Delta S_f \quad \text{or} \quad \Delta S_f = \Delta H_f / T_e$$

Here, $\Delta H_f = H_L - H_S$ is the change in enthalpy during melting, or the volumetric latent heat. ΔS_f is the entropy of fusion (melting). At a temperature lower than T_e:

$$\Delta G_v = \Delta H_f - T\frac{\Delta H_f}{T_e} = \Delta H_f \frac{T_e - T}{T_e} = \Delta S_f \Delta T \tag{2.8}$$

ΔT is the undercooling at which the liquid-to-solid transformation occurs. From this equation the undercooling is defined as:

$$\Delta T = \Delta G_v / \Delta S_f \tag{2.9}$$

Note that if $\Delta T = 0$, $\Delta G_v = 0$. This means that, if there is no undercooling under the equilibrium temperature, the system is at equilibrium, and no transformation can occur.

Thermodynamics does not allow further clarification of the nature of under-cooling. It simply demonstrates that undercooling is necessary for solidification to occur. Kinetics considerations must be introduced to further understand this phenomenon.

This analysis has been conducted under the four simplifying assumptions (*a* to *d*) previously listed. This amounts to stating that the only change in free energy upon solidification is because of the change of a volume of liquid into a solid, ΔG_v. However, when the four assumptions are relaxed the system will increase its free energy. This increase can be described by the sum of the increases resulting from the relaxation of each particular assumption:

$$\Delta F = -\Delta G_v + \Delta G_r + \Delta G_T + \Delta G_c + \Delta F_P \tag{2.10}$$

The four positive right hand terms are the increase in free energy because of curvature, temperature, composition, and pressure variation, respectively. Let us now evaluate the terms in this equation.

2.3   Curvature undercooling

In the evaluation of the equilibrium temperature presented so far, it has been assumed that the liquid-solid interface is planar (flat), *i.e.*, of infinite radius (assumption c). This is seldom the case in real processes, and never the case at the beginning of solidification, because solidification is initiated at discrete points (nuclei) in the liquid, or at the walls of the mold that contains the liquid. As the volume of a solid particle in a liquid decreases, its surface/volume ratio increases and the contribution of the interface energy to the total free enthalpy of the particle increases. Thus, when the particle size decreases in a liquid-solid system, the total free enthalpy of the solid increases. The curve describing the free energy of the solid on Figure 2.2 is moved upward by ΔG_r. This results in a decrease of the melting point (equilibrium temperature) as shown in Figure 2.3.

If solidification begins at a point in the liquid, a spherical particle is assumed to grow in the liquid, and an additional free energy associated with the additional interface, different than ΔG_v, must be considered. This additional energy, results from the formation of a new interface, and is a function the curvature of the interface.

In two dimensions the curvature of a function is the change in slope, $\delta\theta$, over a length of arc, δl, (Figure 2.4) becomes:

$$K = \delta\theta/\delta l = \delta\theta/(r\delta\theta) = 1/r \tag{2.11}$$

In three dimensions, the curvature is the variation in surface area divided by the corresponding variation in volume:

$$K = dA/dv = 1/r_1 + 1/r_2 \tag{2.12}$$

where r_1 and r_2 are the principal radii of curvature (minimum and maximum value for a given surface).

For a sphere $r_1 = r_2$ and thus $K = 2/r$

For a cylinder $\qquad$ $r_1 = \infty$, $r_2 = r$ $\qquad$ and thus $\qquad$ $K = 1/r$.

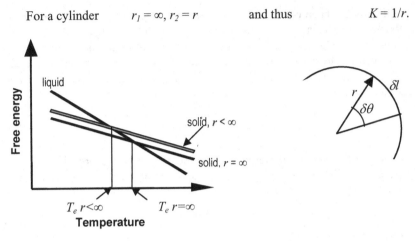

Figure 2.3. Decrease of equilibrium temperature because of the curvature of the S/L interface.   **Figure 2.4.** Definition of curvature.

General definition of curvature. In general, if a curve is represented by $\mathbf{r}(t)$, where t is any parameter, the curvature of that curve is:

$$K(t) = \frac{\sqrt{(\mathbf{r'\cdot r'})(\mathbf{r''\cdot r''}) - (\mathbf{r'\cdot r''})^2}}{(\mathbf{r'\cdot r})^{3/2}}$$

where $r' = dr/dt$ and $r'' = d^2r/dt^2$. In Cartesian coordinates, for a curve $y = y(x)$:

$$K(x) = \frac{|y'|}{\left(1 + y'^2\right)^{3/2}} \qquad \text{where} \quad y' = dy/dx, \text{ etc.}$$

Assuming that the radius of the spherical particle is r, when the particle increases by dr, the work resulting from the formation of a new surface, $d(4\pi r^2 \gamma)/dr$, must be equal to that resulting from the decrease of the free volumetric energy, i.e., $\dfrac{d}{dr}\left(\dfrac{4}{3}\pi r^3 \Delta G_v\right)$. Equating the two, after differentiation, the increase in free energy is:

$$\Delta G_v = 2\gamma/r \qquad \text{or, more general} \quad \Delta G_v = \gamma\, K \qquad\qquad (2.13)$$

where γ is the liquid-solid surface energy, and K is the curvature. Then, from the definition of undercooling, Eq. (2.9), we obtain:

$$\Delta S_f \Delta T_r = \gamma K \quad \text{or} \quad \Delta T_r = T_e - T_e^r = \left(\gamma/\Delta S_f\right)K = \Gamma K \qquad (2.14)$$

where ΔT_r is the curvature undercooling, T_e^r is the equilibrium (melting) temperature for a sphere of radius r, and Γ is the Gibbs-Thomson coefficient. The Gibbs-Thomson coefficient is a measure of the energy required to form a new surface (or expand an existing one). For most metals $\Gamma = 10^{-7}$ K·m. In some calculations molar ΔH_f and ΔS_f are used, for which the units are J·mole^{-1} and J mole^{-1}·K^{-1}, respectively. Then the Gibbs-Thomson coefficient becomes:

$$\Gamma = v_m \gamma / \Delta S_f \tag{2.15}$$

where v_m is the molar volume in m^3/mole.

For a spherical crystal $\Delta T_r = 2 \cdot \Gamma / r$. Using this equation it follows that for $\Delta T_r = 2°C$, $r = 0.1\mu m$, and for $\Delta T_r = 0.2 °C$, $r = 1\mu m$. Thus, the S/L interface energy is important only for morphologies where $r < 10\mu m$, *i.e.*, nuclei, interface perturbations, dendrites and eutectic phases.

2.4 Thermal undercooling

Let us now relax assumption (d), and allow a thermal gradient to exist in the liquid (Figure 2.5). As long as nucleation of solid and subsequent growth of these nuclei is rather fast, the only S/L interface undercoolings for the pure metal are kinetic and curvature. However, if nucleation difficulties are encountered, or if growth of the solid lags heat transport out of the liquid, an additional undercooling, *thermal undercooling*, ΔT_T, occurs. When ignoring kinetic undercooling, this additional undercooling is simply the amount the liquid is under the equilibrium temperature of the pure metal solidifying with a planar interface (no curvature). Thus, the bulk thermal undercooling is:

$$\Delta T_T^{bulk} = T_e - T_{bulk} \tag{2.16}$$

where T_{bulk} is the bulk liquid temperature (temperature far from the interface that can be measured through a thermocouple).

At the S/L interface the rejection of latent heat must also be considered. As shown in Figure 2.6, a boundary layer of height $\Delta H_f/c$, and length δ_T will form at the interface (position $x = 0$), because of heat accumulation at the interface. The interface thermal undercooling can be calculated as:

$$\Delta T_T^* = T^* - T_{bulk} \tag{2.17}$$

The corresponding increase in free energy is $\Delta G_T = \Delta S_f \left(T^* - T_{bulk} \right)$.

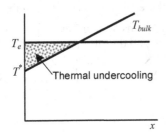

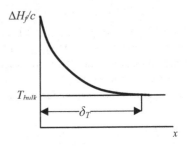

Figure 2.5. Bulk thermal undercooling.   **Figure 2.6.** Interface thermal undercooling.

Sometimes, metals can undercool considerably before solidifying. For example, pure iron can be undercooled under its melting (equilibrium) temperature by 300°C, or even more, under certain controlled conditions.

2.5   Constitutional undercooling

Up to this point, only pure metals have been considered (assumption a). For alloys, the solutal field introduces an additional change in the free energy, which corresponds to an additional undercooling. Figure 2.7 shows the left corner of the phase diagram of a hypothetical alloy solidifying to form a single-phase solid solution. T_L is the liquidus temperature, T^* is the interface temperature at some arbitrary time during solidification, and T_S is the solidus temperature. Note that for alloys, $T_e = T_L$. At temperature T^*, the composition of the solid at the interface is C^*_S, while the composition of the liquid is C^*_L. The bulk composition of the alloy, at the beginning of solidification, is C_o. The ratio between the solid composition and the liquid composition at the interface is called the *equilibrium partition coefficient, k*:

$$k = \left(C^*_S / C^*_L\right)_{T,P} \tag{2.18}$$

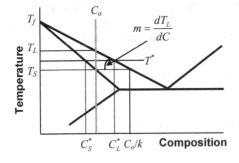

Figure 2.7. Schematic region of a phase diagram for a solid solution alloy.

The indices T and P mean that calculations are made at constant temperature (isotherm) and at constant pressure (isobar). Note that at the end of solidification, T_S, it can be calculated that the last liquid to solidify should be of composition C_o/k. For the particular case described in Figure 2.7, there is more solute in the liquid than in the solid at the interface. This partition is the cause of the occurrence of macro- and micro-segregation in alloys, which will be discussed later in more detail.

The partition coefficient is constant only when the liquidus slope, m, is constant. Since for most alloys m is variable, so is k. Nevertheless, for mathematical simplicity, in most analytical calculations m and k are assumed constant. Note that $k < 1$ when the left hand corner of a phase diagram is considered. However, $k > 1$ when the slopes of the liquidus and solidus lines are positive.

The following relationships exist between the various temperatures and compositions in Figure 2.7:

$$\Delta T_o = T_L - T_S = -m \cdot \Delta C_o \quad \text{and} \quad \Delta C_o = C_o(1-k)/k \tag{2.19}$$

where ΔT_o is the liquidus-solidus temperature interval at C_o, and ΔC_o is the concentration difference between liquid and solid at T_S.

For dilute solutions, the Van't Hoff relation for liquid-solid equilibrium holds and relates k and m as:

$$k = 1 - m\Delta H_f^A \Big/ \left[R\left(T_f^A\right)^2\right] \tag{2.20}$$

where ΔH_f^A and T_f^A are the latent heat and the melting temperature of the pure solvent, A, respectively, and R is the gas constant. The partition coefficient can also be calculated with:

$$k = \exp\left[\left(\Delta H_f^B / R\right)\left(1/T_f^B - 1/T_f^A\right)\right] \tag{2.21}$$

where ΔH_f^B and T_f^B are the latent heat and the melting temperature of the solute B, respectively.

The difference between the solid and liquid solubility of the alloying element is responsible for the occurrence of an additional undercooling, called *constitutional, or compositional, or solutal, undercooling* (ΔT_c). The concept was first introduced by Chalmers (1956). Consider the diagrams in Figure 2.8. The first diagram in the upper left corner is a temperature - composition plot, that is, a phase diagram. C_o is the composition of the solid at temperature T_S, while C_o/k is the composition of the liquid at the same temperature. These compositions have been translated onto the lower diagram, which is a composition - distance (x) diagram. A diffusion boundary layer, δ_c, is shown on the diagram. This layer occurs because at the interface the composition of the liquid is higher (C_o/k) than farther away in the bulk liquid (C_o), and consequently, the composition of the liquid, C_L, decreases from the interface toward the liquid.

The third diagram, on the upper right, is a temperature-distance diagram. It shows that the liquidus temperature in the boundary layer is not constant, but increases from T_S at the interface, to T_L in the bulk liquid. This is a consequence of the change in composition, which varies from C_o/k (at temperature T_S) at the interface, to C_o (at temperature T_L) in the bulk liquid. A liquidus (solutal) temperature gradient, G_L, can now be defined as the derivative of the $T_L(x)$ curve with respect to x at the temperature of the interface, T^* (Figure 2.9).

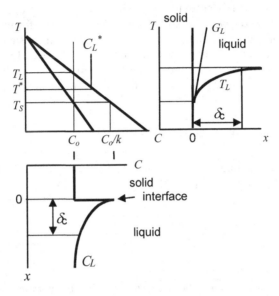

Figure 2.8. The thermal and solutal field in front of the solid/liquid interface.

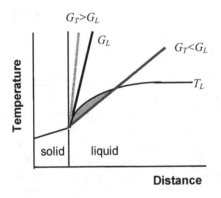

Figure 2.9. Constitutional undercooling diagram comparing thermal (G_T) and liquidus (compositional) (G_L) gradients.

Since heat is flowing out from the liquid through the solid, there is also a thermal gradient in the liquid, G_T, which is determined by the evolution of the thermal field. The two gradients are compared in Figure 2.9. If $G_L < G_T$, the temperature of the liquid ahead of the interface is above the liquidus temperature of the alloy. If on the contrary, $G_L > G_T$, over a certain distance ahead of the interface, the liquid will be at a temperature lower than its liquidus. Thus, while the bulk liquid may be at a temperature above its liquidus, the liquid at the interface may be at a temperature below its liquidus, because of the solute concentration in the diffusion layer. This liquid is constitutionally undercooled. The undercooling associated with this liquid is called *constitutional, or compositional, or solutal, undercooling,* ΔT_c. Based on Figure 2.8 it can be calculated as:

$$\Delta T_c = T_L - T^* = -m\left(C_L^* - C_o\right) \tag{2.22}$$

Note that the sign convention here is that m is negative. The corresponding increase in free energy is:

$$\Delta G_c = -\Delta S_f\, m\left(C_L^* - C_o\right)$$

2.6   Pressure undercooling

Let us now relax assumption (b) and consider that local pressure is applied on the S/L interface or that pressure is applied on the whole system. The change in free energy of the liquid and solid with small changes in pressure and temperature can be calculated from Eq.(2.5) as:

$$\Delta F_L = v_L\,\Delta P - S_L\,\Delta T \qquad \text{and} \qquad \Delta F_S = v_S\,\Delta P - S_S\,\Delta T$$

This is true assuming that the internal energy, the volume and the entropy of the condensed matters (liquid and solid) change little under the proposed conditions. Then, from the equilibrium condition, $\Delta F_L = \Delta F_S$, the change in equilibrium temperature because of the applied pressure is:

$$\Delta T_P = \Delta P\,\Delta v / \Delta S_f \tag{2.23}$$

This equation is known as the Clapeyron equation. During solidification, the change in volume Δv is positive. Thus, an increase in pressure ($\Delta P > 0$) will result in an increase in undercooling.

For metals the pressure undercooling is rather small, of the order of 10^{-2} K/atm. Hence, pressure changes typical for usual processes have little influence on the melting temperature. However, in certain applications, such as particle engulfment by the S/L interface, the local pressure can reach relatively high values, and ΔT_P may become significant.

2.7   Kinetic undercooling

The concept of undercooling can also be understood in terms of atom kinetics at the S/L interface. While this analysis is done at the atomic scale level, and a more in-depth discussion of this subject will be undertaken in Section 14, some concepts will be introduced here for clarity. When a S/L interface moves, the net transfer of atoms at the interface results from the difference between two atomic processes (Verhoeven, 1975):

atoms in solid ⇨ atoms in liquid   melting
atoms in liquid ⇨ atoms in solid   solidification

The rate of these two processes is:

$$Rate\ of\ melting\ (S{\rightarrow}L) \quad = \left(\frac{dn}{dt}\right)_{melt} = p_{melt}\,n_S\,v_S\exp\left(-\frac{\Delta G_{melt}}{k_B T}\right) \tag{2.24}$$

$$Rate\ of\ solidification\ (L{\rightarrow}S) = \left(\frac{dn}{dt}\right)_S = p_S\,n_L\,v_L\exp\left(-\frac{\Delta G_S}{k_B T}\right) \tag{2.25}$$

where n_S, n_L are the number of atoms per unit area of solid and liquid interface respectively, v_S, v_L are the vibration frequencies of solid and liquid atoms respectively, ΔG_{melt}, ΔG_S are the activation energy for an atom jumping through the interface during melting and solidification, respectively, and p_M, p_S are probabilities given by:

$$p_{M,S} = f_{M,S} \cdot A_{M,S} \tag{2.26}$$

Here $f_{M,S}$ is the probability that an atom of sufficient energy is moving toward the interface, and $A_{M,S}$ is the probability that an atom is not kicked back by an elastic collision upon arrival.

At equilibrium, the flux of atoms toward and away from the interface must be equal, that is:

$$(dn/dt)_M = (dn/dt)_S$$

Thus, the two curves must intersect at T_e (Figure 2.10). For solidification to occur, more atoms must jump from L to S than from S to L. Consequently, the solidifying interface must be at lower temperature than T_e by an amount that is called *kinetic undercooling*, ΔT_k.

Another approach to this problem (*e.g.*, Biloni and Boettinger 1996) would be to consider that the overall solidification velocity is simply:

$$V = Rate\ of\ solidificaion - Rate\ of\ melting = V_c - V_c\exp(-\Delta G/RT_i)$$

where ΔG is expressed in J/mole. V_c corresponds to the hypothetical maximum growth velocity at infinite driving force. Then, using series expansion for the exponential term ($1 - e^{-x} \approx x$), neglecting 2nd and higher order terms, and assuming that Eq.(2.9) is valid near equilibrium we obtain:

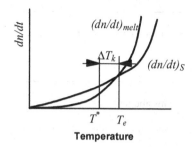

Figure 2.10. Requirement of kinetic undercooling based on atomic kinetics considerations.

$$V = V_c \frac{\Delta H_f \Delta T_k}{RT_e^2} \quad \text{or} \quad \Delta T_k = \frac{RT_e^2}{\Delta H_f} \frac{V}{V_c} \tag{2.27}$$

Two hypotheses have been used to evaluate V_c. The first one (*e.g.*, Turnbull 1962) assumes that the rate of forward movement (atoms incorporation in the solid) is the same as the rate at which atoms can diffuse in the melt. Thus, $V_c = D_L/a_o$, where a_o is the interatomic spacing. The second one, the so-called *collision limited growth model* (Turnbull and Bagley, 1975), assumes that the solidification event may be limited only by the impingement rate of atoms with the crystal surface. Then $V_c = V_o$, where V_o is the speed of sound. Note that V_o is approximately three orders of magnitude higher than D_L/a_o. Experimental analysis of rapidly growing dendrites in pure melts (Coriell and Turnbull, 1982) has confirmed the collision limited growth model. Typically, as will be calculated in Section 15.3, for metals the kinetic undercooling is of the order of 0.01 to 0.05 K.

2.8   Departure from equilibrium

We have demonstrated that for solidification to occur a certain amount of undercooling is necessary. Solidification cannot occur at equilibrium. Depending of the amount of undercooling different degrees of departure from equilibrium may occur, following a well-defined hierarchy. As shown in Table 2.1, as the undercooling or the solidification velocity increases, the liquid-to-solid transformation changes form fully diffusional to non-diffusional.

Global equilibrium, (I), requires uniform chemical potentials and temperature across the system. Under such conditions, no changes occur with time. In solidification processing such conditions exist only when the solidification velocity is much smaller than the diffusion velocity. Such conditions truly exist only when solidification takes place over geological times (Biloni and Boettinger, 1996), or after long time annealing. When global equilibrium exists, the fraction of phases can be calculated with the lever rule, and the phase diagram gives the uniform composition of the liquid and solid phases.

Table 2.1. Hierarchy of equilibrium (Boettinger and Perepezko, 1985).

Increa- sing under- cooling or solidi- fication velocity ↓	I. Full diffusional (global) equilibrium
	    A. No chemical potential gradients (composition of phases are uni- form)
	    B. No temperature gradients
	    C. Lever rule applicable

	II. Local interfacial equilibrium
	    A. Phase diagram gives compositions and temperatures only at liq- uid - solid interface
	    B. Corrections made for interface curvature (Gibbs - Thomson ef- fect)

	III. Metastable local interface equilibrium
	    A. Stable phase cannot nucleate or grow sufficiently fast
	    B. Metastable phase diagram (a true thermodynamic phase diagram missing the stable phase or phases) gives the interface condi- tions

	IV. Interface non-equilibrium
	    A. Phase diagram fails to give temperature and compositions at the interface
	    B. Chemical potentials are not equal at the interface
	    C. Free energy functions of phases still lead to criteria for impossi- ble reactions

During solidification of most castings, both temperature and composition gradients exist across the casting. Nevertheless, in most cases, the overall kinetics can be described with sufficient accuracy by using the mass, energy and species transport equations to express the temperature and composition variation within each phase, and equilibrium phase diagrams to evaluate the temperature and composition of phase boundaries, such as the solid/liquid interface. This is the local equilibrium condition, (II). Most phase transformations, with the exception of massive (parti- tionless) and martensitic transformations can be described with the conditions present under (II).

Metastable equilibrium, (III), can also be used locally at the interface. The most common case is the gray-to-white (metastable-to-stable) transition in cast iron that occurs as the cooling rate increases. The stable eutectic graphite-austenite is gradu- ally substituted by the metastable iron carbide-austenite because on difficulties in the nucleation of graphite and the higher growth velocity of the metastable eutec- tic. Metastable transformation can occur at solidification velocities exceeding 0.01m/s. Usually, solidification occurring at rate above this value is termed rapid solidification.

For both stable and metastable local equilibrium, the chemical potentials of the components across the interface must be equal for the liquid and for the solid. However, at large undercooling, achieved for example when using high solidifica- tion velocities, this condition ceases to be obeyed. The solidification velocity ex- ceeds the diffusive speed of solute atoms in the liquid phase. The solute is trapped into the solid at levels exceeding the equilibrium solubility. These conditions, (IV), correspond to rapid solidification. Typically, for solute trapping to occur, the so- lidification velocity must exceed 5m/s (Boettinger and Coriell, 1986).

The preceding analysis is useful in attempting to classify practical solidification processes based on the degree of equilibrium at which they occur as follows:

- processes occurring with local interface equilibrium: shape casting, continuous casting, ingot casting, welding (arc, resistance), directional solidification;
- processes occurring with interface non-equilibrium: welding (laser), melt spinning, atomization, surface remelting.

2.8.1   Local interface equilibrium

For the time scale (cooling rates) typical for solidification of castings, the assumption of local interface equilibrium holds very well. However, the interface temperature is not only a function of composition alone, as implied by the phase diagram. Interface curvature, as well as heat and solute diffusion affect local undercooling. Accordingly, to express the condition for local equilibrium at the S/L interface all the contributions to the interface undercooling must be considered. The total undercooling at the interface with respect to the bulk temperature, T_{bulk}, is made of the algebraic sum of all the undercoolings derived above (see Figure 2.11):

$$\Delta T = \Delta T_k + \Delta T_r + \Delta T_c + \Delta T_T + \Delta T_P \tag{2.28}$$

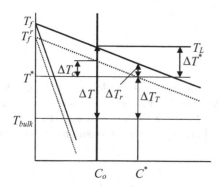

Figure 2.11. The various components of interface undercooling with respect to the bulk temperature under the condition of local interface equilibrium.

Ignoring the kinetic and pressure undercooling, and since $T_L = T_f + m\, C_o$, the interface undercooling under the condition of local equilibrium for castings solidification can be written as:

$$\Delta T = \Delta T_T + \Delta T_c + \Delta T_r = (T^* - T_{bulk}) + (T_L - T^*) + \Gamma K = T_f + m\,C_o + \Gamma K - T_{bulk} \tag{2.29}$$

where T_f is the melting point of the pure metal (see Application 2.2).

In practical metallurgy, the solidification velocity is increased by increasing the cooling rate. As the cooling rate increases the length scale of the microstructure (e.g., dendrite arm spacing) decreases. For cooling rates up to 10^3 K/s local equilibrium with compositional partitioning between the liquid and solid phases at the solidification interface is maintained. The interface undercooling is small. However, when the cooling rate increases above 10^3 K/s non-equilibrium solidification occurs.

Local equilibrium can occur even at remarkable undercooling under the equilibrium temperature if nucleation is avoided. In this case, the liquidus and solidus lines can be extended as metastable lines, as shown in Figure 2.12.

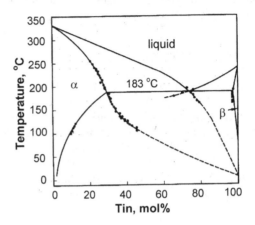

Figure 2.12. The stable Pb-Sn phase diagram (solid line) with superimposed calculated metastable extensions (dotted lines) of the liquidus and solidus lines, and measured data (Fecht and Perepezko, 1989). With kind permission of Springer Science and Business Media.

2.8.2   Interface non-equilibrium

It has been shown that for a multiphase system a condition for equilibrium is that the chemical potential of each component must be the same in all phases, as stated by Eq. (2.4). This is shown graphically in Figure 2.13. It is noticed that, while the chemical potentials in the liquid and solid are equal, the compositions are not. The

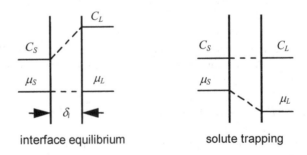

Figure 2.13. Interface composition and chemical potential for equilibrium and difusionless solidification (solute trapping).

necessary condition for interface equilibrium is $V \ll D_i/\delta_i$, where V is the solidification velocity, D_i is the interfacial diffusion coefficient, and δ_i is the atomic jump distance. Note that D_i is smaller than the bulk liquid diffusion coefficient, D_L. The equilibrium partition coefficient is calculated from the phase diagram with Eq. (2.18).

If the ratio between the two velocities is reversed, that is $V \gg D_i/\delta_i$, as shown in Figure 2.13, the equality between the chemical potentials is lost but the composition becomes uniform across the interface. The partition coefficient becomes one. Solute trapping occurs. Using the typical values of $D_i = 2.5 \cdot 10^{-9} \mathrm{m^2/s}$ and $\delta_i = 0.5 \cdot 10^{-9} \mathrm{m}$, the critical velocity for solute trapping is calculated to be 5m/s.

For solute trapping to occur, the interface temperature must be significantly undercooled with respect to T_L. During partitionless solidification ($C_S^* = C_L^*$), a thermodynamic temperature exists which is the highest interface temperature at which partitionless solidification can occur. This temperature is called the T_o temperature, and is the temperature at which the molar free energies of the solid and liquid phases are equal for the given composition. The locus of T_o over a range of compositions constitutes a T_o curve. The liquid and solid phase compositions are equal along the T_o curve.

Some examples of such curves are given in Figure 2.14. They can be used to evaluate the possibility of extension of solubility by rapid melt quenching. If the T_o curves are steep (Figure 2.14a) single phase α or β crystals with compositions beyond their respective T_o cannot form from the melt. The solidification temperature in the vicinity of the eutectic composition can be depressed to the point where an increased liquid viscosity stops crystallization (glass temperature transition, T_g). If the T_o curves are shallow (Figure 2.14b), for composition below both T_o curves, a mixture of α and β crystals could form, each phase having the same composition as the liquid.

Baker and Cahn (1971) formulated the general interface condition for solidification of binary alloys by using two response functions:

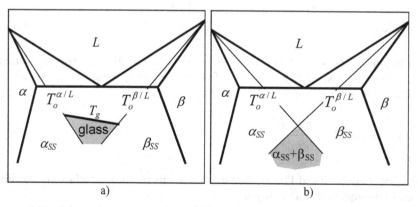

Figure 2.14. Schematic representation of T_o curves for two different eutectic systems (Perepezko and Boettinger, 1983).

$$T^* = T(V, C_L^*) - \Gamma K \tag{2.30}$$
$$C_S^* = C_L^* k^*(V, C_L^*) \tag{2.31}$$

At zero interface velocity (equilibrium), the functions T and k^* are directly related to the phase diagram. Indeed, $T(0, C_L^*)$ describes the liquidus temperature of the phase diagram and $k^*(0, C_L^*)$ is the equation for the equilibrium partition coefficient, Eq. (2.18). The dependence of k^* on interface curvature is ignored.

Several models have been proposed to describe the dependence of the partition coefficient on velocity. The most widely accepted is that one proposed by Aziz (1982). Ignoring the composition dependence of the partition coefficient, its functional dependence for continuous growth is:

$$k^*(V) = \frac{k_e + \delta_i \cdot V/D_i}{1 + \delta_i \cdot V/D_i} \tag{2.32}$$

where k_e is the equilibrium partition coefficient.

Note that for $V = 0$, $k^* = k_e$, and for very large V, $k^* = 1$. D_i is unknown. In some other models liquid diffusivity rather than interfacial diffusivity is used. The atomic diffusion speed $V_i = D_i/\delta_i$, is usually obtained by fitting Eq. (2.32) to experimental curves showing velocity dependence on partition coefficients. Some typical values for V_i are 17m/s for Sn (Hoaglund et al., 1991), 33m/s for Ni-0.6 at% C (Barth et al., 1999), and 5m/s for Ag-5at% Cu (Boettinger and Coriell, 1986). From this analysis it follows that for solute trapping to occur two conditions are necessary: $k^* = 1$ and $T^* < T_o$.

By evaluating the change in free energy and assuming a linear kinetic law for the interface velocity (from Eq. (2.27), Baker and Cahn (1971) calculated the two response functions for a flat interface to be:

$$T^* = T_f + m_L(V)C_L^* + \frac{m_L}{1 - k_e} \frac{V}{V_o} \quad \text{with} \quad m_L(V) = \frac{m_L}{1 - k_e}\left[1 - k^*\left(1 - \ln\frac{k^*}{k_e}\right)\right] \tag{2.33}$$

$$C_S^* = k^* C_L^* \tag{2.34}$$

Boettinger and Coriell (1986) have proposed a slightly different derivation, substituting the last term in Eq.(2.33) for interface temperature with the kinetic undercooling given by Eq. (2.27), to obtain:

$$T^* = T_f + m_L(V)C_L^* - \frac{RT_e^2}{\Delta H_f} \frac{V}{V_o} \tag{2.35}$$

Note that if $D_i/\delta_i = 0$ and $V_o = \infty$, then the conditions for local interface equilibrium revert to the equations previously introduced:

$$T^* = T_f + m_L C_L^* \quad \text{and} \quad C_S^* = k_e C_L^* \tag{2.36}$$

2.9   Applications

Application 2.1

Calculate the time required to solidify unidirectionally a rod having the length l = 10 cm, so that full diffusional equilibrium operates during solidification.

Answer:
Assume D_L = 10^{-9}m/s. For equilibrium solidification to occur diffusion will have to go to completion; that is the solute should be able to diffuse over the entire length of the specimen. The diffusion velocity for complete diffusion over the sample of length l is D_L/l = $10^{-9}/10^{-2}$ = 10^{-7}m/s. The solidification velocity must be much higher than the diffusion velocity, i.e. V_S << D_L/l. Assume V_S = 10^{-10}m/s. Then, the solidification time is t = l/V_S = $10^{-2}/10^{-10}$ = 10^8s = 3.17 years.

Application 2.2

Consider a Cu-10%Sn bronze (phase diagram in Appendix C). Assume solidification with planar S/L interface under local equilibrium conditions. A thermocouple placed far from the interface reads 950°C. What is the interface undercooling at the beginning of solidification? Calculate the change in interface undercooling when the average (bulk) composition has changed from 10% to 12%.

Answer:
The interface undercooling is given by Eq. (2.29). The contribution of curvature is ignored as the interface is planar. From the phase diagram T_f = 1085°C. The liquidus slope can be calculated using values at the temperature of 798°C, as follows: m = $\Delta T/\Delta C$ = (1085 - 798)/(-26) = -11. C_o is given as 10%. Substituting in Eq. (2.29) we obtain the initial interface undercooling to be ΔT = 25°C.

The change in interface undercooling when the bulk composition increases to 12% is simply m (C_o - C_{bulk}) = -11 (10 - 12) = 22°C.

References

Aziz M.J., 1982, *J. Appl. Phys.* **53**:1158.
Baker J.C., Cahn J.W., 1971, in: *Solidification*, ASM Metals Park, OH, p.23.
Barth M., Holland-Moritz D., Herlach D.M., Matson D.M., Flemings M.C., 1999 in: *Solidification 1999*, W.H. Hofmeister *et al.*, eds., The Minerals, Metals and Materials Soc., p.83.
Biloni H, Boettinger W.J., 1996, Solidification, in: *Physical Metallurgy*, R. W. Cahn, P. Haasen, eds., Elsevier Science BV, p.670.
Boettinger W.J., Perepezko J.H., 1985, in: *Rapidly Solidified Crystalline Alloys*, S.K. Das, B.H. Kear, C.M. Adam eds., The Metallurgical Soc., Warrendale PA, p.21.
Boettinger W.J., Coriell S.R., 1986, in: *Science and Technology of the Supercooled Melt*, P.R. Sahm, H. Jones, C.M. Adams, eds., NATO ASI Series E-No. 114, Martinus Nijhoff, Dordrecht, p.81.
Chalmers B. 1956, *Trans. AIME* **200**:519
Coriell S.R., Turnbull D., 1982, *Acta metall.* **30**:2135.
Fecht H.C., Perepezko J.H., 1989, *Metall. Trans.* **20A**:785.
Hoaglund D. E., Aziz M.J., Stiffer S.R., Thomson M.O., Tsao J.Y., Peercy P.S., 1991, *J. Cryst. Growth* **109**:107.

Perepezko J. H., Boettinger W.J., 1983, in: *Mat. Res. Soc. Symp. Proc.*, **19**:223.

Turnbull D., 1962, *J. Phys. Chem.* **66**:609.

Turnbull D., Bagley B.G., 1975, in: *Treatise on Solid State Chemistry*, N.B. Hannay, ed., Plenum, NY, **5**:513.

Verhoeven J.D., 1975, *Fundamentals of Physical Metallurgy*, John Wiley & Sons, New York, p.238.

3

MACRO-SCALE PHENOMENA - GENERAL EQUATIONS

The problem to solve is to describe mathematically casting solidification and associated phenomena such as macrosegregation and macroshrinkage. To build a correct model it is first necessary to understand the physics associated with the solidification of castings. At the macro-scale level the casting is a two-phase system comprised of solid and liquid. Nucleation and growth of phases is ignored at this stage.

Solidification is considered a thermodynamic process driven by:

- diffusion of species and energy
- convection of mass and energy driven by natural convection (thermal and solutal buoyancy) and by solidification contraction

3.1 Relevant Transport Equations

The mathematical problem is to solve the mass, energy and momentum transport equations for the particular geometry and material of the casting. The general transport equation written in its standard form for advection-diffusion is:

$$\frac{\partial}{\partial t}(\rho \cdot \phi) + \nabla \cdot (\rho \cdot \mathbf{V} \cdot \phi) = \nabla \cdot (\rho \cdot \Gamma \cdot \nabla \phi) + S \tag{3.1}$$

where t is time, ρ is the density, ϕ is the phase quantity, $\mathbf{V}$ is the velocity vector, Γ is the general diffusion coefficient, and S is the source term. The *del* operator[1] was used to reduce the number of equations to be written.

---

[1] The *del* operator (or the *gradient*) of a function u is: $\nabla u \equiv \frac{\partial u}{\partial x}\mathbf{i} + \frac{\partial u}{\partial y}\mathbf{j} + \frac{\partial u}{\partial z}\mathbf{k}$. Note that ∇u

is ∇ operating on u, while $\nabla \cdot \mathbf{A}$ is the vector dot product of *del* with $\mathbf{A}$ (or the *divergence*):

The first left hand (LH1) term is the temporal term, the LH2 is the convective term, and the first right hand (RH1) term is the diffusive term. The specific phase quantities and diffusivity for the four basic equations required in solidification modeling are given in Table 3.1. The following notations were used: H is the sensible enthalpy, $\alpha = k/(\rho\, c)$ is the thermal diffusivity, D is the species diffusivity, $v = \mu/\rho$ is the kinematic viscosity, μ is the dynamic viscosity.

Table 3.1. Phase quantities, diffusivities, and origin of the source term.

Quantity	Mass	Energy	Species	Momentum
ϕ	1	H	C	$\mathbf{V}$
Γ	0	$\alpha = k/(\rho\,c)$	D	$v = \mu/\rho$
S	- phase motion	- phase transformation - phase motion	- phase transformation - phase motion	- phase motion - S/L interaction - natural convection - shrinkage

The relevant transport equations can now be obtained from Eq. (3.1) as follows:

Conservation of mass (continuity):

$$\frac{\partial \rho}{\partial t} + \nabla \cdot (\rho \mathbf{V}) = S_m \qquad (3.2)$$

The term $\partial \rho / \partial t$, which expresses the change in density over time, describes the shrinkage-induced flow. The flow is driven toward the volume element when the averaged density of the solid and liquid phases increases with time, as is the case for most alloys.

Conservation of energy:

$$\frac{\partial}{\partial t}(\rho H) + \nabla \cdot (\rho \mathbf{V} H) = \nabla \cdot \left(\frac{k}{c} \nabla H\right) + S_H \qquad (3.3)$$

where k is the thermal conductivity and c is the specific heat.

Conservation of species:

$$\frac{\partial}{\partial t}(\rho C) + \nabla \cdot (\rho \mathbf{V} C) = \nabla \cdot (\rho D \nabla C) + S_C \qquad (3.4)$$

$\nabla \cdot \mathbf{A} = \dfrac{\partial \mathbf{A}_x}{\partial x} + \dfrac{\partial \mathbf{A}_y}{\partial y} + \dfrac{\partial \mathbf{A}_z}{\partial z}$. Furthermore, $\nabla^2 u$ is the product of the *del* operator with itself, or

$\nabla \cdot \nabla = \dfrac{\partial}{\partial x}\left(\dfrac{\partial}{\partial x}\right) + \dfrac{\partial}{\partial y}\left(\dfrac{\partial}{\partial y}\right) + \dfrac{\partial}{\partial z}\left(\dfrac{\partial}{\partial z}\right)$ operating on u.

Conservation of momentum:

$$\frac{\partial}{\partial t}(\rho \mathbf{V}) + \nabla \cdot (\rho \mathbf{V} \cdot \mathbf{V}) = \nabla \cdot (\rho \nu \cdot \nabla \mathbf{V}) + S_m \qquad (3.5)$$

Alternatively, for fluids assumed to have constant density and viscosity, conservation of momentum can be written in the form of the Navier-Stokes equation:

$$\rho \frac{D\mathbf{V}}{Dt} = \mu \nabla^2 \mathbf{V} - \nabla P + \rho g \qquad (3.6)$$

where $D\mathbf{V}/Dt$ is the substantial derivative[2], P is the pressure and g is the gravitational acceleration. It simply states that the total force (left hand term) is equal to the sum of the viscous forces ($\mu \nabla^2 \mathbf{V}$), the pressure forces (∇P) and the gravitational force (ρg).

The source terms of these equations can be complicated. An example of their formulation will be provided later as an application of these equations to macrosegregation modeling.

There are two main difficulties in solving these equations for the problem of interest. Firstly, the application of these equations to a two-phase or multi-phase system, where all the quantities must describe not one but two phases; secondly, the formulation of the source terms for the various types of transport. Two main approaches, based on concepts from Continuum Mechanics, have been developed to solve the complicated problem of a two-phase system (Figure 3.1).

In the *mixture-theory model* proposed by Bennon and Incropera (1987) each phase is regarded as a continuum that occupies the entire domain, and described by a set of variables that are continuous and differentiable functions of space and of time. Any location within the domain can be simultaneously occupied by all phases. The macroscopic transport equations are formulated using the classical mixture theory. Summation over the computational domain is used.

In the *volume-averaged mo*del proposed by Beckermann and Viskanta (1988), Ganesan and Poirier (1990), and Ni and Beckermann (1991) all phases are considered separated. Phase quantities are continuous in one phase but discontinuous over the entire domain. Discontinuities are replaced by phase interaction relationships at interface boundary. Integration of microscopic equations over a finite volume is used.

The volume averaging technique has been utilized extensively to produce solidification models that attempt to describe microstructure evolution in castings by incorporating three phases: liquid, equiaxed solid and columnar solid (e.g. Wang

---

[2] The substantial derivative of a function u is given by:

$$\frac{Du}{Dt} = \frac{\partial u}{\partial t} + u_x \frac{\partial u_x}{\partial x} + u_y \frac{\partial u}{\partial y} + u_z \frac{\partial u}{\partial z}$$

It can be applied to any property of a fluid, the magnitude of which varies with time and position.

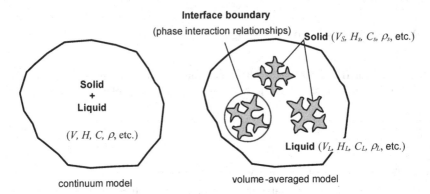

Figure 3.1. Schematic representation of the computational domain and physical quantities for two-phase models.

and Beckermann 1996, Wu and Ludwig 2006). However, the lack of information regarding the microscopic configuration at interface boundaries is a serious complication. While these models will be discussed in more details in Chapter 13, because of space restrictions, in this text we will develop only the mixture-theory model.

The typical mixture theory relationships used in the continuum mixture-theory model are derived based on the spatial average:

$$\bar{\phi} = \frac{1}{v} \int_v \phi \, dv$$

where $\bar{\phi}$ is the averaged value of ϕ over both the solid and liquid phases, and v is the volume. In particular, the following relationships are postulated:

$$g_L + g_S = 1 \qquad f_L + f_S = 1 \qquad f_L = \frac{g_L \rho_L}{\rho} \qquad f_S = \frac{g_S \rho_S}{\rho}$$

$$\rho = g_L \rho_L + g_S \rho_S \qquad k = g_L k_L + g_S k_S$$

$$\mathbf{V} = f_L \mathbf{V_L} + f_S \mathbf{V_S} \qquad \mathbf{V} = V^x \cdot i + V^y \cdot j + V^z \cdot k$$

$$D = f_L D_L + f_S D_S \qquad H = g_L H_L + g_S H_S \tag{3.7}$$

where the subscripts S and L denote solid and liquid, respectively, g is the volume fraction, f is the mass fraction, and k is the thermal conductivity. Additional assumptions must be made to describe the relative movement of the solid and liquid phases during solidification (see for example Chang and Stefanescu, 1996).

The simplified macroscopic transport equations as derived by Bennon and In-cropera (1987) based on the continuum model and the mixture theory relationships are in principle similar to those introduced above. The source term in the continuity equation was considered zero. The conservation of momentum equation has been modified to describe flow in the mushy zone, assuming equiaxed grains floating in the liquid. For the x direction:

$$\frac{\partial}{\partial t}(\rho V^x) + \nabla \cdot (\rho V V^x) = \nabla \cdot \nabla \left(\mu^* \frac{\rho}{\rho_L} V^x \right) + S_m \qquad (3.8)$$

where μ^* is the relative viscosity (the viscosity of the solid-liquid mixture, which is a function of the solid fraction) and V^x is the velocity component in the x direction.

3.2   Introduction to diffusive transport

3.2.1   Flux laws

Diffusive transport of energy, mass and momentum can be described through flux laws whose fundamental form is:

$$flux = \frac{flow\,rate}{area} = transport\,property \cdot potential\,gradient$$

The three laws describing diffusive transport are:

Energy: $q = -k\nabla T$ Fourier's law   (3.9a)

Mass (species): $J_A = -D_{AB}\nabla C_A$ Fick's law   (3.9b)

Momentum: $\tau_{xy} = -\mu\frac{\partial V^y}{\partial x}$ Newton's law of viscosity   (3.9c)

where q, J_A and τ_{xy} are the heat, mass and momentum flux, respectively, and V_y is the fluid velocity along the y axis. Analogous to mass diffusivity we can define the thermal diffusivity, $\alpha = k/(\rho\,c)$ and momentum diffusivity $\nu = \mu/\rho$. Hence, the flux laws can be written in their diffusion form as follows:

flux = - diffusivity · concentration gradient

Now the three flux laws for energy, mass and momentum transport can be written as:

Energy: $q = -\alpha \nabla(\rho c T)$ (3.10a)

Mass (species): $J_{Ax} = -D_{AB} \nabla C_A$ (3.10b)

Momentum: $\tau_{xy} = -v \dfrac{\partial(\rho V^y)}{\partial x}$ (3.10c)

Note that the quantities at the numerators on the right hand term represent energy concentration (energy/volume), mass concentration (mass/volume), and momentum concentration (momentum/volume), respectively.

The general transport equation can also describe diffusive processes by recognizing that when there is no relative movement of phases, *i.e.* $\mathbf{V} = 0$, the advective term in Eq. (3.1) disappears:

$$\frac{\partial}{\partial t}(\rho \cdot \phi) = \nabla \cdot (\Gamma \cdot \rho \cdot \nabla \phi) + S \qquad (3.11)$$

While further discussion could be conducted for the general transport equation, it is believed that a treatment of the heat diffusion equation with appropriate examples will be easier to follow. However, the discussion is equally applicable to the other forms of diffusive transport.

3.2.2   The differential equation for macroscopic heat transport

The macroscopic heat flow equation in terms of temperature rather than enthalpy can be obtained from the conservation of energy Eq. (3.3). Indeed, for $\mathbf{V} = 0$, assuming constant ρ and c, and since the enthalpy is $H = c \cdot T$:

$$\partial T / \partial t = \alpha \nabla^2 T + S_H / \rho c \qquad (3.12)$$

Further assuming that the source term is the heat flow rate resulting from the latent heat of solidification, $S_H = \dot{Q}_{gen}$, the heat flow equation becomes:

$$\frac{\partial T}{\partial t} = \alpha \nabla^2 T + \frac{\dot{Q}_{gen}}{\rho c} \qquad (3.13)$$

To solve the partial differential equation (PDE) Eq. (3.13), we need an initial condition (because it has one time derivative) and two boundary conditions (because it has two spatial derivative). The initial condition at $t = 0$ is the initial temperature distribution:

$$T(x,0) = f(x) \qquad (3.14)$$

The following types of boundary conditions may be used:

prescribed temperature (Dirichlet problem): $T(0,t)=T_I$ (3.15)

insulated boundary (Neumann problem): $\frac{\partial T}{\partial x}(0,t) = 0$ (3.16)

known heat flux (Newton's law of cooling): $-k\frac{\partial T}{\partial x}(0,t) = h(T(t) - T(0))$ (3.17)

where h is the heat transfer coefficient.

A number of simplifications of Eq. (3.13) are possible. If there is no phase transformation or heat generation (no source term):

$$\partial T/\partial t = \alpha \nabla^2 T$$ (3.18)

The steady state solution, when boundary conditions and the source term are independent of time, is:

$$\nabla^2 T + \dot{Q}_{gen}/k = 0 \qquad \text{Poisson's equation}$$ (3.19)

This equation gives the equilibrium temperature distribution. If there is no heat generation this equation is reduced to:

$$\nabla^2 T = 0 \qquad \text{Laplace's equation}$$ (3.20)

References

Beckermann C. and Viskanta R., 1988, *PhysicoChem. Hydrodyn.* **10**:195
Bennon W. D. and Incropera F. P., 1987, *Int. J. Heat Mass Transfer* **30**:2161, 2171
Chang S. and Stefanescu D. M., 1996, *Metall. Mater. Trans.* **27A**:2708
Ganesan S. and Poirier D.R., 1990, *Metall. Trans.* **21B**:173
Ni J. and Beckermann C., 1991, *Metall. Trans.* **22B**:349
Wang C.Y. and Beckermann C., 1996, *Metall. Mater. Trans.* **27A**:2754
Wu M.G. and Ludwig A., 2006, *Metall. Mater. Trans.* **37A**:1613-1631

MACRO-MASS TRANSPORT

The mechanisms of mass transport include species diffusion and momentum transfer (fluid convection). The effects of these mechanisms on casting solidification will be discussed in this chapter.

4.1 Solute diffusion controlled segregation

When alloys having a partition coefficient $k < 1$ solidify, solute atoms are rejected from the first region to solidify into the liquid. These atoms build up in the liquid just ahead of the solid/liquid (S/L) interface, forming a boundary layer, which has a content of solute than that of the bulk liquid. When $k > 1$, a boundary layer depleted in solute is formed. Thus, three zones for *mass transfer* can be defined:

- the solid: mass transfer occurs only by chemical diffusion
- the interface: there is a boundary layer of thickness δ_c in the liquid at the S/L interface, where the solute is transported through diffusion (*diffusion boundary layer*)
- the bulk liquid: mass transport is done by diffusion and convection.

The schematic evolution of composition across the interface (compositional profile) is shown in Figure 4.1 for $k < 1$. The amount of solute rejected at the S/L interface because of partitioning is $C_L^* = C_S^*/k$. It will diffuse down the concentration gradient until the composition will be that of the bulk liquid, C_o. A diffusion boundary layer, in which the concentration is above C_o will exist. Assuming no convection in the liquid, the diffusion boundary layer can be defined as the distance from the interface, at which the diffusion rate becomes equal to the solidification rate, *i.e.*, $D_L/\delta_c = V$. Thus, the thickness of the boundary layer is:

$$\delta_c = D_L/V \tag{4.1}$$

Let us derive an equation for the shape of the diffusion (solutal) boundary layer. We will assume $k < 1$, no source term, constant density, and constant diffu-

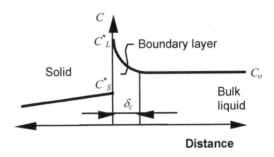

Figure 4.1. Composition profile resulting from mass transfer.

sivity. Under these assumptions the equation governing the diffusion process , Eq. (3.4), becomes:

$$D\nabla^2 C = \frac{\partial C}{\partial t} + \nabla \cdot (VC) \tag{4.2}$$

Assuming directional solidification, this equation can be used in one-dimensional (1D) form. Further assuming constant solidification velocity, the equation becomes:

$$\frac{\partial^2 C}{\partial x^2} - \frac{V}{D}\frac{\partial C}{\partial x} = \frac{1}{D}\frac{\partial C}{\partial t} \tag{4.3}$$

Setting the reference point at the interface, the velocity in the advection term is the liquid velocity that compensates for shrinkage. It is equal to the solidification velocity but has opposite sign: $V = V_L - V_{ref} \rightarrow -V$. Consequently, the advection term is subtracted from the diffusion flux. For steady state, $\partial C/\partial t = 0$ and thus:

$$\frac{\partial^2 C}{\partial x^2} + \frac{V}{D}\frac{\partial C}{\partial x} = 0 \tag{4.4}$$

This equation is known as the time independent form of the directional growth equation, or as the characteristic partial differential equation of the quasi-steady state. Its solution is (see inset for derivation):

$$C_L = C_o + \Delta C_o \exp\left(-\frac{V}{D_L}x\right) = C_o\left[1 + \frac{1-k}{k}\exp\left(-\frac{V}{D_L}x\right)\right] \tag{4.5}$$

Derivation of the equation for the composition in the boundary layer, Eq. (4.5).
The solution of Eq. (4.4) is $C_L = A + B\exp(bx)$, where b must be a solution of the equation $b^2 + Vb/D = 0$. The solutions of this "auxiliary" equation are $b = 0$ and $b = -V/D$. There-fore, the general solution of Eq. (4.4) is:

$$C_L = A + B \exp\left(-\frac{V}{D}x\right) \tag{4.6}$$

The following boundary conditions are used:

BC1 far-field condition: for $x = \infty$ $C = C_o$

BC2 flux at the interface: for $x = 0$ $V(C_L^* - C_S^*) = -D(\partial C/\partial x)$

The second boundary condition simply states that:

rate of solute rejection at the interface = diffusional flux in the liquid

Introducing BC1 in Eq. (4.6) gives $A = C_o$. Then, since from Eq. (4.6), for $x = 0$ and $A = C_o$, we have $C_L^* = C_o + B$, the left hand side term in BC2 is:

$$V(C_L^* - C_S^*) = V(1-k)C_L^* = V(1-k)(C_o + B) \tag{4.7}$$

The right hand side term in BC2 is:

$$-D\left(\frac{\partial C}{\partial x}\right)_{x=0} = -D \cdot B \frac{\partial}{\partial x}\left(\exp\left(-\frac{Vx}{D}\right)\right)_{x=0} = B \cdot V \tag{4.8}$$

Equating these last two equations gives $B = C_o(1-k)/k = \Delta C_o$ (see Eq. (2.19)). After substituting the constants A and B, Eq. (4.6) becomes Eq. (4.5).

The boundary layer given by this equation is of infinite extent, since $C_L = C_o$ at $x = \infty$. However, typically, the thickness of the boundary layer is taken to be as calculated by Eq. (4.1). To obtain a convenient practical estimate of its thickness, Kurz and Fisher (1989) have defined an equivalent boundary layer, δ_e. The equivalent boundary layer contains the same total solute amount as the infinite layer, and has constant concentration gradient G_c^e across its thickness. It is easily demonstrated that $\delta_e = 2D_L/V$.

Since the composition of the liquid and solid during solidification are different, chemical diffusion will be active during and after solidification. Thus, it is important to explore the solute redistribution resulting from this diffusion. Assuming only diffusive transport, the final composition in a solidifying casting depends on the liquid and solid diffusivity and on the partition coefficient. In our analysis, we will consider the simple case of directional solidification. This means that energy transport is only in the x-direction (Figure 4.2). There is no heat flux in the y- and z- directions. By controlling the end temperatures, the solidification velocity and the temperature gradient at the S/L interface can be maintained constant.

A rigorous complete solution of this problem can be obtained by solving the diffusion equation in three-dimensions (3D). However, this can only be done numerically, unless some simplifying assumptions are used.

Analytical solutions to the solute redistribution problem during directional solidification can be obtained on the basis of the following assumptions:

a) equilibrium solidification: $D_S = \infty$, $D_L = \infty$

b) no diffusion in solid, complete diffusion in liquid: $D_S = 0$, $D_L = \infty$
c) no diffusion in solid, limited diffusion in liquid: $D_S = 0$, $0 < D_L < \infty$
d) partial (back) diffusion in solid, complete diffusion in liquid: $D_S > 0$, $D_L = \infty$
e) limited diffusion in solid and liquid: $0 < D_S \neq D_L < \infty$
f) no diffusion in solid, partial mixing in liquid: $D_S = 0$, convection in liquid

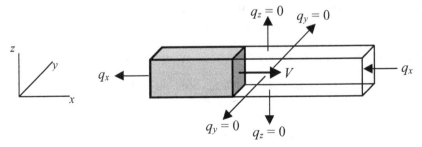

Figure 4.2. Directional solidification.

4.1.1   Equilibrium solidification

If there is enough time for solute diffusion to proceed to completion in both the solid and the liquid the composition becomes uniform throughout the sample. This is equilibrium solidification, where either enough time is available for the solid- and liquid-solutions to become completely homogeneous from the chemical stand-point ($V_S \to 0$), or diffusion is very rapid in both the solid and the liquid ($D_S = \infty$, $D_L = \infty$). Such conditions require solidification times that are much higher than encountered in practical metallurgy (see Application 2.1). As solidification proceeds under equilibrium conditions, the solute composition in the solid, C_S, and in the liquid, C_L, vary along the solidus and the liquidus line of the phase diagram, respectively (Figure 4.3).

Figure 4.4 shows the composition profile of the solute in the solid and in the liquid, in a directionally solidified sample, at three different stages: immediately after the beginning of solidification, at an intermediate time, and at the end of so-

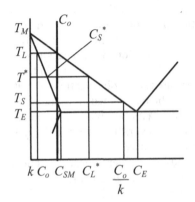

Figure 4.3. Schematic phase diagram.

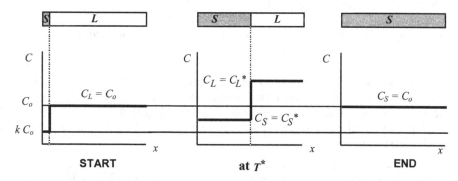

Figure 4.4. Equilibrium solute redistribution in a directionally solidified casting.

lidification. The initial composition of the alloy is C_o, i.e., the alloy contains C_o % solute. Thus the liquid composition at the beginning of solidification is $C_L = C_o$. As imposed by the partition coefficient, k, the first amount of solid to form will have the composition $C_S = k\,C_o$. This means that the solid will have less solute than the liquid, and thus the solidifying solid will reject some solute in the liquid. At an intermediate time during solidification, when the interface temperature is T^*, the composition of the liquid has risen to $C_L^* > Co$, and that of the solid to $C_S^* > k\,C_o$. Writing a material balance (conservation of solute atoms) at T^* gives:

$$C_S f_S + C_L f_L = C_o \quad\quad \text{with} \quad\quad f_S + f_L = 1$$

where f_S and f_L are the mass fraction of solid and the fraction of liquid, respectively. Knowing that $C_S = k\,C_L$, the composition of the solid is:

$$C_S = \frac{kC_o}{1-(1-k)f_S} \quad\quad\quad\quad (4.9)$$

This equation is called *the equilibrium lever rule*. It is valid assuming $\rho_S = \rho_L$.

   At the end of solidification, because of rapid solid diffusion, the composition of the solid is uniform across the volume element, and equal to the initial composition of the liquid. Note that in spite of the equilibrium nature of solidification, substantial solute redistribution occurs during solidification. The material is homogeneous only before and after solidification.

   For the more general case, when $\rho_S \neq \rho_L$, a similar equation can be obtained by using volume fractions, g_S and g_L, rather than mass fractions. The relationships between volume and mass fractions are defined in Eq. (3.7). Then, when substituting f_S with $g_S \rho_S / \rho$ in Eq. (4.9), the equilibrium lever rule becomes:

$$C_S = \frac{1-(1-\rho_S/\rho_L)g_S}{1-(1-k\,\rho_S/\rho_L)g_S}kC_o \qu\quad\quad (4.10)$$

Note that, if $\rho_S = \rho_L$, this equation reduces to the previous one.

4.1.2   No diffusion in solid, complete diffusion in liquid (the *Gulliver-Scheil* model)

The basic assumption is that diffusion is very rapid ($D_L = \infty$), or there is complete mixing (convection) in the liquid, but there is no diffusion in the solid ($D_S = 0$). The graphical representation of this case is given in Figure 4.5. Note that since complete mixing in the liquid is assumed there is no diffusion boundary layer ahead of the solidifying interface.

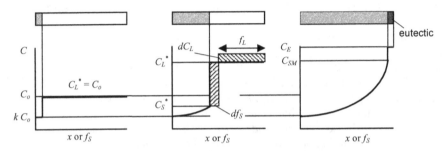

Figure 4.5. Solute redistribution for non-equilibrium solidification for $D_S = 0$ and $D_L = \infty$.

At the beginning of solidification, the situation is identical with that for equilibrium solidification. Then, since there is no diffusion in the solid, as the solidifying liquid has increased solute, a concentration gradient will be established between the initial solid composition $k \cdot C_o$ and the solid composition at the intermediate time (or temperature, T^*), which is C_S^*. In the liquid, the composition is homogeneous and equal to $C_L^* > C_o$, since diffusion is very rapid. The composition of the solid will continue to grow to the end of solidification, and will finally reach the maximum solubility in the solid solution on the phase diagram, C_{SM}.

    To find an equation for the solid composition as a function of the solid fraction, material balance equations must be used. In the original derivation by Gulliver (1913) and Scheil (1942) a material balance at the interface was written:

solute rejected when df_S is formed = solute increase in liquid

that is:

$$\left(C_L^* - C_S^*\right)df_S = \left(1 - f_S\right)dC_L \tag{4.11}$$

Since $C_L = C_S/k$, and $dC_L = dC_S/k$, on integration this equation becomes:

$$\int_0^{f_S} \frac{df_S}{1-f_S} = \frac{1}{1-k}\int_{kC_o}^{C_S} \frac{dC_S}{C_S} \quad \text{or} \quad C_S = kC_o\left(1 - f_S\right)^{k-1} \tag{4.12}$$

This is known as the *Scheil* (more recently *Gulliver-Scheil*) equation, or the non-equilibrium lever rule. Note that, for $f_S = 1$ this equation calculates $C_S = \infty$. This is of course absurd for alloy solidification. The composition of the solid can only increase to the maximum solid solubility, C_{SM}, and that of the liquid to the eutectic composition, C_E. As solidification proceeds, the solid composition follows the solidus line from $k\,C_o$ to C_{SM} and then to C_E (Figure 4.3). The Gulliver-Scheil equation can also be derived from overall mass balance (Rappaz and Voller, 1990).

4.1.3   No diffusion in solid, limited diffusion in liquid

The basic assumptions are: $D_S = 0$ and $0 < D_L < \infty$. The solute redistribution for this case is shown in Figure 4.6. A diffusion layer will exist ahead of the interface, and equations that are more complicated are used to calculate the liquid and solid composition. Three distinctive zones are seen:

I.   the initial transient, between T_L and T_S: because of the boundary layer
II. the steady state, at T_S
III. the final transient, between T_S and T_E: buildup of solute occurs because the boundary layer reaches the end of the crucible

The initial and final transient represent chemical segregation. The shaded areas in Figure 4.6 must be equal to conserve mass balance, so that the average composition remains C_o.

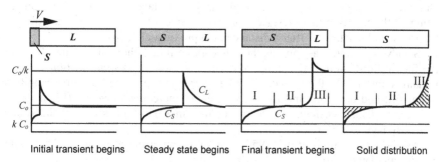

Figure 4.6. Solute redistribution for non-equilibrium solidification for $D_S = 0$ and $0 < D_L < \infty$.

Calculation of solute redistribution during the *initial transient* can be done by using the time dependent form of the diffusion Eq. (4.3). The solution of this equation is:

$$C_S = C_o\left[1-(1-k)\exp\left(-k\frac{V}{D}x\right)\right] \tag{4.13}$$

Alternatively, solute flux balance can be used to derive the equation for the initial transient, as demonstrated in the inset.

Derivation of the initial transient equation from solute flux balance (Kurz and Fisher, 1989). During non-steady state solute flux balance at the interface gives:

solute accumulation in boundary layer = solute rejected by solid - solute diffusing in liquid

or: $\dfrac{dC_L^*}{dt}\delta_c = V\left(C_L^* - C_S^*\right) - \left[-D\left(\dfrac{\partial C}{\partial x}\right)_{x=0}\right]$

During steady state there is no accumulation of solute in the boundary layer. The compositional gradient can be calculated from Eq. (4.5), or it can be approximated as:

$\left(\dfrac{\partial C}{\partial x}\right)_{x=0} = \dfrac{C_o - C_L^*}{\delta_c} = \left(C_o - C_L^*\right)\dfrac{V}{D}$ Substituting in the flux balance equation:

$\dfrac{dC_L^*}{dt}\delta_c = V\left(C_o - kC_L^*\right)$ and, since $\delta_c = D/V$ $\dfrac{dC_L^*}{C_o - kC_L^*} = \dfrac{V^2}{D}dt$

or, since $V = dx/dt$: $\displaystyle\int_{C_o}^{C_L^*}\dfrac{dC_L^*}{C_o - kC_L^*} = \dfrac{V}{D}\int_0^x dx$. Then, after integration: $\ln\left[\dfrac{C_o(1-k)}{C_o - kC_L^*}\right]^{1/k} = \dfrac{V}{D}x$

Rearranging, we obtain Eq.(4.13).

During *steady state* solidification, the planar S/L interface grows at T_S. The composition of the liquid at the interface is $C_L^* = C_o/k$ and then decreases according to Eq. (4.5), and reaches C_o after a distance of approximately $2D_L/V$.

Steady state exists as long as there is enough liquid ahead of the interface for the forward diffusion of the solute to occur, and as long as solidification velocity remains constant. As the boundary of the sample is approached the first condition is not fulfilled anymore and the solute content increases above C_o (Figure 4.7). This is the *final transient*. The length of the final transient is that of the solute boundary layer, D_L/V.

As shown by Smith *et al.* (1955), the solid composition in the final transient can be calculated with:

$$\dfrac{C_S}{C_o} = 1 + 3\dfrac{1-k}{1+k}\exp\left(-\dfrac{2Vx}{D}\right) + 5\dfrac{(1-k)(2-k)}{(1+k)(2+k)}\exp\left(-\dfrac{6Vx}{D}\right) +$$

$$+ \ldots + (2n+1)\dfrac{(1-k)(2-k)..(n-k)}{(1+k)(2+k)..(n+k)}\exp\left[-\dfrac{n(n+1)Vx}{D}\right] \tag{4.14}$$

where $x = 0$ at the end of the specimen.

So far, it was assumed that the solidification velocity is constant during solidification. However, this condition does not hold in most solidification processes. Notable exceptions are controlled directional solidification and crystal growth. If the solidification velocity V is suddenly increased ($V > V_o$), the diffusion layer decreases, which means that the amount of solute transported forward decreases. Conservation of solute atoms requires then an increase in the composition of the solid, and a band rich in solute (positive segregation) is formed, as shown in Figure 4.8.

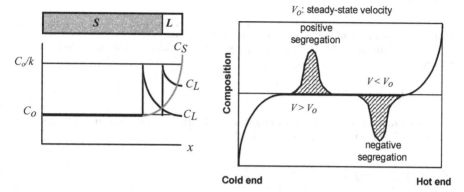

Figure 4.7. Solute accumulation during the final transient.

Figure 4.8. Formation of positive and negative segregation when solidification velocity is different from steady state velocity.

On the contrary, when the solidification velocity is suddenly decreased ($V < V_o$), a band poor in solute (negative segregation) is formed. If V or δ_c varies periodically, then periodical composition changes are produced. They are called *banding*.

4.1.4   Limited diffusion in solid, complete diffusion in liquid

The first model that attempted to describe this problem was proposed by Brody and Flemings (1966). The basic assumptions of the model included $D_S > 0$, $D_L \to \infty$, and some back-diffusion of solute in the solid occurs at the interface. To solve the mass balance equation an additional assumption was necessary, *i.e.*, $V = ct$. Two cases can be considered:

- linear growth: $f_S = t/t_f = x/l$
- parabolic growth: $f_S = \sqrt{t/t_f}$

where t_f is the final solidification time. Unidirectional solidification typically imposes linear growth in the specimen. Solidification of dendrites is commonly assumed to follow parabolic growth.

For linear growth, it was shown that:

$$C_S = kC_o\left(1 - \frac{f_S}{1 + k\alpha}\right)^{k-1} \tag{4.15}$$

For parabolic growth the equation is (see inset for derivation):

$$C_S = kC_o\left[1 - (1 - 2\alpha k)f_S\right]^{(k-1)/(1-2\alpha k)} \tag{4.16}$$

In these equations α is the dimensionless back-diffusion coefficient:

$$\alpha = 2D_S t_f / l \tag{4.17}$$

Simplified derivation of the *Brody-Flemings* equation (Kurz and Fisher, 1989).
When examining the shaded areas in Figure 4.9, it is seen that mass balance at the interface requires:

$$A_1 = A_2 + A_3 \qquad A_1 = (C_L - C_S)\, dx \qquad A_2 = (l - x)\, dC_L \qquad A_3 = (\delta_S / 2)\, dC_S$$

The approximation that allows writing this last equation is shown in Figure 4.9. Since $x/l = f_S$ and $dx/l = df_S$, after substituting and dividing by l:

$$(C_L - C_S)df_S = (1 - f_S)dC_L + \frac{\delta_S}{2l}dC_S \tag{4.18}$$

where $\delta_S = 2D_S/V = 2D_S/(dx/dt)$. If linear growth is considered, since $dx/dt = l/t_f$, we have $\delta_S = 2D_S t_f / l$. Then, substituting in Eq. (4.18):

$$C_S(1 - k)df_S = (1 - f_S)dC_S + kdC_S \frac{D_S t_f}{l^2}$$

A dimensionless solid-state back-diffusion coefficient is defined as $\alpha = D_S t_f / l^2$. Rearranging and integrating:

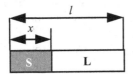

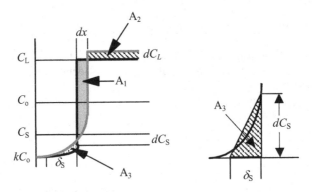

Figure 4.9. Mass balance at the interface when complete diffusion in liquid and partial diffusion in solid.

$$\frac{1}{1-k}\int_{kC_o}^{C_S}\frac{dC_S}{C_S}=\int_0^{f_S}\frac{df_S}{1+k\alpha-f_S}$$

Further manipulations produce Eq. (4.15).

Eqs. (4.15) and (4.16) have been obtained without solving the 'Fickian' diffusion. Because of that, when significant solid-state diffusion occurs, mass balance is violated. This can be understood by examining Figure 4.9. Mass balance for the boundary layer δ_S is correctly described by the equation given for A_3 only as long as the boundary layer is smaller than the solidified region. Consequently, the application of these equations is limited to slow diffusion when the boundary layer is small.

Another problem is the solutal profile shown in Figure 4.9. If diffusion in solid is finite, the solutal profile should be intermediate between that predicted by Scheil and equilibrium, as shown in Figure 4.10. C_S^* should decrease which in turn will determine a lower C_L^*.

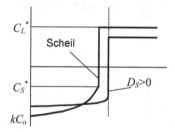

Figure 4.10. Comparison between Scheil and finite diffusion in solid profiles.

Indeed, in Eq. (4.15) for $\alpha = 0$ we have $l^2 >> D_S\, t_f$, and this equation reduces to the Scheil equation. However, for the equilibrium condition which is $\alpha = \infty$, $l^2 << D_S\, t_f$. This gives $C_S = k\, C_o$, which is the interface equilibrium condition but not the equation for equilibrium solidification. Similarly, Eq. (4.16) reduces to the Scheil equation for $\alpha = 0$, and the lever rule is obtained for $\alpha = 0.5$. Unfortunately, $\alpha = 0.5$ does not describes the physics of equilibrium.

Clyne and Kurz (1981) have used the Brody-Flemings model and added a spline fit to match predictions by Scheil equation and the equilibrium equation for infinitesimal and infinite diffusion coefficient, respectively. This relation has no physical basis.

Kobayashi (1988) obtained an exact solution (Laguerre polynomial). Solidification rate and physical properties were considered constant. Parabolic solidification was assumed for the planar geometry. A large number of terms (20,000 for a Fo = 0.05) is required for convergence. However, calculations with the second order approximate solution were very close to the exact solution. This approximation is:

$$C_S = kC_o\xi^{\frac{k-1}{1-\beta k}}\left\{1+\Gamma\left[\frac{1}{2}\left(\xi^{-2}-1\right)-2\left(\xi^{-1}-1\right)-\ln\xi\right]\right\} \tag{4.19}$$

where $\gamma = 2\,\alpha$ for planar geometry and:

$$\xi = 1 - (1 - \beta k)f_S \qquad \beta = \frac{2\gamma}{1 + 2\gamma} \qquad \Gamma = \beta^3 k(k-1)[(1+\beta)k - 2](4\gamma)^{-1}(1 - \beta k)^{-3}$$

Note that, for $D_S = 0$ we have $\gamma = 0$, $\beta = 0$, $\Gamma = 0$, and $\xi = 1 - f_S$ and this equation reduces to the Scheil equation. In addition, for $D_S \rightarrow \infty$ it gives $C_S = k\,C_o$, which upon integration gives equilibrium composition. Kobayashi has also demonstrated that the Brody-Flemings and Clyne-Kurz solutions underestimate segregation by overestimating the effect of D_S, and are particularly inaccurate for low values of k and α.

Himemiya and Umeda (1998) developed an integral profile method that can consider all significant diffusion cases. For finite diffusion in solid and complete diffusion in liquid, a second order differential equation was obtained.

For linear growth the equation is:

$$f_S^2(1 - f_S)\frac{d^2 C_L}{df_S^2} + \left[(k-4)f_S^2 + (3k\,\alpha - 3\alpha + 2)f_S + 3\alpha\right]\frac{dC_L}{df_S} = (1-k)(2f_S + 3\alpha)C_L\,C_o$$

For parabolic growth the equation is:

$$f_S^2(1 - f_S)\frac{d^2 C_L}{df_S^2} + \left[(k-4-6\alpha)f_S + (3k\,\alpha + 6\alpha + 2)\right]\frac{dC_L}{df_S} = (1-k)(2 + 6\alpha)C_L\,C_o$$

The Runge-Kuta method was used to solve these equations.

4.1.5   Limited diffusion in solid and liquid

The Himemiya-Umeda model is applicable to this problem. However, complicated equations describing an initial value problem must be solved. A simpler analytical model proposed by Nastac and Stefanescu (1993) is only valid at the micro-scale because of some of the assumptions made during derivation. This model is described in detail in Section 7.2.

4.1.6   Partial mixing in liquid, no diffusion in solid

The segregation measured in solids is, in most cases, intermediate between that for complete mixing and no mixing. When a temperature gradient exists in the liquid, thermal convection will occur, because of the difference in density between the cold and hot metal. Therefore, mass transport is not only by diffusion but also by fluid flow. A more complicated situation must be considered, and an additional assumption is necessary.

As discussed earlier, within the diffusion layer of thickness δ (Figure 4.11), mass transport is by diffusion only, while outside it convection insures homogeneity within the liquid. In terms of hydrodynamics, the diffusion layer is stagnant.

The diffusion layer is treated by using an *effective distribution coefficient, k_{ef}*. It can be shown from boundary layer theory (Burton *et al.*, 1953) that k_{ef} is related to k by the equation:

$$k_{ef} = \frac{k}{k + (1-k)\exp(-V\delta/D_L)} \qquad \text{with} \qquad 1 \geq k_{ef} \geq k \qquad (4.20)$$

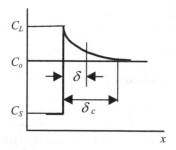

Figure 4.11. Boundary layer when convection in the liquid is assumed.

An equation similar to the Scheil equation is derived for the calculation of the solid composition as a function of fraction of solid:

$$C_S = k_{ef} C_o (1 - f_S)^{k_{ef}-1} \qquad (4.21)$$

Note that for $D_L \to \infty$, $k_{ef} = k$, and the equation for complete mixing (Scheil) is obtained. For $D_L \to 0$, $k_{ef} = 1$, $C_S = C_o$, which means that no mass transport occurs.

For limited diffusion in liquid ($0 < D_L < \infty$) and no diffusion in solid, Nastac (2004) derived the following equation for the liquid concentration profile:

$$
\begin{aligned}
C_L = C_o &+ \frac{C_o(1-k)}{2k} \exp\left(-\frac{Vx}{D_L}\right) erfc\left(\frac{x}{2\sqrt{D_L t}} - \frac{V}{2}\sqrt{\frac{t}{D_L}}\right) - \frac{C_o}{2} erfc\left(\frac{x}{2\sqrt{D_L t}} + \frac{V}{2}\sqrt{\frac{t}{D_L}}\right) \\
&- \frac{C_o(1-2k)}{2k} \exp\left[(k-1)\left(k\frac{V^2 t}{D_L} + \frac{Vx}{D_L}\right)\right] erfc\left(\frac{x}{2\sqrt{D_L t}} - V\left(\frac{1}{2} - k\right)\sqrt{\frac{t}{D_L}}\right)
\end{aligned}
$$

Finally, a graphic summary of the various solute redistribution analytical models discussed in this chapter is presented in Figure 4.12. Note that complete mixing (Scheil model) occurs when considerable convection exists in the liquid. This is the case for most directional solidification experiments performed in the earth's gravitational field. For experiments conducted in a micro-gravity environment the no-mixing model gives a more realistic description of reality.

It must be noted that, when using analytical models to evaluate segregation, it must be assumed that all physical properties are constant. The solid-state concentration can only be calculated at the interface, and cannot be modified by subsequent solid diffusion. In other words, only the trace of the solid-state concentration can be plotted. Thus, the equilibrium and Scheil model predict the solid concentra-

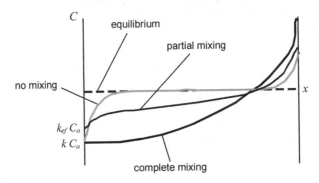

Figure 4.12 Summary of solute redistribution.

tion across the whole length as well as at the interface, while the Brody-Flemings and the Kobayashi models can only calculate the interface solid composition. The applicability of various models presented in this section can be understood by studying Application 4.4.

Many other analytical and numerical models have been proposed. Some of them will be reviewed as part of the discussion on micro-segregation.

The study of Figure 4.12 reveals that the last part of the casting to solidify is richer in solute than the initial one. This difference in composition at the macroscopic level is called *macrosegregation*. It can alter mechanical properties dramatically. It can be easily understood to occur in single crystal castings that are solidified unidirectionally. When considering three-dimensional solidification of castings with columnar structure, the middle of the casting will be richer in solute than the skin that has solidified in contact with the mold walls.

Macrosegregation is significant in large castings, but can become a factor also in small or medium size castings, when the partition coefficient is relatively high. Typical alloys that will exhibit such a behavior are some aluminum and copper alloys.

When the occurrence of macrosegregation is governed by a law such as the Scheil equation, it is termed *normal segregation*. Note that normal segregation is in fact a positive segregation. The degree of normal segregation increases as the solidification velocity, V, or the solute boundary layer, δ_c, decrease.

The occurrence of macrosegregation is more complicated than the solute diffusion models previously discussed. Other effects must be considered in order to obtain an accurate description of the process, as follows (Ohnaka, 1992):

- gravity effect on density differences caused by phase, compositional or thermal variations (natural convection)
- solidification contraction
- capillary forces
- external centrifugal or electromagnetic forces
- deformation of solid phases due to thermal stress and static pressure

Most of these effects are related to the fluid flow and will be discussed later.

4.1.7   Zone melting

The understanding of segregation phenomena has led to the development of solidi-fication techniques for metal purification. Metal purification through solidification processing can be performed in two ways:

- successive directional solidification of the alloy, and rejecting the last part after each cycle; it is not practical, but it is possible; a typical set-up is the Bridgman-type furnace in Figure 4.13a;
- moving a short molten zone along a solid bar; this is zone melting; a typical set-up is illustrate in Figure 4.13b; the liquid is held in place by a crucible or by surface tension (floating zone).

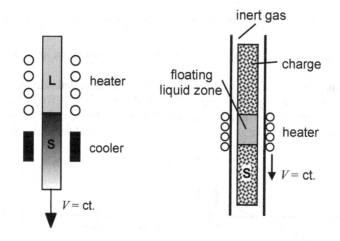

a) the Bridgman method   b) zone melting (floating zone)

Figure 4.13. Single crystal growth processes.

Let us consider a molten zone of length ℓ that is moved along the crucible (Figure 4.14). Assuming no convection in the liquid, flux balance at the interface gives $(C_o - C_S)\,dx = \ell\,dC_L$, where ℓ is the length of the molten zone, assumed to be constant and small. Integrating between kC_o and C_S, and 0 and x, we obtain the equation for the solid concentration after a single path:

$$C_S = C_o\left[1 - (1 - k)\exp\left(-\frac{k}{\ell}x\right)\right] \qquad (4.22)$$

This equation was originally derived by Pfann (1952). If convection is present, k_{ef} rather than k are used in the above equation. This distribution is true with the exception of the terminal transient region where a rapid increase in concentration occurs. The net result of the process is transport of solute from one end to the other of the crucible (Figure 4.14). The composition in the final transient can be calculated with Scheil's equation.

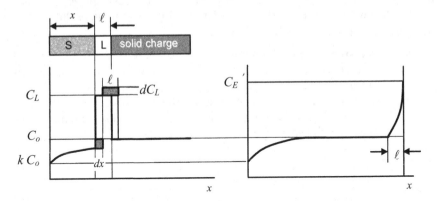

Figure 4.14. Solute redistribution during zone melting after a single path.

If several passes (directional solidification cycles) are executed, further purification becomes possible, as shown in Figure 4.15. If the partition coefficient is low, concentration is reduced fast in a few passes. On the contrary, if the partition coefficient is close to unity, a large number of passes are required for purification.

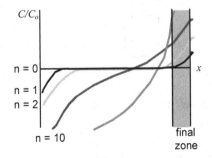

Figure 4.15. Metal purification as a result of successive passes.

Commercial applications of zone melting include refining of metals, crystal growth, and fabrication of superconductors. Depending on the goal, zone melting can be used in two different ways:

- to achieve maximum uniformity; this is called zone leveling, and one pass is sufficient;
- to achieve maximum purity; this is called zone refining; since maximum transport of solute is needed, a large number of passes is used (see Application 4.5).

4.2   Fluid dynamics during mold filling

4.2.1   Fluidity of molten metals

It is generally understood that metal flow through the mold improves as fluidity increases. But what is fluidity? For mold filling, fluidity cannot be considered simply as the inverse of viscosity, since as it flows, the metal cools and may even solidify. Consequently, the solidification pattern of the metal will greatly influence its fluidity. As shown in Figure 4.16, for pure metals the maximum length covered by the flowing metal, L_f, is attained upon complete solidification of the metal. For alloys, L_f is typically smaller, because flow will stop before complete solidification, when the viscosity of the mushy zone becomes too high.

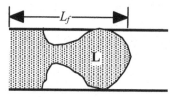
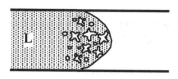

a) Pure metals and eutectic alloys   b) Single phase alloys

Figure 4.16. Solidification fronts of various alloys (after Campbell, 1991).

For pure metals or eutectic alloys, fluidity as measured by the length covered by the metal before solidifying can be calculated as $L_f = V\,t_f$. For single phase alloys the fluidity is given by $L_f = f_S^{cr} V t_f$, where t_f is the solidification time, and f_S^{cr} is the critical fraction of solid at which flow stops. The critical fraction solid corresponds to dendrite coherency and is typically at 0.2 to 0.4.

The influence of the solidification front morphology on fluidity is demonstrated through Figure 4.17. It is seen that, indeed, fluidity is maximum for the pure metal and the eutectic alloy that have very small solidification intervals, and decreases for the solid solution alloy that have large mushy zones, as predicted by the simple preceding mathematical analysis.

A correct mathematical description of fluid flow during mold filling requires solving the momentum and energy conservation equations. Mold filling models that approach this problem are out of the scope of this text. However, some comments on capillary flow are in order, since this type of flow is relevant not only for flow through the gating system, but also for flow in the mushy zone. In addition a succinct discussion on the design of gating systems will be provided.

4.2.2   Capillary flow

Young's equation describes the surface energy balance at the trijunction between a liquid droplet, its solid and its vapor phase (Figure 4.18):

$$\gamma_{SV} - \gamma_{SL} = \lambda_{LV} \cos\theta \tag{4.23}$$

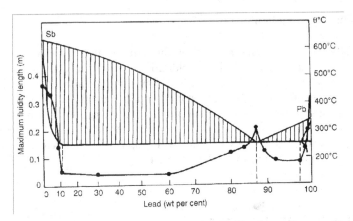

Figure 4.17. Fluidity curves superimposed on the Pb-Sb phase diagram (Portevin and Bastien, 1934).

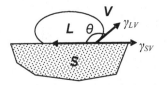

Figure 4.18. Surface energy balance for a liquid droplet on its solid.

Based on Young's equation it can be demonstrated (see Stefanescu 2002) that the pressure-drop along the mold (resistance to flow) because of metal-mold interaction, that is the resistance to flow through a channel of radius r, can be calculated with:

$$\Delta P_{\gamma} = -\frac{2}{r}\gamma_{LV}\cos\theta \tag{4.24}$$

where γ_{LV} is the surface energy between the liquid metal and its vapor phase, and θ is the contact angle between the metal and the mold. This equation can be used to describe the pressure drop during flow through capillary channels.

When the melt wets the mold (*i.e.*, $\theta < 90°$), $\Delta P_{\gamma} < 0$, which means that flow occurs spontaneously. In other words, no outside pressure is required. On the contrary, if the melt does not wet the mold (*i.e.*, $\theta > 90°$), $\Delta P_{\gamma} > 0$, and outside pressure is required for flowing.

In general, for the metal to be able to flow through the mold it is necessary that the external pressure acting on the metal be higher than ΔP_{γ}, *i.e.* $\Delta P_{ext} \geq \Delta P_{\gamma}$. The external pressure is:

$$\Delta P_{ext} = P_{met} - P_{mold} = \left(P_{atm} + P_{st} + P_{dyn}\right) - \left(P_{atm} + P_{gas}\right) = \rho g h + \rho V^2 - P_{gas} \quad (4.25)$$

where P_{met} and P_{mold} are the pressures on the liquid/air interface on the metal and mold side respectively, $P_{st} = \rho g h$ is the metallostatic pressure, $P_{dyn} = \rho V^2$ is the dynamic pressure, and P_{gas} is the pressure exercised by the gas in the mold. Thus, to have flow it is necessary that:

$$\rho g h + \rho V^2 - P_{gas} \geq -\frac{2\gamma}{r}\cos\theta \quad (4.26)$$

It is apparent from this equation that if the metal wets the mold $(\theta < 90^o)$, $\cos\theta < 1$ and the metal will flow even if $\Delta P_{ext} = 0$ (spontaneous flow). However, in most cases, the metal does not wet the mold and external pressure is required for flow. To maximize ΔP_{ext} one needs to minimize P_{gas}. Thus, mold venting is good practice.

4.2.3   Gating systems for castings

Manufacture of a sound casting is highly dependent on the correct design of the gating system. This is not a trivial task, as simple calculations with Eq. (4.40) show that the Reynolds numbers for casting alloys are considerably above 2000, which indicates turbulent flow. A schematic casting-gating system assembly is presented in Figure 4.19. Traditionally gating systems have been designed on the basis of two fundamental laws, the law of continuity and Bernoulli's theorem.

A simple analytical equation for the law of continuity can be derived from the conservation of mass, assuming steady flow (local acceleration is zero) in a system with incompressible walls. The mass entering the system is equal to the mass leaving the system, i.e., $dm/dt = \rho_1 \cdot A_1 \cdot V_1 = \rho_2 \cdot A_2 \cdot V_2$, where m is the mass, A is the cross sectional area and V is the average velocity taken normal to the cross sectional area. If it is further assumed that the flow is incompressible $(\rho = 0)$, the continuity equation becomes:

$$Q = \rho_1 \cdot A_1 \cdot V_1 = \rho_2 \cdot A_2 \cdot V_2 = ct. \quad (4.27)$$

where Q is the volumetric flow rate. The differential form of this equation that holds for both steady and unsteady flow is the conservation equation, Eq. (3.2).

Bernoulli's theorem states that the total energy of unit mass of fluid is constant throughout the system, i.e. potential energy + pressure energy + kinetic energy + frictional energy = const., or:

$$g \cdot h + P/\rho + V^2/(2g) + E_f = ct. \quad (4.28)$$

where h is the metallostatic head, P is the external pressure(atmospheric, metallostatic, applied pressure), and E_f is the friction energy.

The sequence of calculations in the design of the gating system involves the following steps:

- establish the optimum pouring time;
- calculate the choke area;
- select the gating ratio.

The pouring time is calculated with empirical equations that can be expressed in a general form as:

$$t_{pour} = k_{size} \cdot m_{cast}^{n} \tag{4.29}$$

where k_{size} is a factor depending on section size or casting weight and on the nature of the mold, m_{cast} is the mass of metal to be poured, and n is a coefficient. The factor and coefficient in this equation are empirical values available in tables or graphs.

The choke area is the area that most restricts the rate of pouring. To calculate the choke area it is first necessary to express the filling time of the mold as a function of the choke area. This will depend on the position of the casting with respect to the parting line (see Figure 4.19 and Figure 4.20). The filling time will be longer if more of the casting is in the cope because of the reduced head pressure.

If the choke area is at the base of the sprue and the part is in the drag (Figure 4.20a), the pouring time can be calculated as:

$$t_{pour} = \frac{v}{Q} = \frac{m_{cast}/\rho}{AV} = \frac{m_{cast}}{A\rho C\sqrt{2gh}}$$

where v is the volume of metal to be poured, A is the cross sectional area of the base of the sprue (choke area), and C is the efficiency factor of the gating system.

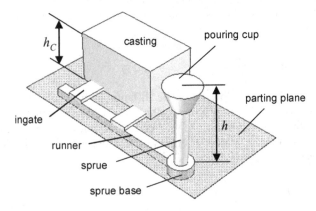

Figure 4.19. Basic components of a gating system for a horizontally parted mold. Casting is in the cope mold.

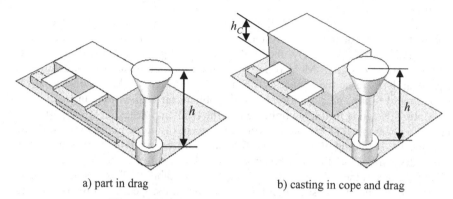

a) part in drag   b) casting in cope and drag

Figure 4.20. Possible positions of casting in mold.

When friction in the gating system is ignored $C = 1$. Then, the choke area is calculated as:

$$A_{ch} = \frac{m_{cast}}{\rho t_{pour} \sqrt{2gh}} \qquad (4.30)$$

More complicated equations are derived for the case of a casting that is positioned in the cope or partly in the cope and partly in the drag (*e.g.*, Wukovici and Metevelis 1989, Upadhya and Paul 1993), as follows:

$$A = \frac{1.5h_c m_{cope}}{\rho t_{pour} \sqrt{2g} \left[\sqrt{h^3} - \sqrt{(h - h_c)^3} \right]} \qquad \text{for part in cope} \qquad (4.31)$$

$$A = \frac{1}{\rho t_{pour} \sqrt{2g}} \left[\frac{m_{drag}}{\sqrt{h}} + \frac{1.5\, h_c\, m_{cope}}{\sqrt{h^3} - \sqrt{(h - h_c)^3}} \right] \qquad \text{for part in cope and drag} \qquad (4.32)$$

where h is the cope height, h_c is the height of the casting above the parting line (height in the cope), and m_{cope} and m_{drag} are the masses of the casting in the cope and drag respectively.

The next step is the selection of the gating ratio, which is the ratio between the area of the sprue, that of the runner, and that of the ingates, that is: $A_{sp} : A_{run} : A_{ing}$. Based on the relative values of these three components, gating systems can be classified into two categories: (i) pressurized systems with $A_{sp} > A_{run} > A_{ing}$ where the metal enters the mold at high velocity; and (ii) un-pressurized systems with $A_{sp} < A_{run} < A_{ing}$, in which passages are incompletely filled and turbulence and aspiration effects are possible. Some examples of typical gating ratios are as follows: for steel: 1:2:1.5 or 1:3:3; for gray iron: 1:4:4; for ductile iron: 4:8:3; for aluminum: 1:2:4.

A correct calculation of the gating system does not guarantee a defect-free casting, as the design of the gating system can contribute significantly to the solution of the problem. Unfortunately, the quality of the design is proportional to the experience of the designer. Mold filling simulation through numerical modeling can significantly increase the chances for a successful design, as they allow visualization of mold filling and correction of local problems through redesigning.

Gating design is further complicated by the oxidation behavior of the particular cast alloy. Campbell (1991) introduced the concept of folded oxide films (bifilms) which is the enfolding of the oxidized film on liquid surface into the bulk melt. The outer, dry surfaces of the film become opposed in the folding action, and do not bond. A crack is thereby formed in the liquid. The stability of oxides, nitrides and other film compounds is such that bifilms can remain suspended in melts for long periods. Thus, according to Campbell (2005) most metallurgical alloys are suspensions of bifilms. A more detailed discussion on this subject is found in section 12.2. Drawing on extensive practical experience and applying the bifilms concept, Campbell (2004) developed the "10 rules of casting". The rules related to fluid dynamics during mold filling are as follows:

- Good quality melts.
- Prevent liquid front damage: Maximum meniscus velocity <0.5m/sec. No top gating.
- Avoid liquid front arrests: Liquid should not stop at any point of the front, progressing only uphill in a continuous, uninterrupted advance.
- No bubble damage: Bubbles of air entrained by the filling system should not pass through the liquid metal into the mold cavity. Design the sprue and runner to fill in one pass and possibly use ceramic filters and bubble traps. Avoid the use of wells.
- No core blows: Bubbles from the out-gassing of cores should not pass through the liquid metal into the mold cavity. Further, control out-gassing of cores.
- Avoid convection damage: Avoid convective loops in the geometry of the casting; eliminate convection by rolling the mold over after filling.

4.3   Fluid dynamics during solidification

During casting solidification, significant flow of the molten metal will affect the local composition. The driving forces for fluid flow can be internal or external. The internal sources of fluid flow include shrinkage (solidification contraction) flow, natural convection, capillary forces, formation of gas bubbles, and deformation of solid phases because of thermal stress and static pressure. The external driving forces may include centrifugal and/or electromagnetic forces.

4.3.1   Shrinkage flow

During solidification the vast majority of metals and alloys shrink. Solidification contraction can be calculated as:

$$\beta = \frac{v_L - v_S}{v_L} = \frac{\rho_S - \rho_L}{\rho_S} \qquad (4.33)$$

where v_L and v_S are the liquid and solid specific volume, respectively.

The macroscopic transport equation that describes the shrinkage flow is the mass conservation (continuity) Eq. (3.2). Note that, if in this equation the change of density over time is ignored, the equation simply states that the velocity gradient must be constant. Thus, any model that attempts to describe shrinkage flow should use the continuity equation and assume different liquid and solid densities.

4.3.2   Natural convection

Natural convection is the flow that results from the effect of gravity on density differences caused by phase or solute variations in the liquid. As the liquid is colder next to the interface (thus denser) than in the bulk, a downward flow driven by the temperature gradient will occur next to the interface (Figure 4.21). As a consequence, an upward flow will occur in the bulk. When the solute rejected at the interface is denser than the solvent a downward flow, driven by the composition gradient (more solute at the interface than in the bulk liquid), will also occur next to the interface (Figure 4.21). In the opposite case, more complex flow patterns will result.

In a first simplified analysis, it can be stated that a necessary condition for stability is that the liquid has throughout a negative gradient of liquid density upward (Flemings, 1974). Thermal expansion is produced by changes in temperature and/or density. Thus, a thermal (β_T) and solutal (compositional) (β_C) expansion coefficient can be defined:

$$\beta_T = -\frac{1}{\rho} \frac{\partial \rho}{\partial T}\bigg|_C \quad \text{and} \quad \beta_C = -\frac{1}{\rho} \frac{\partial \rho}{\partial C}\bigg|_T \qquad (4.34)$$

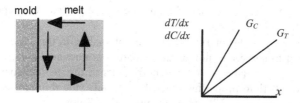

Figure 4.21. Natural convection induced by the flow of denser metal at the mold wall.

Then, since $\beta_T = -\dfrac{1}{\rho}\dfrac{\partial \rho}{\partial x}\dfrac{\partial x}{\partial T}$ and $\beta_C = -\dfrac{1}{\rho}\dfrac{\partial \rho}{\partial x}\dfrac{\partial x}{\partial C}$ the gradient of liquid density is:

$$\frac{\partial \rho_L}{\partial x} = -\rho_L\left(\beta_T\,\frac{\partial T}{\partial x} + \beta_C\,\frac{\partial C_L}{\partial x}\right)$$

A more complete analysis requires expansion of the momentum equation, Eq. (3.5). The source term S_m will include the hydrostatic pressure gradient, ∇P, an additional body force term $\rho\,g$ induced by gravitational acceleration, and an additional viscous term S_{visc}:

$$S = -\nabla P - \rho g - S_{visc} \tag{4.35}$$

For the time being, we will ignore the viscous term, S_{visc}. Then, since $\rho = \rho_L$, and assuming that the body force is oriented in the y direction, the momentum equation, Eq. (3.8) in the y direction becomes:

$$\frac{\partial}{\partial t}(\rho V^y) + \nabla\cdot(\rho \mathbf{V}V^y) = \nabla\cdot\nabla(\mu V^y) - \nabla P - \rho g \tag{4.36}$$

For a static fluid this equation simplifies to $\nabla P_0 + \rho_0\,g = 0$, where P_0 is the hydrostatic pressure corresponding to density ρ_0 and temperature T_o (in this case the liquidus temperature T_L). Combining these last two equations:

$$\frac{\partial}{\partial t}(\rho V^y) + \nabla\cdot(\rho \mathbf{V}V^y) = \nabla\cdot\nabla(\mu V^y) - \nabla(P - P_o) - (\rho - \rho_o)g \tag{4.37}$$

To solve this equation it is now necessary to formulate the pressure gradient, ∇P, and the density as a function of composition and temperature.

Several methods have been proposed for the calculation of the unknown pressure field. We will only introduce the principles of the *Semi-Implicit Method for Pressure-Linked Equations*, SIMPLE (for details see Patankar, 1980). In this method, the correct pressure, P, is assumed to be described by $P = P^* + P'$, where P^* is the guessed pressure and P' is the pressure correction. Then, a sequence of calculation is performed to evaluate in order guessed values for V^x, V^y, V^z, then P', then P, then the corrected values for V^x, V^y, V^z. The corrected pressure P is then treated as the new guessed pressure P^*, and the whole procedure is repeated until a converged solution is obtained.

For certain alloys data for $\rho = f(C,T)$ are obtainable. In this case, all the information required to solve the momentum equation is now available. If this data is not available, we can make use of the coefficients of thermal expansion. Ignoring the variation of all fluid properties other than density (the *Boussinesq approximation*) the last term of Eq. (4.37) can be formulated as follows:

$$d\rho = \frac{\partial \rho}{\partial C}\bigg|_T dC + \frac{\partial \rho}{\partial T}\bigg|_C dT = \rho\left[-\beta_C(C - C_o) - \beta_T(T - T_o)\right] \qquad (4.38)$$

where C_o is the initial composition. In general, β_C and β_T vary with both density and temperature. Assuming the coefficients of thermal expansion to be constant over certain ranges of temperature and composition, the previous equation can be integrated to $\rho = \rho_o \exp\left[-\beta_C^o(C - C_o) - \beta_T^o(T - T_o)\right]$, where T_O is the liquidus temperature, and C_o is the liquid concentration at the liquidus temperature. Using only the first two terms of the series expansion, we have $\rho = \rho_o\left[1 - \beta_C^o(C - C_o) - \beta_T^o(T - T_o)\right]$, where β_C^o and β_T^o have the same expressions as above with ρ substituted by ρ_o. Both are assumed constant throughout solidification. Then, the buoyancy source term in Eq. (4.37) becomes ((the *Boussinesq approximation for natural convection*):

$$S_b = \rho_o g\left[\beta_C^o(C - C_o) + \beta_T^o(T - T_o)\right] \qquad (4.39)$$

A characteristic temperature difference, ΔT, and a characteristic length scale, l_b, can be defined to characterize the horizontal thermal gradient, and consequently the horizontal density gradient. In addition, a characteristic concentration difference, ΔC, can be defined. The resulting nondimensional parameters are:

Thermal Grashof number: $\mathrm{Gr}_T = g\,\beta_T\,\Delta T\,l_b^3\,v^{-2}$

Solutal Grashof number: $\mathrm{Gr}_C = g\,\beta_C\,\Delta C\,l_b^3\,v^{-2}$

Prandtl number: $\mathrm{Pr} = v\,\alpha^{-1}$

Schmidt number: $\mathrm{Sc} = v\,D^{-1}$

where α is the thermal diffusivity, μ is the dynamic viscosity, and v is the kinematic viscosity. The Rayleigh number can be defined as the sum of the thermal and the solutal Rayleigh numbers as follows:

$$\mathrm{Ra} = \mathrm{Ra}_T + \mathrm{Ra}_C = \mathrm{Gr}_T\cdot\mathrm{Pr} + \mathrm{Gr}_C\cdot\mathrm{Sc} \qquad (4.40)$$

The Rayleigh number compares the buoyancy forces to the viscous forces. For small Rayleigh numbers the viscous forces dominate. The conduction regime is maintained with negligible thermosolutal convection effects on macrosegregation. Above a critical Rayleigh number (~2000) buoyancy forces become important, a convection regime is established, and the effect of thermosolutal convection on macrosegregation becomes significant (Nastac 2004).

From this analysis it is clear that a model that attempts to include the effect of fluid on segregation must include at least the density difference in the continuity equation and the body force term in the momentum equation.

4.3.3   Surface tension driven (Marangoni) convection

When free or deformable interfaces exist in the system (*e.g.* liquid-liquid, or liquid-gas interfaces) the temperature and compositional gradient will impose a surface tension gradient along the interface, which exerts a shear stress on the fluid. This induces a flow toward regions with higher values of γ termed *Marangoni convection*. The force balance can be written as (Shy *et al.* 1996):

$$\nabla(\mathbf{V}\cdot\mathbf{t})\cdot\mathbf{n} = \frac{\partial\gamma}{\partial T}\nabla T\cdot\mathbf{t} \tag{4.41}$$

where $\mathbf{n}$ is the unit normal vector to the free surface, and $\mathbf{t}$ is the unit tangent vector to the free surface. This equation states that the surface tension gradient is proportional to the temperature gradient and gives rise to the normal derivative of the tangential velocity at the free surface. To describe surface convection strength the nondimensional Marangoni number can be used:

$$\mathrm{Ma} = \left|\frac{d\gamma}{dT}\right|\frac{\Delta T\, l_b}{\mu\alpha} \tag{4.42}$$

The governing equation describing both the thermal and solutal Marangoni convection is:

$$\tau = \mu\left(\frac{\partial V_y}{\partial y}\right) = \left(\frac{\partial\gamma}{\partial T}\right)\left(\frac{\partial T}{\partial x}\right) + \sum_i\left(\frac{\partial\gamma}{\partial a_i}\right)\left(\frac{\partial a_i}{\partial x}\right)$$

where τ is the shear stress caused by the surface tension gradients, V_y is the velocity component parallel to the surface, x and y are the coordinates parallel and perpendicular to the surface, and a_i is the thermodynamic activity of alloying element i.

4.3.4   Flow through the mushy zone

The mushy zone is the region where solid and liquid coexist as a mixture. As long as the equiaxed dendrites are free to flow with the liquid, it may be assumed that the flow velocity is affected only by the change in viscosity. However, when dendrite coherency is reached and a fixed solid network is formed, or if the dendrites are columnar, the flow will be considerably influenced by the morphology of the mushy zone. In most solidification models, the flow through the mushy zone is treated as flow through porous media and *Darcy's* law is used for its mathematical description. The standard form of Darcy's law as applied to flow through a fixed dendritic network is (*e.g.*, Poirier 1987):

$$V_L = - (K/\mu g_L)(\nabla P - \rho g) \qquad (4.43)$$

where V_L is the velocity of the interdendritic liquid, K is the specific permeability of the mushy zone, and g_L is the volume fraction of interdendritic liquid. Darcy's law is valid under the following assumptions (Ganesan and Poirier, 1990):

- slow flow ($V_L \rightarrow 0$); this allows ignoring inertial effects
- steady flow
- uniform and constant volume fraction of liquid
- negligible liquid-liquid interaction forces

The permeability must be now defined. Two models may be used to this purpose: the *Hagen-Poiseuille* model or the *Blake-Kozeny* model.

The Hagen-Poiseuille Model

Following the analysis by Poirier (1987), the Hagen-Poiseuille law gives the velocity of an incompressible fluid under laminar flow conditions through a tube as:

$$V = -\frac{r^2}{8\mu}\left(\frac{\partial P}{\partial y} - \rho g\right) \qquad (4.44)$$

where the flow is in the direction of gravity (the y-direction), and r is the tube radius. Applying this law to the flow through the interdendritic network we derive:

$$V = -\left(g_L \lambda_I^2/8\pi \mu\right)\left(dP/dy - \rho g\right)$$

where λ_I is the primary dendrite arm spacing (DAS). Comparing this equation with Eq. (4.43) the permeability is derived to be:

$$K = C_1 \lambda_I^2 g_L^2 \qquad (4.45)$$

where C_1 is a parameter that depends on the geometry (tortuosity) of the flow channels. For Pb-Sn alloys, experimental data give $3.75 \cdot 10^{-4}$ for this parameter.

The Blake-Kozeny Model

For flow through porous media in the vertical direction the Blake-Kozeny equation gives:

$$V = -C_2 \frac{d^2 g_L^3}{\mu(1 - g_L)^2}\left(\frac{dP}{dy} - \rho g\right) \qquad (4.46)$$

where d is a characteristic dimension of the solid phase. For spheres, d is the diameter of the sphere.

By comparing Eqs. (4.43) and (4.46) the permeability is (Poirier, 1987):

$$K = C_2 d^2 g_L^3 / (1 - g_L)^2$$

Assuming further that the characteristic dimension is related to the volume fraction of liquid by $(1 - g_L) \propto (D/\lambda_l)^2$, the final equation for permeability is:

$$K = C_2 \lambda_l^2 g_L^3 / (1 - g_L) \tag{4.47}$$

For the Pb-Sn system, C_2 is of the order of $1.43 \cdot 10^{-3}$.

4.4   Macrosegregation

Gravity plays an important role in the formation of segregation. Settling or flotation of liquid or solid phases having a different composition, and therefore a different density than the bulk liquid, will produce *gravity segregation*. Typical examples are dendrites settling at the bottom of ingots, and coarse primary lamellar graphite (kish) or spheroidal graphite floating on top of large cast iron castings. Centrifugal forces enhance the gravitational forces applied on the casting, and have significant effects on segregation.

Figure 4.22 shows the types of macrosegregation formed in a killed steel ingot (an ingot procured from a melt which has been deoxidized). The "+" and "-" signs denote positive and negative segregation, respectively. The type of segregation has also been traditionally defined in terms of the shape or the location of segregation. The streaks arranged in a V-pattern at the center of the casting are called *channel*, *centerline*, or *V-type* segregation. The *A-type* segregation, also called *freckles*, refers to the streaks oriented almost vertically in an A-pattern at the upper and outer regions of the ingot. These are all positive segregations. Negative segregation is distributed in a cone at the base of the ingot.

The main driving force of the fluid flow during solidification is *solidification contraction*. Interdendritic liquid flow can cause solute concentration to be higher than the average concentration in the earlier solidified regions. This is opposite to the distribution shown on Figure 4.12 for the beginning of solidification. This is termed *inverse segregation*. Such complex segregation patterns cannot be explained through solute diffusion alone.

The physics of macrosegregation formation can be summarized as follows. Segregation starts at the microscopic level as solidification proceeds. During solidification, solute is rejected ($k < 1$) or depleted ($k > 1$) continuously from the precipitated solid and the composition of the surrounding liquid is consequently affected. If significant concentration gradients are developed at the interface, the interdendritic liquid can be driven simultaneously by thermal and solutal buoyancy, as well as by solidification contraction. The induced flow will wash away the liquid next to the interface, resulting in segregation at the macroscopic level (macrosegregation).

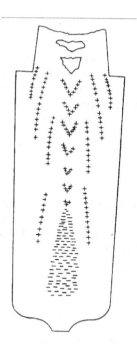

Figure 4.22. Macrosegregation in a killed steel ingot (Derge, 1964).

4.4.1   Fluid flow controlled segregation

Let us first discuss the case where segregation arises mainly from fluid flow rather than from solute diffusion at the macroscopic scale. This particular problem is of interest because the solute diffusion layer is much smaller than the typical dimension of the volume element. In other words, the volume element is open to fluid flow but closed to diffusion. However, diffusion is active within the volume element. The governing equation is that of conservation of species, *i.e.* (3.4). Assuming no source term, fast solute diffusion ($\nabla C = 0$) and constant density, the conservation equation reduces to:

$$\frac{\partial C}{\partial t} + \nabla \cdot (\mathbf{V}C) = 0 \qquad\qquad (4.48)$$

The second term represents the solute transport that is associated with fluid flow. Assuming that the solid does not move (*e.g.*, fixed dendrite skeleton) the average velocity within the volume element can be obtained from the real velocity of the liquid phase, V_L: $V = (1 - f_S) V_L$. Note that:

- for $f_S = 0$ $V = V_L$
- for $f_S = 1$ $V = 0$

The average composition over the volume element is:

$$C = \int_v C_L \, df_L + \int_v C_S \, df_S \tag{4.49}$$

The assumptions made on the type of solute redistribution will determine the form of C in Eq.) and the form of the governing fluid flow controlled segregation (macrosegregation) equation obtained when substituting C in Eq. (4.48) (Rappaz and Voller, 1990). Assuming, for example, complete mixing in both liquid and solid, *i.e.* the lever rule, we obtain the macrosegregation equation:

$$\nabla(VC_L) + \frac{\partial C_L}{\partial t} - \frac{\partial[f_S(1-k)C_L]}{\partial t} = 0 \tag{4.50}$$

Coupling between the thermal and solutal field is done through the liquidus line in the phase diagram, $C_L(t)$. Since V is unknown, this equation must be coupled with the momentum equation. If complete mixing in the liquid and no diffusion in solid are assumed (Scheil model), Eq. (4.49) takes the form:

$$C = C_L f_L + \int_v C_S \, df_S = C_L(1 - f_S) + \int_0^{f_S} C_S(f_S) \, df_S$$

Substituting in Eq. (4.48) and after derivation with respect to time we obtain:

$$\nabla(VC_L) + \frac{\partial[(1-f_S)C_L]}{\partial t} + C_S^* \frac{\partial f_S}{\partial t} = 0 \tag{4.51}$$

Eliminating C_S^* by using the equilibrium condition is no longer trivial for an open system.

4.4.2   Fluid flow /solute diffusion controlled segregation

From the preceding discussion, it follows that macro-scale segregation is controlled simultaneously by fluid flow and solute diffusion. Accordingly, macrosegregation models must describe both phenomena. This is a rather complicated problem, since fluid flow through the liquid /solid mixture must be described.

Analytical models cannot tackle the intricacies of flow through the mushy zone. Nevertheless, they contribute to the understanding of the physics of macrosegregation. A first analytical model for macrosegregation was proposed by Flemings and Nereo (1967). First, Eq. (4.48) was written as:

$$\nabla(\rho_L g_L VC_L) + \partial(\bar{\rho}\bar{C})/\partial t = 0$$

where $\bar{\rho}$ and $\bar{C}$ are the average density and composition, respectively, in the volume element. Manipulation of this equation yields:

$$\frac{\partial C_L}{\partial t} = -\left(\frac{1-k}{1-\beta}\right)\frac{C_L}{g_L}\frac{\partial g_L}{\partial t} - \mathbf{V}\cdot\nabla C_L \tag{4.52}$$

where β is given by Eq.(4.33). This is the local solute redistribution equation used to calculate macrosegregation (Flemings, 1974). It describes the influence of shrinkage flow on the composition of the solid at each fraction solid. However, it does not include the effect of natural convection.

A slightly modified form of this equation was suggested by Ohnaka (1992):

$$\frac{\partial C_L}{\partial t} = \left(\frac{1-k}{1-\beta}\right)\left(1 + A - \frac{V_n}{dx/dt}\right)^{-1}\frac{C_L}{1-g_S}\frac{\partial g_S}{\partial t} \tag{4.53}$$

where V_n is the flow velocity normal to the isotherms (equivalent to V_L) and dx/dt is the velocity of the isotherms. Also $A = 0$ for no diffusion in solid and $A = k\,g_S\,(1 - \beta)^{-1}\,(1 - g_S)^{-1}$ for complete diffusion in solid. Since:

$$\mathbf{V}\cdot\nabla C_L = V_L \frac{\partial C_L}{\partial t}\frac{\partial t}{\partial x} = -\frac{V_L}{V_S}\frac{\partial C_L}{\partial t} \quad \text{(for 1D)}$$

where V_S is the solidification velocity, Eq. (4.52) can be integrated to give:

$$C_S = kC_o\left(1-g_S\right)^{\frac{k-1}{(1-\beta)\left[1-V_L/V_S\right]}} = kC_o\left(1-g_S\right)^{\frac{k-1}{\xi}} \tag{4.54}$$

Analysis of this equation suggests three possible scenarios:

a) $\xi = 1$, e.g., no solidification shrinkage ($\beta = 0$) and no flow ($V_L = 0$); the equation becomes Scheil equation, and negative macrosegregation ($C_S < C_o$) is calculated for the first part of solidification, while positive segregation ($C_S > C_o$) is predicted for the second part of solidification;

b) $0 < \xi < 1$, e.g., no flow ($V_L = 0$), C_S is larger than predicted by the Scheil equation, which means that a strong tendency toward positive segregation exists. For example, since at the mold wall $V_L = 0$, positive segregation, i.e. inverse segregation, can occur next to the mold wall. This is illustrated Figure 4.23 for the case of directional solidification against a chill;

c) $\xi > 1$ or $V_L/VS < -\beta/(1-\beta)$, C_S is smaller than predicted by the Scheil equation, which means that a strong tendency toward negative segregation exists.

The morphology of the mushy zone will influence the flow during solidification in a complex manner. Two main cases may be considered. The first one involves the flow of the liquid through a fixed solid network ($V_S = 0$). This will be a good approximation for columnar solidification, or for equiaxed solidification after dendrite coherency is reached. This flow can be treated as flow through a porous medium (Darcy flow).

The second case involves the flow through a solid /liquid mixture when the solid can move with the liquid ($V_S = V_L$). This will be true for equiaxed solidifica-

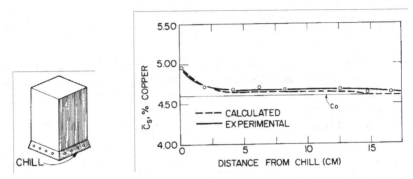

Figure 4.23. Macrosegregation in a directionally solidified Al-4.6% Cu alloy (Flemings, 1974).

tion *before coherency*. Here, the concept of *relative viscosity* is applied to describe the viscosity of the liquid /solid mixture. The relative viscosity can be calculated with:

$$\mu^* = \mu\,(1 + \varphi) \tag{4.55}$$

where φ is a correction factor. Numerous expressions were proposed for the correction factor (see review by Kaptay, 2000). Selected expressions for the correction factor are given in Table 4.1. In this table, f_S is the fraction solid, and f_S^{cr} is the critical fraction solid at which coherency is reached.

Table 4.1. Equations for calculation of the correction factor in the relative viscosity equation

Correction factor φ	Reference
$2.5\,f_S$	Einstein
$2.5 \cdot f_S + 10.05 \cdot f_S^2 + 0.00273 \cdot \exp(16.6 \cdot f_S)$	Thomas (1965)
$\left(1 - f_S / f_S^{cr}\right)^{-2.5 f_S^{cr}} - 1$	Krieger (1972)

The complex macrosegregation problem can only be solved numerically. This will be discussed in some detail in Chapter 6.

4.5   Fluid dynamics during casting solidification - macroshrinkage formation

During cooling and solidification in the mold most metals and alloys shrink. The combined effect of metal shrinkage and mold behavior during casting solidification dictates casting soundness. Improper management of heat flow may result in casting defects such as cold shuts or shrinkage defects. These defects are responsible for considerable financial loss in the metal casting industry.

4.5.1   Metal shrinkage and feeding

It is convenient to distinguish three types of shrinkage: liquid, solidification, and solid shrinkage (Figure 4.24). *Liquid shrinkage*, occurring from the pouring temperature to the liquidus temperature, is usually compensated by flow of liquid from the gating systems and the risers. *Solidification shrinkage* can also be compensated through liquid feeding from the risers. However, since feeding channels may be interrupted during solidification before all parts of the casting are fully solid, local shrinkage cavities may occur. *Solid shrinkage* (also called *patternmaker shrinkage*) is accommodated by allowing suitable corrections to the dimensions of the pattern used for making the mold. It is apparent that because of liquid and solidification shrinkage a mass deficit may result in certain regions of the casting. This mass deficit translates into shrinkage cavities that may cause rejection of the castings. Therefore, it is important to understand and control the feeding of regions of mass deficit. The feeding mechanisms are summarized in Figure 4.25 as suggested by Campbell (1969), for alloys with freezing range.

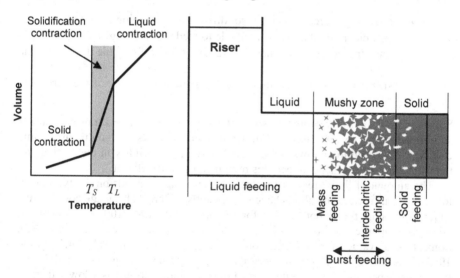

Figure 4.24. Shrinkage regimes.   **Figure 4.25.** Mechanisms of feeding.

Let us try to evaluate the feeding velocity, V_f, which is the average velocity of the mass moving to fill the mass deficit. As explained earlier (see Eq. (3.7) it can be expressed as $V_f = f_L V_L + f_S V_S$, where V_S and V_L are the velocity of the solid and liquid, respectively.

During *liquid feeding*, which occurs before the beginning of solidification, $f_S = 0$ and thus $V_f = V_L$. When solidification starts, solid particles (grains) form in the liquid. As long as these particles are not in contact with one another, that is when $f_S < f_S^{cr}$, it may be assumed that the solid moves with the liquid ($V_S = V_L$), and the metal behaves like a slurry (semisolid). Its relative viscosity is increased (fluidity is decreased). Because of this increased viscosity, during *mass* (semisolid) *feeding* the flow velocity decreases to $V_f < V_L$.

As solidification proceeds, dendrite coherency (i.e. a rigid network of contiguous dendrites) will occur when $f_s < f_s^{cr}$, and a fixed solid network will form. Then, since $V_S = 0$, the feeding velocity becomes $V_f = f_L \cdot V_L$, meaning a further decrease in feeding. Only *interdendritic feeding* is possible at this point.

It has been suggested that the dendritic network collapses during solidification, causing a redistribution of liquid and solid, which has been termed *burst feeding* (Campbell, 1969). More recent research by Dahle *et al.* (1997, 1999) seems to confirm this hypothesis. Indeed, their measurements suggest that interdendritic fluid flow can develop stresses in the mushy region that are of similar magnitude to the shear strength of the interdendritic network.

Once solidification is complete and $f_S = 1$, only limited solid feeding through elastic and plastic deformation of the metal is possible.

For effective feeding to occur during solidification four main requirements must be satisfied:

- a feeding source (riser) that solidifies after the region to be fed;
- sufficient liquid must be available to feed the shrinkage;
- unrestricted feeding channels (path of flow from the feeder to the shrinkage);
- sufficient pressure on the liquid to make it flow toward the shrinkage region.

Satisfaction of the first requirements is dictated by the overall heat transport during solidification and will be discussed in detail in chapter 5 as the solidification time criterion. This criterion allows finding the last region to solidify in the casting and then attaching a riser that can feed this region and solidify after it.

The amount of liquid required to feed the shrinkage depends on the type of alloy, as various metals and alloys have significantly different shrinkage coefficients. Liquid shrinkage for carbon steel is for example 1.6 to 1.8%/100°C of superheat, while for graphitic cast irons it is 0.68 to 1.8%/100°C (Plutshak and Suschil, 1988). Some typical values for solidification shrinkage are given in Table 4.2 (Flinn, 1963). It is seen that for graphitic cast iron expansion may occur during solidification. This is because the graphite formed during solidification has a lower density than the liquid from which it is formed.

Table 4.2. Solidification contraction of various metals and alloys

Material	Volumetric solidification contraction, %	Material	Volumetric solidification contraction, %
carbon steel	2.5 to 3	Cu-30%Zn	4.5
1% carbon steel	4	Cu-10%Al	4
white iron	4 to5.5	aluminum	6.6
gray iron	-2.5 (expansion) to 1.6	Al-4.5%Cu	6.3
ductile iron	-4.5 (expansion) to 2.7	Al-12%Si	3.8
copper	4.9	magnesium	4.2
		zinc	6.5

Gray and ductile iron expand during solidification because of graphite precipitation, and when poured in non - rigid green sand molds an additional 15% feed metal requirement above that needed to satisfy the calculated liquid and solidification shrinkage may be required (Plutshack, 1988). In copper-base alloys an additional 1% volumetric shrinkage may be expected under similar conditions.

The third requirement amounts to efficient feeding channels. The efficiency of the feeding channels is affected by the type of alloy as well as by the geometry of the casting. The type of alloy influences the width of the mushy zone (the solidification interval). Wide mushy zone alloys ($T_L - T_S > 110°C$) that solidify typically with equiaxed grains, rely heavily on semisolid and interdendritic feeding. Thus their feeding velocity is small and significant difficulties are experienced in feeding the numerous tortuous channels. The resistance to flow is relatively high. Alloys with narrow mushy zone ($T_L - T_S < 50°C$), that exhibit columnar structure, rely mostly on liquid feeding, and therefore their feeding velocity is high. They are called *skin forming alloys*.

The local geometry of the solidifying volume can also affect significantly the feeding efficiency. Consider the solidifying plate in Figure 4.26a. There is no temperature gradient along the plate, as heat is only conducted perpendicular to the plate sides. Parallel solidification fronts will move from the mold wall to the center of the casting. The flow of liquid metal is gradually restricted, because liquid feeding is gradually replaced by interdendritic feeding. Eventually the feeding channel will be closed and porosity, known as dispersed centerline shrinkage, will occur between the dendrites. In the case of a solidifying wedge (Figure 4.26b), a steep temperature gradient from the center to the edge of the casting exists. Liquid feeding is possible until the end of solidification.

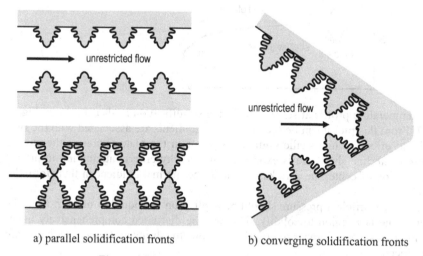

a) parallel solidification fronts   b) converging solidification fronts

Figure 4.26. Plate- and wedge-type solidification.

Consider now the case of the L-shaped casting presented in Figure 4.27. At the corners of the casting either convergent or divergent heat flow may occur. When the heat flow is divergent solidification will occur at a faster rate, since heat is lost faster (see also Figure 5.7). The contrary is true for convergent heat flow which results in the formation of hot spots. From the study of the drawing it can be seen that at the extremity of the plate-casting, as well as at the bottom of the riser, wedge type solidification occurs. Therefore, dispersed shrinkage is unlikely to appear in these regions. The casting exhibits an *end effect* and a *riser effect*, respectively. In the long, horizontal part of the casting, parallel solidification fronts converge toward the center of the plate, and center-line dispersed shrinkage is to be expected.

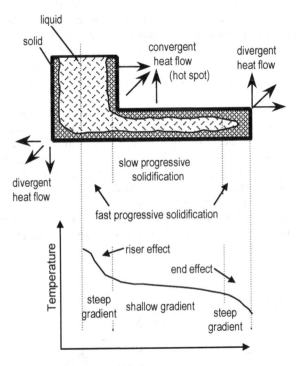

Figure 4.27. Directional (progressive) solidification in an L - shaped casting resulting from increasing temperature gradient from the extremity of the casting to the riser.

The temperature profile at some time during solidification is also shown on Figure 4.27. From this plot it can be seen that steep gradients are associated with the riser and end effects, while shallow gradients occur in the plate-region of the casting where centerline shrinkage is expected. This is why criteria attempting to predict the position of centerline shrinkage include the thermal gradient in their formulation.

Finally, sufficient pressure should be applied on the liquid to move it from the riser to the last region to solidify. This will be discussed comprehensively under section 6.5 where a number of criteria functions for shrinkage defect prediction are introduced.

4.5.2   Shrinkage defects

The soundness of the casting depends of uninterrupted flow of liquid metal to the region that solidifies to feed the mass deficit resulting from solidification contraction. Failure to feed the mass deficit will produce shrinkage defects. Since the terminology is rather ambiguous, in this text we will use the classification and definitions presented in Figure 4.28. Shrinkage defects that are open to the atmosphere (also called shrinkage cavities) are a consequence of metal contraction while cooling in liquid state and during solidification. This defect is a macro-scale defect that can be also termed *macroshrinkage*. The mass deficit produced by shrinkage is compensated by atmospheric gasses, a process that is independent of the gas content of the metal and which does not require gas pores nucleation and growth. On the contrary, closed shrinkage defects correlate well with pores nucleation and growth in the mushy zone or the amount of bifilms, and thus seem to depend on the impurity level and the amount of gas dissolved in the metal. They can be either *macroporosity* or *microshrinkage* (microporosity) defects. In summary, shrinkage cavities are driven only by metal contraction (shrinkage flow) while shrinkage porosity is driven by both metal contraction and pore nucleation and growth or bifilms growth.

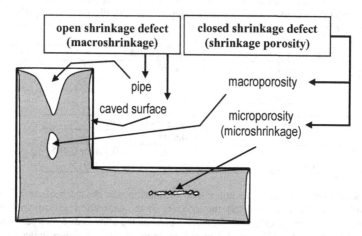

Figure 4.28. Definition and classification of shrinkage defects (Stefanescu 2005). With permission of Maney Publishing, http://www.ingentaconnect.com/content/maney/ijcmr.

From this analysis it is apparent that the accurate prediction of shrinkage cavity formation must be based on three-phase (liquid, solid, and gas) mass conservation, coupled with energy conservation. To include casting distortion stress must be also modeled. This is a problem that has not yet found a satisfactory answer. The state of the art in modeling of macroshrinkage will be presented in chapter 6, while that of shrinkage porosity in chapter 14.

4.6   Applications

Application 4.1

Calculate the amount of eutectic at the end of solidification in an Al-51wt%Zn alloy. Hint: use Scheil equation and the maximum solubility in solid, 82.2wt%.

Answer:
From the Al-Zn phase diagram, at the eutectic temperature, we obtain the partition coefficient: $k = 0.822/0.95 = 0.865$. We further know that $C_o = 0.51$, and that $C_{SM} = 0.822$. The governing equation is the Scheil equation. At the eutectic temperature $C_s = C_{SM}$. Substituting into the Scheil equation gives: $C_{SM} = kC_o(1-f_s)^{k-1} = kC_o(f_L)^{k-1}$. Noting that $f_L = f_E$ we now solve for the fraction of eutectic:

$$f_E = (C_{SM}/kC_o)^{\frac{1}{k-1}} = (0.822/(0.865 \cdot 0.51))^{\frac{1}{0.865-1}} = 0.0099.$$

Application 4.2

Cylindrical steel rods, 60 mm in length, were solidified in a Bridgman directional solidification furnace at a velocity of 2.8μm/s. An induction coil was used for heating. Consequently, complete mixing in the liquid can be assumed. The composition of the steel was as follows: 0.25%C, 0.57%Cr. Solidification was interrupted by quenching the samples in water and the interface solid composition was measured using a Scanning Electron Microprobe. The following results were obtained (Pershing, 1997):

Solidified length, m	Fraction of solid	%C	%Cr
0.01	0.167	0.12	0.52
0.03	0.5	0.135	0.56
0.06	1	0.23	0.72

Element	Materials constants	
	D_S in austenite	k
C	$1 \cdot 10^{-9}$	0.196 or 0.495
Cr	$1.2 \cdot 10^{-12}$	0.915

The solid diffusivity in austenite rather than in δ-ferrite was chosen because it is the rate controlling value for diffusion near the interface in the peritectic transformation in steel. There are two partition coefficients for carbon because the alloy is of hyper-peritectic composition. The partition coefficient will change as the temperature deceases under the peritectic temperature. Calculate the solid composition using the equilibrium, Scheil, Brody-Flemings and Kobayashi models. Compare with the experimental results.

Answer:
Calculations for the equilibrium, Scheil and Brody-Flemings models are straightforward. For the Kobayashi model the local solidification time is calculated as $t_f = l/V = 2.14 \cdot 10^6$ s. Then, some terms of the equation are calculated on the Excel spreadsheet as follows:

	D_S	k	C_o	α	β	Γ
	data	data	data	Eq. (4.17)	$4\alpha/(1+4\alpha)$	from Eq. (4.19)
for C	1.00E-09	0.43	0.25	5.95E-03	2.3E-02	6.9E-05
for Cr	1.20E-12	0.915	0.57	7.14E-06	2.9E-05	3.4E-11

These terms are then used to calculate the change in composition as a function of fraction solid. Different values for the fraction of solid are used as input on the Excel spreadsheet.

A	B	C	D	H	J	N
f_S	0.01	0.100	0.167	0.500	0.700	0.980
ξ	Eq.(4.19)	Eq.(4.19)	Eq.(4.19)	Eq.(4.19)	Eq.(4.19)	Eq.(4.19)
ξ_C	0.990	0.900	0.834	0.502	0.303	0.024
ξ_{Cr}	0.990	0.900	0.833	0.500	0.300	0.020
C_S	Eq.(4.19)	Eq.(4.19)	Eq.(4.19)	Eq.(4.19)	Eq.(4.19)	Eq.(4.19)
C_S for C	0.049	0.053	0.057	0.085	0.226	0.751
C_S for Cr	0.522	0.526	0.530	0.553	0.578	0.727

Note that as the carbon content increase above 0.1% (max. solubility in the δ phase) the peritectic temperature the partition coefficient of carbon changes (see the Fe-C diagram in Appendix C). Thus, and IF statement must be used when calculating ξ or C_S. For example, for column C in the calculation of C_S: IF(C_S in column B < 0.1, Eq.(4.19) with $k = 0.196$, Eq.(4.19) with $k = 0.495$). Similar calculations are performed for equilibrium, Eq. (4.9), Scheil, Eq. (4.12) and Brody-Flemings, Eq.(4.16) models.

The calculation and experimental results are presented in Figure 4.29 for carbon and chromium, respectively. It is seen that for carbon, that has a relatively high solid diffusivity (interstitial solution), the equilibrium assumption gives the closest results. Yet, the experimental value is only predicted at 0.98 fraction solid. This is because the equilibrium equation calculates the carbon concentration at the interface only, while the experimental data include some significant solid diffusion. Also, note the change in slope on the curves, as the carbon content increases above 0.1%. The Brody-Flemings and the Kobayashi equations predict close results to the equilibrium equation.

For chromium, that has a much lower solid diffusivity (substitutional solution), the Scheil equation works well. There is no difference between the prediction of Scheil, Brody-Flemings and Kobayashi equations. This means that the solid diffusivity of chromium is negligible.

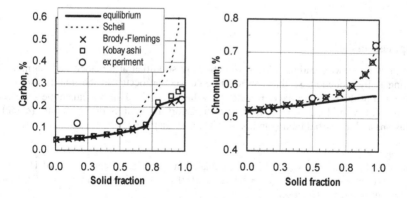

Figure 4.29. Comparison of segregation calculations with various analytical models for $V_S = 2.8 \cdot 10^{-6}$ m/s.

Application 4.3

Consider the problem in Application 4.2. Compare calculation results for equilibrium, Scheil, Brody-Flemings and Kobayashi assuming a solidification velocity of 0.028 μm/s. Discuss the differences.

Answer:
Using the same calculation procedure as in Application 4.2 we obtain the results plotted in Figure 4.29. It is seen that for carbon the Brody-Flemings and Kobayashi models are very close to the Scheil calculation. This is because the solidification velocity is now very close to the diffusion velocity ($D_S/l = 1.67 \cdot 10^{-8}$) and carbon diffusivity can not be considered infinite. However, for chromium there are no differences between the three non-equilibrium models. This is because for chromium the diffusion velocity ($2 \cdot 10^{-11}$) is still much smaller than the solidification velocity.

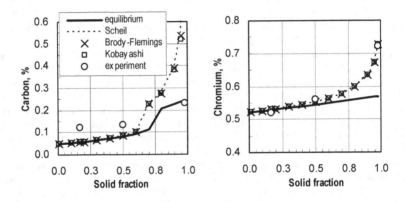

Figure 4.30. Comparison of segregation calculations with various analytical models for $V_S = 2.8 \cdot 10^{-8}$ m/s.

Application 4.4

A vanadium – 8wt% carbon alloy is directionally solidified at a velocity of 10μm/s. The heating is by induction, so that considerable convection occurs in the liquid. Microprobe evaluation of the boundary layer produced a value of $\delta = 2 \cdot 10^{-4}$m. Calculate the solid composition at 0.5 solid fraction for the case of transport in liquid by diffusion only (complete diffusion) and for partial mixing in liquid. The phase diagram is provided in Appendix C.

Answer:
From the phase diagram $k_{max} = 1.88$ is calculated at the eutectic temperature of 1653°C. A $k_{min} = 1.16$ is calculated at the temperature of 2180°C. The average value, $k = 1.52$, will be used in subsequent calculation.

For the case of complete diffusion in liquid, no diffusion in solid Scheil Eq. (4.12) is used. For $f_S = 0.5$ it is obtained $C_S = 8.48\%$. For the case of partial mixing first, the effective partition coefficient must be calculated with Eq. (4.20). Taking $D_L = 10^{-8}$ m²/s we obtain $k_{ef} = 1.39$. Then, equation Eq. (4.21) is used and we obtain $C_S = 8.49\%$. This is a slight increase over the Scheil value.

Application 4.5

Consider zone melting. Calculate the composition profile (*C*-x graph) for a steel rod of length 60 mm having the composition 0.21% C and 0.9% Si, for one and two passes. Assume the length of the molten zone to be 3 mm.
Note: for the last cell composition is calculated from mass balance.

Answer:
The steel is of hypereutectic composition. This means that as the carbon content of the solid increases above 0.1% the partition coefficient will change from 0.196 to 0.495. This will require the implementation of an IF statement.

Select the computational grid so that the total sample length is a multiple of the length of the molten zone. The Excel spreadsheet could be organized as shown in the following table. Equation (4.22) is used in cell B3 with $\ell = 3$ and $x =$ value in cell A3. For all other cells the following IF statement must be implemented: IF(%C in preceding cell < 0.1, Eq. (4.22) with $k = 0.196$, Eq. (4.22) with $k = 0.495$). In all cells in row 24 a mass balance equation is implemented.

	A	B	C	D	E
1		**First Pass**		**Second Pass**	
2	**Length (mm)**	**%C**	**%Si**	**%C**	**%Si**
3	3	0.0712	0.8042	0.0242	0.7185
4	6	0.0959	0.8556	0.0260	0.8134
5	9	0.1163	0.8795	0.0665	0.8594
6	12	0.1954	0.8905	0.1083	0.8811
23	57	0.2100	0.9000	0.1067	0.9000
24	60	0.5940	1.0785	2.3523	1.2440

Results of calculations with this scheme are presented in Figure 4.31. This calculation can be run until the eutectic composition is reached. Then the composition in the section next to the last one should be allowed to increase. Note that for Si that has a relatively high partition coefficient a large zone leveling region occurs..

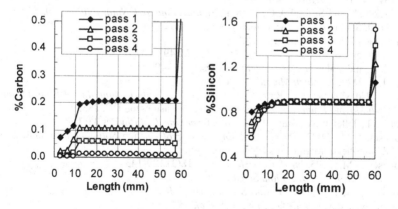

Figure 4.31. Composition evolution during zone refining.

References

Brody H.D., Flemings M.C., 1966, *Trans. TMS-AIME* **236**:615

Burton J.A., Prim R.C., Slichter W.P., 1953, *J. Chem. Phys.* **21**:1987

Campbell J., 1969, *AFS Cast Metals Research Journal*, March:1.

Campbell J., 1991, *Castings*, Butterworth-Heinemann, Oxford

Campbell J., 2004, *Casting Practice, The 10 Rules of Castings* Elsevier-Butterworth-Heinemann

Campbell J., 2005, in: *Shape Casting: The John Campbell Symposium*, M. Tiryakioglu and P.N. Crepeau editors, The Minerals, Metals, & Materials Society, Warrendale Pa. p.3

Clyne T.W., Kurz W., 1981, *Metall. Trans.* **12A**:965

Dahle A.K., Arnberg L., Apelian D., 1997, *AFS Trans.* **105**:963

Dahle A.K., Thevik H.J., Arnberg L., StJohn D.H., 1999, *Met. Mater. Trans. B*, **30B**:287

Derge G. ed., 1964, *Basic Open Hearth Steel Making*, 3d ed., AIME, New York

Flemings M.C., 1974, *Solidification Processing*, McGraw-Hill

Flemings M.C. and Nereo G.E., 1967, *Trans. AIME* **239**:1449

Flinn R.A., 1963, *Fundamentals of Metal Casting*, Addison - Wesley

Ganesan S. and Poirier D.R., 1990, *Metall. Trans.* **21B**:173

Gulliver G.H., 1913, *J. Inst. Met.* **9**:120

Himemiya T. and Umeda T., 1998, *ISIJ Intern.* **38**:730

Kaptay G. and Kelemen K.K., 2000, in: *State of the Art in Cast Metal Matrix Composites in the Next Millenium*, P.K. Rohatgi ed., TMS, Warrendale, Pa. p.45

Kobayashi S., 1988, *Trans. Iron Steel Inst. Jpn.* **28**:728

Krieger I.M., 1972, *Adv. in Coll. Interf. Sci.* **3**:111

Kurz W. and Fisher D.J., 1989, *Fundamentals of Solidification*, Trans Tech Publications, Switzerland

Nastac L., 2004, *Modeling and Simulation of Microstructure Evolution in Solidifying Alloys*, Kluwer Academic Publishers, Boston, 285p

Nastac L. and Stefanescu D.M., 1993, *Metall. Trans.* **24A**:2107

Ohnaka I., 1992, in: *ASM Handbook vol.15 Casting*, D.M. Stefanescu ed., ASM International, Metals Park, OH

Patankar, S.V., 1980, *Numerical Heat Transfer and Fluid Flow*, Hemisphere Publ. Corp., New York

Pershing M.A., 1997, An assessment of some models for micro and macrosegregation as applied to cast steel, *MS Thesis*, The University of Alabama, Tuscaloosa

Pfann W. G., 1952, *Trans. AIME* **194**:747

Plutshack L.A. and Suschil A.L., 1988, in *Metals Handbook Ninth Edition, vol. 15 Casting*, D.M. Stefanescu ed., ASM International, Ohio p576

Poirier D.R., 1987, *Metall. Trans.* **18B**:245

Portevin A. and Bastien P., 1934, *J. Inst. Metals* **54**:49

Rappaz M. and Voller V., 1990, *Metall. Trans.* **21A**:749

Scheil E., 1942, *Zeitschrift Metallkde.* **34**:70

Shyy W., Udaykumar H.S., Rao M.M., Smith R.W., 1996, *Computational Fluid Dynamics with Moving Boundaries*, Taylor & Francis, Washington DC, 285p

Smith V.G., Tiller W.A., Rutter J.W., 1955, *Canadian J. of Physics* **33**:723

Stefanescu D.M., 2002, *Science and Engineering of Casting Solidification*, Kluwer Academic/Plenum Publishers, New York, 342p

Stefanescu D.M., 2005, *Int. J. Cast Metals Res.* **18**(3):129-143

Thomas D.G., 1965, *J. of Colloid Science* **20**:267

Upadhya G. and Paul A.J., 1993, *AFS Trans.* **101**:919

Wukovici N. and Metevelis G., 1989, *AFS Trans.* **97**:285

5

MACRO-ENERGY TRANSPORT

There are three forms of energy transport: conduction (diffusive transport), convection (heat transmitted by the mechanical motion of the fluid) and/or radiation (through space). All three are active during solidification of a casting. Energy diffusion and convection occurs within the casting, at the metal/mold interface, and within the mold. Energy is transported by radiation from the mold to its environment, which is typically the air.

The corresponding equations for the heat flux are:

conduction: $\quad q = -\alpha \dfrac{\partial(\rho c_p T)}{\partial x} \quad$ Fourier's law

convection: $\quad q = h(T(t) - T(0)) \quad$ Newton's law

radiation: $\quad q = \varepsilon\sigma(T_1^4 - T_2^4) \quad$ Stefan-Boltzman's law

where α is the thermal diffusivity, h is the heat transfer coefficient, σ is the Stefan-Boltzman constant, and ε is the emissivity factor.

Steady state equations can be derived and solved based on the above equations, for the case of no phase change. However, solidification of castings is a process that can be either steady state or non-steady state, and involves phase transformation. Thus, it is a transient problem involving partial differential equations. In turn, partial differential equations can be solved either analytically or through numerical approximation methods. Specific boundary conditions must be used to describe various casting processes. In the following sections we will discuss the governing equation with the source term and the most common boundary conditions.

5.1   Governing equation for energy transport

The governing equation for solidification of a casting is the conservation of energy equation written in its advection-diffusion form, Eq.(3.3), repeated here for convenience:

$$\frac{\partial}{\partial t}(\rho H) + \nabla \cdot (\rho \mathbf{V} H) = \nabla \cdot \left(\frac{k}{c} \nabla H \right) + S \qquad (5.1)$$

The source term as derived by Benon and Incropera (1987) for a two-phase system is:

$$S = \nabla \cdot \left[\frac{k}{c_S} \nabla (H_S - H) \right] - \nabla \cdot \left[\rho f_S (\mathbf{V} - \mathbf{V}_S)(H_L - H_S) \right] \qquad (5.2)$$

The first right hand term (RH1) in the source term equation is the energy flux associated with phase transformation, and the second right hand term (RH2) is the energy flux associated with phase motion.

The energy equation for a two-phase system can be expressed in terms of temperature instead of enthalpy on the basis of the following relationships:

$$H = g_L H_L + g_S H_S$$

$$H_S = c_S T$$

$$H_L = c_L T + (c_S - c_L) T_e + \Delta H_f$$

Assuming that $c_S = c_L = c_p$, the energy equation becomes:

$$\frac{\partial}{\partial t}(\rho c_p T) + \nabla \cdot (\rho c_p \mathbf{V} T) = \nabla \cdot (k \nabla T) + S \qquad (5.3)$$

where $S = -\frac{\partial}{\partial t}(\Delta H_f \rho_L g_L) - \nabla \cdot [\Delta H_f \rho g_L (\mathbf{V} - \mathbf{V}_S)] - \nabla \cdot (\Delta H_f \rho_L g_L \mathbf{V})$

In this equation, ΔH_f is the latent heat of fusion of the alloy. In the source term equation, RH1 is the latent heat from generation of solid fraction and RH2 and RH3 are the energy flux associated with phase motion. Note that when the advective terms are ignored this equation reduces to:

$$\frac{\partial}{\partial t}(\rho c_p T) = \nabla \cdot (k \nabla T) - \frac{\partial}{\partial t}(\Delta H_f \rho_L g_L) \quad \text{or} \quad \frac{\partial T}{\partial t} = \nabla \cdot (\alpha \nabla T) + \frac{\Delta H_f}{\rho c_p} \frac{\partial}{\partial t}(\rho_S g_S - \rho)$$

If it is assumed that $\rho_S = \rho_L = \rho$, then $g_S = f_S$, $\partial\rho/\partial t = 0$, and the source term in Eq. (3.13) is:

$$\dot{Q}_{gen} = \rho_S \Delta H_f \frac{\partial f_S}{\partial t} \tag{5.4}$$

For the first stage of equiaxed solidification, when the grains move freely with the liquid, $V = V_S = V_L$, and Eq. (5.3) can be simplified to:

$$\frac{\partial}{\partial t}\left(\rho c_p T\right) + \nabla \cdot \left(\rho c_p V T\right) = \nabla \cdot \left(k \nabla T\right) - \frac{\partial}{\partial t}\left(\Delta H_f \rho_L g_L\right) - \nabla \cdot \left(\Delta H_f \rho_L g_L V\right) \tag{5.5}$$

where RH3 is the convected latent heat.

In the second stage of equiaxed solidification, when after dendrite coherency the solid is fixed, $V_S = 0$. The energy equation can then be written as:

$$\frac{\partial}{\partial t}\left(\rho c_p T\right) + \nabla \cdot \left(\rho c_p V T\right) = \nabla \cdot \left(k \nabla T\right) + \frac{\partial}{\partial t}\left(\Delta H_f \rho_S g_S\right) \tag{5.6}$$

Since g_S and g_L are function of time and temperature, appropriate expressions must be provided for these quantities in order to solve the energy equation. The energy equations derived above are second order partial differential equations. The complete equations cannot be solved analytically without further simplifying assumptions. Thus, numerical solutions have been proposed. Regardless of the solution, boundary conditions and initial conditions are necessary. They will be discussed in the following section.

5.2   Boundary conditions

The classic boundary conditions used in diffusive transport have been discussed in Section 3.2.1. However, for the specific case of castings further discussion is needed. Typical boundary conditions (BC) used for computational modeling of casting solidification are shown in Figure 5.1. These BCs can be expressed as follows:

known heat flux (convective BC): $\quad -k\dfrac{\partial T}{\partial x}(0,t) = h\big(T(t) - T(0)\big) \tag{5.7a}$

insulated boundary: $\quad\quad\quad\quad\quad -k\dfrac{\partial T}{\partial x}(0,t) = 0 \tag{5.7b}$

The radiation loss requires additional discussion. As summarized by Upadhya and Paul (1994) heat loss through a surface in the general case is by radiation and convection, and the heat transfer coefficient becomes:

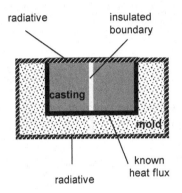

Figure 5.1. Typical boundary conditions for castings.

$$h = h_r + h_c = \sigma \varepsilon F_{m-a}\left(T_S^2 + T_o^2\right)\left(T_S + T_o\right) + c\left(T_S + T_o\right)^{1/3} \tag{5.7c}$$

where h_r and h_c are the radiative and convective heat transfer coefficients, respectively, F_{m-a} is the view factor of the mold with respect to air, and c is a constant dependent on the surface geometry. The view factor is defined as the fraction of the radiation that leaves surface i in all directions and is intercepted by surface j. When two surfaces, dA_1 and dA_2, undergo radiation exchange, the view factor can be mathematically expressed as (Siegell and Howell, 1981):

$$F_{1-2} = \iint\limits_{A_1 A_2} \frac{\cos\theta \cos\phi}{R_{1-2}^2} dA_1 dA_2$$

where R_{1-2} is the distance between the two surfaces, and θ and ϕ are the angles of the two surface normals with the line joining the two surfaces.

The calculation of the view factor may become rather complicated when multiple surfaces are involved in the radiation process. A code for view factor calculation has been developed by Lawrence Livermore National Laboratory (Shapiro, 1983).

For sand and die casting the first two boundary conditions are generally used. However, for investment casting the third one is necessary, so that the extension of solidification models for sand castings to investment castings is not trivial.

In the following sections we will discuss some analytical solutions of the energy transport equation applied to particular cases for castings. It is convenient to discuss these solutions first for steady-state and then for non-steady-state energy transport.

5.3   Analytical solutions for steady-state solidification of castings

Steady-state solidification occurs during controlled directional solidification. Directional solidification is widely used for industrial applications. Steady state solidification is used mainly for the production of single crystals.

The processes used for growth of single crystals include (Figure 5.2):

- normal freezing: the boat method (Figure 5.2a) or the Bridgman method (Figure 4.13a)
- crystal pulling (Czochralski) (Figure 5.2b)
- zone melting and zone freezing: with crucible or crucibles (surface held by surface tension and/or magnetic forces) (Figure 4.13b)

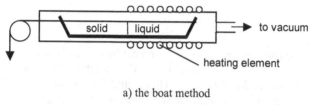

a) the boat method

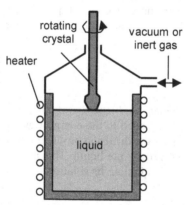

c) crystal pulling

Figure 5.2. Single crystal growth processes.

Two solutions of the steady-state heat conduction equation (Laplace equation) are given in Appendix A. Unfortunately they are not useful for the study of directional solidification because they are not applicable to the moving boundary problem that is directional solidification. Thus, other solutions must be developed.

The basic heat flow objectives are to obtain a constant thermal gradient across the S/L interface and to move the temperature gradient at a controlled rate. Let us consider an S/L interface. Heat flux balance at the interface is:

$$q_S = q_L \qquad \text{(in J m}^{-2}\text{s}^{-1} = \text{W m}^{-2}) \tag{5.8}$$

with $q_S = -k_S\, G_S$ and $q_L = -k_L\, G_L - \Delta H_f\, V$, where k is the thermal conductivity (J m^{-1} s^{-1} K^{-1}) and G is the gradient (K m^{-1}). Note the heat generation term in the second equation, where the latent heat is expressed in J m^{-3}. Solving for V we obtain:

$$V = \frac{k_S\, G_S - k_L\, G_L}{\Delta H_f} \tag{5.9}$$

The maximum solidification velocity can then be calculated for $G_L \to 0$:

$$V = \frac{k_S\, G_S}{\Delta H_f}$$

G_S can be evaluated from experiments or from heat flow calculations.

Let us try to calculate G_S assuming that in the directionally solidified sample energy transport is by conduction alone. This would be a good approximation for Bridgman-type solidification. The problem is one of steady state heat flow with moving boundary. The governing equation is Eq. (5.3) simplified for the case of steady state and no source terms. In one-dimensional form it becomes:

$$\frac{\partial^2 T}{\partial x^2} - \frac{V}{\alpha}\frac{\partial T}{\partial x} = 0 \tag{5.10}$$

Again, this is a moving boundary problem. Setting the reference system at the S/L interface, the advective term is added to the energy flow (V is positive). Note that for the case of energy transport the diffusion problem is solved for the solid, while for solute transport it was solved for the liquid. This is a linear, homogeneous second-order ordinary differential equation. Its solution is:

$$T = A + B\exp\left(\frac{V}{\alpha_S}x\right) \tag{5.11}$$

where A and B are constants. The boundary conditions are as follows:

BC1   at $x = 0$ $T = T_f$
BC2   at $x \to -\infty$ $T = T_o$

From BC1, $T_f = A + B$, and from BC2, $T_o = A$. Substituting in Eq. (5.11):

$$T = T_o + (T_f - T_o)\exp(V\,x/\alpha_S) \tag{5.12}$$

The interface thermal gradient in the solid is:

$$G_S = (dT/dx)_{x=0} = (T_f - T_o) V/\alpha_S \qquad (5.13)$$

Typical experimental gradients are 1 K/cm for high melting point crystals, and 5 K/cm for low melting point crystals (Flemings, 1974).

5.4   Analytical solutions for non-steady-state solidification of castings

Non-steady state heat transport is typical for some DS processes such as chill-casting or controlled directional solidification of superalloys, as well as for the vast majority of casting processes, including sand casting, die casting, investment casting, continuous casting, ingot casting, etc. Typical non-steady-state directional solidification processes are presented schematically in Figure 5.3. The main difference between the two processes is that in chill casting the solidification velocity decreases continuously during solidification, while it is typically maintained constant during DS. In all non-steady-state processes the solidification velocity and/or the temperature gradient at the interface change continuously during solidification. Accordingly, steady state energy transport can no longer be assumed.

The PDE that must be solved is Eq. (5.3). This equation cannot be solved analytically without further simplifying assumptions. The general simplifying assumptions include:

- the metal is poured without superheat, that is at T_f
- pure metal (has a melting temperature not a melting range)
- mold is semi-infinite (outside temperature is that of the ambient, T_o)
- no heat generation

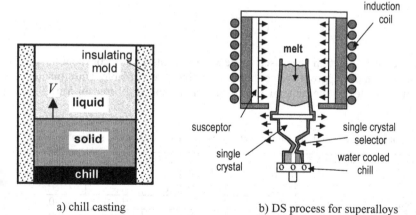

a) chill casting   b) DS process for superalloys

Figure 5.3. Non-steady state directional solidification processes.

The temperature profile in the mold is shown in Figure 5.4 for the general case. The temperature decrease within a domain, or at the interface between two domains, can be considered an interface resistance. Additional assumptions are made on the relative values of different thermal resistances, as summarized in Table 5.1. The temperature profile across a metal - mold section following these assumptions is shown in Figure 5.5.

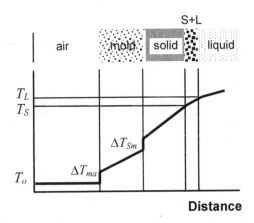

Figure 5.4. Temperature profile in a casting - mold assembly.

The governing equation derived from Eq. (5.3) assuming $\mathbf{V} = 0$ and $S = 0$ is:

$$\partial T/\partial t = \alpha\left(\partial^2 T/\partial x^2\right) \tag{5.14}$$

Several methods may be used to solve this equation. Three Fourier series solutions for different boundary conditions are given in Appendix A. When solving with the method of combination of variables the solution is:

$$T = C_1 + C_2 \, \mathrm{erf}(u) \quad \text{where} \quad u = x/\left(2\sqrt{\alpha t}\right) \tag{5.15}$$

$\mathrm{erf}(u)$ is the error function discussed in detail in the inset. The constants in Eq. (5.15) can be found if appropriate boundary conditions are available. Unfortunately this is not possible for the general case. The simplifying BC listed in Table 5.1 must be used.

Table 5.1. Assumptions on thermal resistance in a casting-mold system

Casting process	Figure 5.5	Resistance			
		In solid	At solid/mold interface	In mold	At mold/air interface
insulating molds	a	0	0	high	0
permanent molds	b	0	high	0	0
ingot molds	c	high	0	0	0
ingot molds	d	high	0	high	0

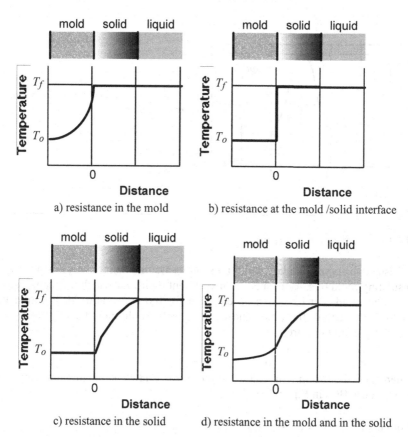

Figure 5.5. Temperature profile in the mold and in the casting for different assumptions.

The error function. The error function is derived from the "bell" curve shown in Figure 5.6. The curve is generated by the equation $y = \exp(-x^2)$. The dashed area under the curve has the property $\int_0^\infty \exp(-x^2)\,dx = \sqrt{\pi}/2$. The error function of x, erf(x), is defined as the ratio between the area under the curve $y|_{x=0}^{x=x}$ and $y|_{x=0}^{x=\infty}$. That is $erf(x) = \dfrac{2}{\sqrt{\pi}} \int_0^x \exp(-x^2)\,dx$

Hence, some properties of the error function are:

  erf(0) = 0   erf(1) = 0.842   erf(2) = 1
  erf(∞) = 1   erf(-∞) = - erf(∞)

The series representation of the error function is $erf(x) = \dfrac{2}{\sqrt{\pi}}\left(x - \dfrac{x^3}{3\cdot 1!} + \dfrac{x^5}{5\cdot 2!} - \dfrac{x^7}{7\cdot 3!}\cdots \right)$.

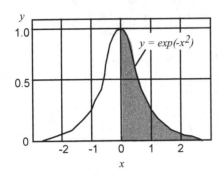

Figure 5.6. Bell curve used for definition of error function.

5.4.1   Resistance in the mold

To establish some boundary conditions, let us assume that the metal is poured in an insulating mold (*e.g.* sand, shell, or investment mold) for which the main resistance is in the mold. Let us derive an equation for the temperature distribution in the mold. If the zero of the *x*-axis chosen at the solid/mold interface (Figure 5.5a), the boundary conditions are:

BC1	at	$x = -\infty$	$T = T_o$
BC2	at	$x = 0$	$T = T_f$

where T_f is the fusion temperature of the metal, and T_o is the ambient temperature. Using these BCs in Eq. (5.15):

from BC1: $T_o = C_1 + C_2 erf(-\infty) = C_1 - C_2 erf(\infty) = C_1 - C_2$

from BC2: $T_f = C_1 + C_2 erf(0) = C_1$

Substituting in Eq. (5.15), the temperature distribution in the mold can be calculated with:

$$T = T_f + (T_f - T_o) erf\left(x/(2\sqrt{\alpha t})\right)$$ (5.16)

Let us now try to evaluate the thickness of casting, *L*, solidified over some time *t*. Heat flux balance at the solid /mold interface gives:

$$-k_m \left(\frac{\partial T}{\partial x}\right)_{x=0} = -\rho_S \Delta H_f \frac{dx}{dt}$$

From Eq. (5.16) the diffusive heat transport through the mold is calculated as:

$$-k_m \left(\frac{\partial T}{\partial x}\right)_{x=0} = -k_m (T_f - T_o)\left(\frac{\partial}{\partial x} erf \frac{x}{2\sqrt{\alpha t}}\right)_{x=0} = -\frac{k_m (T_f - T_o)}{\sqrt{\pi \alpha t}} = -\sqrt{\frac{k_m \rho_m c_m}{\pi t}}(T_f - T_o)$$

By analogy with Ohm's law the resistance in the mold can be defined from $q = k \cdot \Delta T / \Delta x = \Delta T / (\Delta x / k)$ as:

$$R_m = \frac{1}{A} \sqrt{\frac{\pi t}{k_m \, \rho_m \, c_m}}$$

Substituting the diffusive heat transport through the mold in the heat flux balance equation, rearranging and integrating between $t = 0$ and $t = t$, and $x = 0$ and $x = L$ results in:

$$L = \frac{2}{\pi} \left(\frac{T_f - T_o}{\rho_S \, \Delta H_f} \right) \sqrt{k_m \, \rho_m \, c_m} \; t^{1/2} \tag{5.17}$$

Note that the terms in parenthesis are constants referring to the metal, while those under the square root are constants referring to the mold. This equation does not take into account the superheating required for pouring. However, the heat generated during solidification has been included in the heat flux balance. This equation can be used to calculate the thickness solidified for non-divergent heat flow (Figure 5.7b).

In the preceding equations, L is the conductive path length, and is the characteristic linear dimension. In general, a characteristic linear dimension may be obtained by dividing the volume of the solid by its surface area, $L = v/A$. This is exact for plate-type castings, and approximate within 5% error for cylinders and spheres. Taking the characteristic length of the casting to be $L = v/A$, where v is the volume of the casting and A is its cooling surface area, Eq. (5.17) can be rewritten to give the final solidification time for a casting poured into an insulating mold:

$$t_f = \frac{\pi}{4} \left(\frac{1}{k_m \, \rho_m \, c_m} \right) \left(\frac{\rho_S \, \Delta H_f}{T_f - T_o} \right)^2 \left(\frac{v}{A} \right)^2 = ct. \cdot \left(\frac{v}{A} \right)^2 = ct. \cdot M^2 \tag{5.18}$$

where M is the *casting modulus*. This equation is known as the *Chvorinov* equation (Chvorinov, 1940).

The effect of the superheating temperature, ΔT_{super}, on the final solidification time can also be included by writing the heat to be removed from the casting as:

$$Q = \rho_S \, \Delta H_f \, v + \rho_L \, c_L \, \Delta T_{super} \, v$$

Assuming that $\rho_S = \rho_L = \rho$ the heat to be removed is written as:

$$Q = \rho v \left(\Delta H_f + c_L \, \Delta T_{super} \right) = \rho v \, \Delta H_{eff}$$

where ΔH_{eff} is the effective heat of fu-

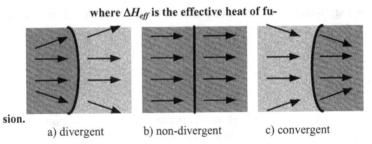

sion.

a) divergent   b) non-divergent   c) convergent

Figure 5.7. Types of heat flow. Divergent heat flow results in heat dissipation, while convergent heat flow results in heat accumulation.

Then, Eq. (5.18) becomes:

$$t_f = \frac{\pi}{4}\left(\frac{1}{k_m\,\rho_m\,c_m}\right)\left(\frac{\rho\,\Delta H_{eff}}{T_f-T_o}\right)^2\left(\frac{v}{A}\right)^2 \tag{5.19}$$

An example of the use of the *Chvorinov* equations (5.18) and (5.19) to calculate the final solidification time of a casting is given in Application 5.1.

The two Chvorinov equations do not fit exactly the experimental data, in particular at the beginning of solidification. Experimental data are better described by an equation having the shape:

$$L = C_1\sqrt{t} - C_2$$

where C_1 is a constant made of the first three terms of Eq. (5.18), and C_2 is a constant resulting from convection in the liquid that removes superheat and delays solidification, which introduces a finite mold - metal resistance to heat transfer.

A graphical representation of this equation is given in Figure 5.8 for experiments in which aluminum was poured in steel and dry sand molds. It can be seen that the slope of the curve increases from the sand mold to the steel mold. This is a result of the increase in the cooling rate, which is a consequence of higher thermal conductivity of the steel mold as compared with the dry sand mold. The dotted line represents schematically the correlation calculated with Eq. (5.17).

More accurate calculation with the Chvorinov equation can be performed using a time-stepping procedure. An example is given in Application 5.2.

5.4.2   Resistance at the mold/solid interface

A rather simple case to analyze is that when the resistance is at the mold/solid (*mS*) interface (Figure 5.5b). The basic assumption is that heat transfer from the solid to the mold is by convection. This is particularly valid when an air gap forms at the interface in such processes as permanent molding or die-casting. Other processes such as splat cooling and atomization can also be described under such an assumption.

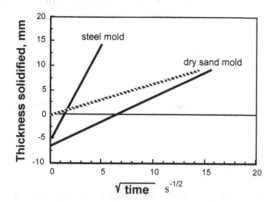

Figure 5.8. Influence of time and cooling rate (mold material) on the solidified thickness.

Heat flux balance at the mold/solid (mS) interface can be written as $h(T_f - T_o) = -\rho_S \Delta H_f \, dx/dt$. The interface resistance is: $R_{mS} = 1/(A \cdot h)$. After the integration of the flux balance equation, between $t = 0$ and $t = t$, and between $x = 0$ and $x = l$, it is obtained that $l = (T_f - T_o) h t /(\rho_S \Delta H_f)$. Then, if non-divergent heat transfer is assumed, and since $l = v/A$:

$$t_f = \frac{\rho_S \Delta H_f}{h(T_f - T_o)} \frac{v}{A} \tag{5.20}$$

This equation is valid when ΔT_{mS} is large compared with ΔT_S or ΔT_m. That is $\Delta T_{mS} \gg \Delta T_S$ when $1/h \gg L/k_S$, or $\Delta T_{mS} \gg \Delta T_m$ when $1/h \gg [\pi t/(k_m \rho_m c_m)]^{1/2}$.

A more complicated case in which $T_S \neq T_f (\Delta T_S > 0)$ can be considered (Figure 5.9). The flux into the mold/solid interface is $q|_{x=0} = -k_S(T_f - T_S)/L$. The flux out of the mold /solid interface is $q|_{x=0} = -h(T_S - T_o)$. Eliminating T_S between these two equations and recognizing that the flux at $x = 0$ must be equal to the flux at $x = L$ to satisfy energy conservation, as well as that the flux into the S/L interface

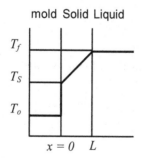

Figure 5.9. Temperature profile assuming solid resistance and interface temperature lower than the melting point.

from the liquid is the result of the latent heat evolved (remember that there is no superheat) we have:

$$q|_{x=L} = \frac{T_f - T_o}{1/h + L/k_S} = \rho_S \Delta H_f \frac{dL}{dt}$$

Integrating between 0 and L, and between 0 and t it is found that:

$$L + \frac{h}{2k_S} L^2 = \frac{h(T_f - T_o)}{\rho_S \Delta H_f} t \qquad (5.21)$$

The problem has been solved by Adams in a more rigorous manner, by assuming that the temperature profile is not necessarily linear. The following equation was obtained:

$$L + \frac{h}{2k_S} L^2 = \frac{h(T_f - T_o)}{\rho_S \Delta H_f a} t \quad \text{where} \quad a = \frac{1}{2} + \sqrt{\frac{1}{4} + \frac{c_S(T_f - T_o)}{3\Delta H_f}} \qquad (5.22)$$

Heat flux balance at the interface can be used to evaluate the temperature gradient during casting solidification, as shown in Application 5.3..

Another approach is the so-called *lump analysis*. If the Biot number (Bi $= \bar{h} L/k$, where $\bar{h}$ is the average heat transfer coefficient for convection for the entire surface, and L is the conduction length) is small, e.g., Bi < 0.1, the internal temperature gradients are also small. The object of analysis is considered to have a single mass averaged temperature and the transient problem can be treated as follows. Consider a macro-volume element of volume v and surface area A. Heat flow rate balance for this element requires:

$$\bar{h} A[T(t) - T_o] = v \dot{Q}_{gen} - v \rho c \frac{dT}{dt} \qquad (5.23)$$

In turn, from Eq. (5.4) $\dot{Q}_{gen} = \rho \Delta H_f (df_S/dt)$, and thus:

$$\bar{h} A[T(t) - T_o] = v \rho \Delta H_f \frac{df_S}{dt} - v \rho c \frac{dT}{dt} \qquad (5.24)$$

This equation can be integrated only if f_S is a function of temperature, e.g. $f_S(T) = a + b \cdot T$. Rearranging:

$$\frac{dT - (\Delta H_f/c) df_S}{T(t) - T_o} = -\frac{\bar{h} A}{v \rho c} dt$$

Integrating and applying the initial condition gives:

$$T(t) = T_o + (T_i - T_o)\exp\left(-\frac{\overline{h}\,At}{v\,\rho\,c(1 - (b/c)\Delta H_f)}\right)$$

(5.25)

From this equation the solidification time of a casting of volume v is:

$$t_f = -\frac{\rho c}{\overline{h}}\frac{v}{A}\left(1 - \frac{b}{c}\Delta H_f\right)\ln\left(\frac{T_f - T_o}{T_i - T_o}\right)$$

(5.26)

where T_i is the initial (superheat) temperature. This equation is valid in the solidification interval. Above T_L and under T_S, $\Delta H_f = 0$.

Eq. (5.24) can be rewritten to describe the cooling rate in the casting, which is the cooling curve:

$$\frac{dT}{dt} = -\frac{\overline{h}\,A[T(t) - T_o]}{\rho c v} + \frac{\Delta H_f}{c}\frac{df_S}{dt}$$

(5.27)

If in this cooling curve equation (RH2) > (RH1), heating occurs that shows up on the cooling curve as recalescence. An example of the use of lump analysis is provided in Application 5.4.

When solidification of an alloy must be described, since the fraction of solid is a function of temperature, it can be expressed as $df_S/dt = (df_S/dT)(dT/dt)$, and the cooling rate is:

$$\frac{dT}{dt} = -\frac{\overline{h}\,A}{\rho c v}[T(t) - T_o]\left(1 - \frac{\Delta H_f}{c}\frac{df_S}{dt}\right)^{-1}$$

From this equation, it is seen that solidification decreases the cooling rate since $df_S/dT < 0$. If lump analysis is not acceptable, or f_S as a function of temperature is not available, numerical methods must be used.

5.4.3   The heat transfer coefficient

The treatment of the solidification problem in the previous section is based on the simple assumption that the heat transfer is either constant or a continuous inverse function of time. However, the heat transfer coefficient at the mold/metal interface can vary over a wide range, depending whether an air gap is formed at the interface or not.

The issue of the numbers to use for the heat transfer coefficient is an open one, since it depends on a number of factors including temperature, casting geometry, and gap formation. In principle, the value of the heat transfer coefficient depends on the mechanism of heat transport at the solid/mold interface, which in turn is a function of the metal-mold contact. Three stages can be rationalized (Ho and Pehlke 1984, Trovant and Argyropoulos 2000) as summarized in Figure 5.10.

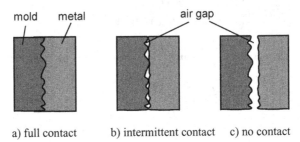

a) full contact   b) intermittent contact   c) no contact

Figure 5.10. Possible metal/mold contacts at the metal/mold interface.

In stage I, at the beginning of solidification, the contact between the liquid metal and the mold can be assumed good. Heat transport is through conduction from liquid metal to the mold wall.

However, as a solid layer forms, the metal will shrink away from the mold and a discontinuous air gap will result (stage II). The mold and solid metal will have partial contact at the asperities of the surfaces. Heat transport is now through mold/solid metal conduction at the points of contact and through gas conduction and radiation through the metal/mold gap.

In stage III the metal will pull away completely from the mold and heat transport is only through the gap. Ho and Pehlke (1984) have demonstrated that it is possible to calculate the interfacial heat transfer for this case by a simple superposition of gas conduction and radiation via the quasi steady state approximation.

Thus, the heat transfer coefficient across the interface may be written as the sum of three components:

$$h = h_{mc} + h_{gc} + h_{gr} \tag{5.28}$$

where h_{mc} accounts for metal-to-mold conduction, h_{gc} for conduction through the air gap, and h_{gr} for radiation through the air gap. The first term can be formulated straightforward for normal casting conditions, assuming good solid contact (see for example Application 5.4). However, it is a function of pressure. Thus, for squeeze casting, large heat transfer coefficients of the order of 50,000Wm^{-2}K^{-1} were reported (Nishida and Matsubara, 1976). Note that the calculation of h_{mc} in stage II must take into account the reduced surface area of contact, which is not trivial. The second term can be calculated as $h_{gc} = k_{air}/l$, where k_{air} is the thermal conductivity of air and l is the gap thickness.

Finally, the third term of Eq. (5.28) can be calculated with the analytical equation:

$$h_{gr} = \frac{\sigma\left(T_{Si}^2 + T_{mi}^2\right)\left(T_{Si} + T_{mi}\right)}{\varepsilon_{Si}^{-1} + \varepsilon_{mi}^{-1} - 1}$$

where σ is Stefan-Boltzman constant, ε is emmisivity, and the subscripts Si and mi, mean solid metal interface and mold interface, respectively.

Some examples of the dependency of the heat transfer coefficient on the gap width for pure aluminum and for the aluminum alloy A356 poured in graphite molds are given in Figure 5.11. Typical average values of heat transfer coefficients for particular solidification processes are given in Appendix B, Table B6.

5.4.4   Resistance in the solid

The typical example for this assumption (Figure 5.5c) is an ingot mold where the mold is water-cooled. The governing equation is again partial differential equation Eq. (5.14) whose solution is Eq. (5.15). The following boundary conditions apply:

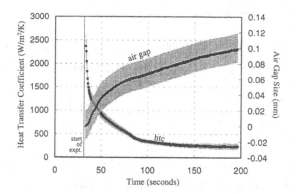

a) aluminum-graphite system

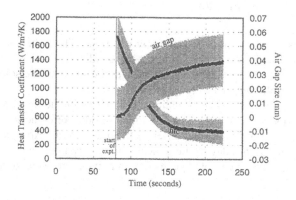

b) A356-graphite system

Figure 5.11. Heat transfer coefficient and air gap width measurement for pure aluminum and A356 poured in graphite molds (Trovant and Argyropoulos, 2000). With kind permission of Springer Science and Business Media.

$$\text{BC1} \quad \text{at } x = 0 \qquad T = T_o$$
$$\text{BC2} \quad \text{at } x = L \qquad T = T_f$$

From BC1 we derive $C_1 = T_o$. From BC2 we have $C_2 = \left(T_f - T_o\right) \big/ erf\left(L/2\sqrt{\alpha_s t}\right)$.

Using the notation $\gamma = L/2\sqrt{\alpha_s t}$ it is seen that, since C_2, T_f, T_o, are constant, γ is also constant. Thus the distance solidified is:

$$L = 2\sqrt{\alpha_s t} \tag{5.29}$$

with γ constant and unknown. When substituting the constants C_1, C_2 in the solution, the temperature distribution in the solid is obtained as:

$$T = T_o + \frac{T_f - T_o}{erf\gamma} erf \frac{x}{2\sqrt{\alpha_s t}} \tag{5.30}$$

The heat flux at the liquid/solid interface $k_s \left(\partial T/\partial t\right)_{x=L} = \rho_s \, \Delta H_f \, \partial L/\partial t$ is used to find γ. The temperature gradient can be calculated from Eq. (5.30):

$$\left(\frac{\partial T}{\partial x}\right)_{x=L} = \frac{\partial}{\partial x}\left[\frac{T_f - T_o}{erf\gamma} erf \frac{x}{2\sqrt{\alpha_s t}}\right]_{x=L} = \frac{T_f - T_o}{erf\gamma} \frac{\exp\left(-\gamma^2\right)}{\sqrt{\pi \alpha_s t}}$$

The solidification velocity can be calculated from Eq. (5.29): $\partial L/\partial t = \gamma\sqrt{\alpha_s/t}$. Substituting in the heat flux balance equation it is found that:

$$\gamma \, erf\gamma \exp\gamma^2 = \left(T_f - T_o\right)\frac{c_s}{\Delta H_f \sqrt{\pi}} \tag{5.31}$$

To obtain γ this equation must be solved by iterations. Then, the solidified thickness is calculated with Eq. (5.29). Note that the solidified thickness is a parabolic function of time ($L \sim t^{1/2}$). This is consistent with the shape of the solidified thickness curve in the solid region of Figure 5.5c.

Analytical solutions to more complicated scenarios including resistance in the solid, the mold and at the interface can be found in other references such as Flemings (1974), Stefanescu (2002) or Poirier and Geiger (1994). However, analytical solutions are rarely used today, as numerical solutions can deal with more complicated problems, including complex geometry, and give more accurate results.

5.5   Applications

Application 5.1

Consider a simple 0.6% C steel casting in the shape of a cube, having a volume of $0.001 \mathrm{m}^3$, poured into a silica sand mold. Neglecting corner effects, calculate:

    a) the solidification time of this casting assuming no superheating;

    b) the solidification time of the casting assuming a superheating temperature of 1550 °C;

    c) the average solidification velocity for the two cases.

Answer:

The governing equations are:

    for case (a): Chvorinov's equation, Eq. (5.18)

    for case (b): Chvorinov's equation, Eq. (5.19)

    for case (c): $V = (v/A)^2/t_f$

    The required data are found in Appendix B. It is calculated that the solidification time without superheat is of 444 s, and with superheat of 619 s. The average solidification velocity is $3.8 \cdot 10^{-5}$ m/s for no-superheat, and $2.7 \cdot 10^{-5}$ m/s with superheat.

Application 5.2

Calculate the solidification time of a 0.6% C steel cube of volume $0.001 \mathrm{m}^3$ poured into a silica sand mold from a superheating temperature of 1550°C. Use a time-stepping analysis assuming resistance in the mold and ignoring the corner effects.

Answer:

Equating the heat flow rate into the mold on the mold side at the mold /metal interface from Eq. (5.16) with the flux coming from the metal (see Eq. (5.4) for heat generation) we have:

$$\sqrt{\frac{k_m \, \rho_m \, c_m}{\pi \, t}}\left(T - T_m\right)A = -v\rho c\,\frac{dT}{dt} + v\rho\,\Delta H_f\,\frac{df_S}{dt} \tag{5.32}$$

Rearranging and writing the equation in time-stepping format, that is $dT = \Delta T = T^{n+1}-T^n$, $dt = \Delta t$, etc.:

$$T^{n+1} = T^n - \frac{A}{v\rho c}\sqrt{\frac{k_m \, \rho_m \, c_m}{\pi \, t^{n+1}}}\left(T^n - T_m\right)\Delta t + \frac{\Delta H_f}{c}\,\Delta f_S \tag{5.33}$$

where T^n and T^{n+1} are the temperatures at times n and $n+1$ respectively, and T_m is the mold temperature. To find an expression for Δf_S, linear evolution of the fraction of solid over the solidification interval ΔT_o is assumed, that is $f_S = a + b \cdot T$.

Since   at: $T = T_L - \Delta T_o$ $f_S = 1$

and   at: $T = T_L$ $f_S = 0$

we have $f_S = \left(T_L - T\right)/\Delta T_o$. Then, $f_S^n = \left(T_L - T^n\right)/\Delta T_o$, and $f_S^{n+1} = \left(T_L - T^{n+1}\right)/\Delta T_o$.

Thus, $\Delta f_S = \left(T^n - T^{n+1}\right)/\Delta T_o$. Substituting in the temperature equation, the final equation is:

$$T^{n+1} = T^n - \frac{A}{v}\,\frac{T^n - T_m}{\rho\left(\Delta H_f/\Delta T_o + c\right)}\sqrt{\frac{k_m \, \rho_m \, c_m}{\pi \, t^{n+1}}}\,\Delta t \tag{5.34}$$

The Excel spreadsheet is organized as shown in the Table 5.2, where the data in Appendix B have been used. The temperature in cell D2 is the initial temperature. Eq. (5.34) was implemented in cell D3. Note that an IF statement must be used since for $f_S = 0$ and $f_S = 1$ the term containing the latent heat evolution is zero. The form of the IF statement is:

IF(OR($T^n > T_L$, $T^n < (T_L - \Delta T_o)$)),Eq.(5.34) with $\Delta H_f = 0$,Eq.(5.34))

Table 5.2. Organization of spreadsheet.

	A	B	C	D
	Data		Calculations	
1	T_{init}	1550	Time	Temp.
2	v	0.001	0	1550
3	A	0.06	10	1522
4	ρ_S	7210	20	1502
5	c_S	794	30	1487
6	ΔH_f	2.72E+05	40	1484
7	ΔT_o	72	50	1482
8	k_m	0.52	60	1480
9	ρ_m	1600	70	1479
10	c_m	1170	80	1477
11	Δt	10	90	1476
12	T_L	1490	100	1474

The cooling curve presented in Figure 5.12 is obtained. From the graph, the solidification time is approximately 880s. This is significantly different from Chvorinov's equation results (619s with superheat).

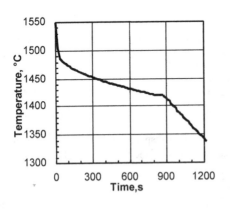

Figure 5.12. Cooling curve of 0.6% C steel casting.

Application 5.3

Calculate the average temperature gradient at the end of solidification in the steel casting described in Application 5.1.

Answer:
A flux balance at the L/S interface is written as:

$$\rho_S \Delta H_f \frac{dL}{dt} = k_S \frac{T_f - T_S}{L}$$

The solidification velocity was calculated in Application 5.1 to be $V = dL/dt = 3.8 \cdot 10^{-5}$ m/s. Then, calculating the gradient at the time when the center of the casting has solidified, and using the data in Appendix B:

$$G_T = (T_f - T_S)/L = (\rho_S \Delta H_f V)/k_S = (7210)(2.72 \cdot 10^5)(3.8 \cdot 10^{-5})/40 = 1863 \text{ K/m}$$

Application 5.4

Consider the casting in Application 5.2. Using the assumption of thermal resistance at the mold /metal interface (small Bi number, no gradient in the casting) calculate the heat transfer coefficient at the interface.

Answer:
We will use a time-stepping procedure in conjunction with lump analysis. The governing equation is flux balance at the interface, Eq. (5.24). In time-stepping format this equation can be written as:

$$h A(T^n - T_o) \Delta t = -\rho c v(T^{n+1} - T^n) + \rho \Delta H_f v \Delta f_S \tag{5.35}$$

Assuming a linear dependence of the solid fraction on temperature as in Application 5.2, $\Delta f_S = (T^n - T^{n+1})/\Delta T_o$, and this equation becomes:

$$T^{n+1} = T^n - \frac{h A}{\rho v} \frac{T^n - T_o}{\Delta H_f / T_o + c} \Delta t$$

Note that this equation is similar to that developed in Application 5.2, except for h. Implementing this equation in an Excel spreadsheet and using the data in Appendix B, the cooling curve given in Figure 5.13a is obtained. A heat transfer coefficient of 315 J/m²·K·s was used (the value for cast iron since no value for steel was available). Note that the solidification time is about 70 s, which is much smaller than that calculated in Application 5.2. This is because of the faster heat transfer for resistance at the interface as compared with resistance in the mold..

Comparing the time-stepping equation in this application with that in Application 5.2, it is found that for the case of a sand mold the heat transfer coefficient can be calculated as:

$$h = \sqrt{(k_m \rho_m c_m)/(\pi t^{n+1})}$$

The time dependency of h is shown in the Figure 5.13b, where the cooling curve from Application 5.2 is also included. The average value of the coefficient from pouring to the end of solidification is $\bar{h} = 34.7$ J/(m²·K·s). When this value is used, the same solidification time of 880 s as in Application 5.2 is obtained.

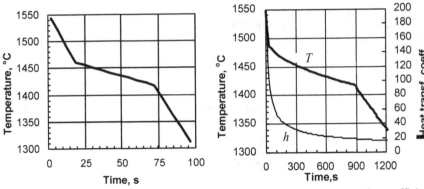

a) cooling curve of 0.6% *C* steel assuming resistance at the mold /metal interface

b) cooling curve and heat transfer coefficient of 0.6% *C* steel assuming resistance in the sand mold

Figure 5.13. Calculated cooling curves for 0.6% *C* steel under different cooling conditions.

Application 5.5

Examine the influence of the time step, Δt, on the results of Applications 5.2 and 5.4.

References

Bennon W. D. and Incropera F. P., 1987, *Int. J. Heat Mass Transfer* **30**:2161, 2171

Chvorinov N., 1940, *Giesserei* **27**:177

Flemings M.C., 1974, *Solidification Processing*, McGraw-Hill

Ho K. and Pehlke R.D., 1984, *AFS Trans.* **92**:587

Nishida Y. and Matsubara H., 1976, *British Foundryman* **69**:274

Poirier D.R. and Geiger G.H., 1994, Transport Phenomena in Materials Processing, TMS, Warrendale PA

Shapiro A., 1983, *FACET-A radiation View Factor Computer Code for Axy-symmetric 2D and 3D Geometries with Shadowing*, Lawrence Livermore National Lab., California

Siegel A. and Howell J.R., 1981, *Thermal Radiation Heat Transfer*, Hemisphere Publ. Co., New York

Stefanescu D.M., 2002, *Science and Engineering of Casting Solidification*, Kluwer Academic/Plenum Publishers, New York, 342p

Trovant M. and Argyropoulos S., 2000, *Metall. and Mater. Trans.* **31B**:75

Upadhya G. and Paul A.J., 1994, *Trans. AFS.* **102**:69

6

NUMERICAL MACRO-MODELING OF SOLIDIFICATION

From the analysis of solidification based on the energy transport equation presented in the previous section, it was seen that analytical solutions of this equation are not always available. Significant simplifying assumptions must be used, assumptions that are many times debilitating to the point that the solution is of little engineering interest. Fortunately, with the development of numerical methods and their application to the solution of partial differential equations, the most complicated equations can be solved numerically. Numerical solutions rely on replacing the continuous information contained in the exact solution of the differential equation with discrete values. Discretization equations are derived from the governing differential equation.

Process modeling has become possible in a much larger extent than allowed by the use of analytical solutions. Process modeling has emerged as a practical industrial tool for the design of manufacturing processes, troubleshooting, and identifying the dependent and independent variables of the process.

Computer simulation of solidification is based on numerical solutions of energy, mass and momentum transport. Its main computational purpose is calculation of the evolution of the thermal and compositional field throughout the casting. To produce a solidification model the following steps are necessary:

- problem formulation
- discretization of governing equations
- solving of the system of algebraic equations

6.1 Problem formulation

Heat transfer (HT) modeling for a given casting - mold combination requires solving of the energy conservation equation for heat conduction with heat generation. Ignoring for the time being the convective term of the energy transport equation, the governing equation is:

$$\frac{\partial T}{\partial t} = \alpha \nabla^2 T + \frac{\dot{Q}_{gen}}{\rho c} \tag{6.1}$$

The source term associated with the phase change, which describes the rate of latent heat evolution during the liquid - solid transformation given by Eq. 5.4 is:

$$\dot{Q}_{gen} = \rho \Delta H_f \frac{\partial f_s(x,t)}{\partial t} \tag{6.2}$$

To solve Eq. (6.1) an appropriate expression for $f_S(x,t)$ must be found.

In one approach, the solution of the discretized energy transport equation at all the nodes or elements of the computational domain, is found by prescribing a solidification path, *i.e.*, by assuming a relationship between f_S and T. The fraction of solid is rewritten as $\partial f_s/\partial t = (\partial f_s/\partial T)(\partial T/\partial t)$. Then, some functional dependency of the fraction of solid on temperature is assumed. Such typical assumptions include linear dependency for eutectics, and equilibrium or Scheil equation for dendritic alloys. For example, assuming that the composition field is governed by Scheil-type diffusion an equation for the fraction solid evolution can be derived as follows. The interface temperature depends on composition according to the relationships $T_f - T_L = -m C_o$ or $T_f - T_L = -m C_L^*$. The liquid composition is given by Scheil equation, $C_L^* = C_o(1 - f_s)^{1-k}$. Then, $f_S = 1 - \left(C_L^*/C_o\right)^{1/(1-k)}$ and finally:

$$f_S = 1 - \left(\frac{T_f - T^*}{T_f - T_L}\right)^{1/(k-1)} \tag{6.3}$$

With these assumptions, Eq. (6.1) can be rewritten in several ways and solved by numerical techniques.

6.1.1   The Enthalpy Method

In the enthalpy method (Pham, 1986) for 1-D Eq. (6.1) is rewritten as:

$$k\frac{\partial^2 T}{\partial x^2} = \left(\rho c - \rho \Delta H_f \frac{\partial f_s}{\partial T}\right)\frac{\partial T}{\partial t} \tag{6.4}$$

Defining the enthalpy as:

$$H(T) = \int_0^T \rho c dT + \rho \Delta H_f \left[1 - f_s(T)\right]$$

and substituting in the previous equation, we obtain:

$$k\left(\partial^2 T/\partial x^2\right) = \rho\,\partial H/\partial t \tag{6.5}$$

The enthalpy discontinuity at the solidification temperature is usually circumvented by the arbitrary selection of a solidification interval.

6.1.2   The Specific Heat Method

In the specific heat method, an effective specific heat, c^*, is defined from Eq. (6.4):

$$c^*(T) = \frac{\partial H}{\partial T} = \rho\,c(T) - \rho\,\Delta H_f\,\frac{\partial f_s}{\partial t}\,\frac{\partial t}{\partial T}$$

which, when reintroduced in the same equation gives:

$$k\frac{\partial^2 T}{\partial x^2} = c^*(T)\frac{\partial T}{\partial t} \tag{6.6}$$

The specific heat method has several drawbacks. Firstly, because of the strong variation of the specific heat close to the liquidus or eutectic temperature, it is difficult to assure energy conservation while solving the equations numerically. Secondly, it requires that $\left(\partial f_s/\partial t\right)\left(\partial t/\partial T\right) \le 0$ so that $c^* \ge c$. This requirement precludes the possibility that $\partial f_s/\partial t$ takes both positive and negative values required to predict recalescence.

6.1.3   The Temperature Recovery Method

For molds having high Biot numbers, that is higher cooling rates, the previously described methods may omit the effect of the latent heat evolved during solidification. The temperature recovery method described graphically in Figure 6.1 correctly recovers the latent heat. In this approach the temperature T_2 at time step $t+\Delta t$ is calculated from Eq. (6.1) without including the latent heat generation term. Then the increased temperature because of latent heat generation, T_2', is calculated from the integral energy balance for the volume element:

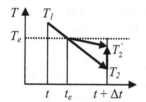

Figure 6.1. Schematic illustration of the temperature recovery method.

$$\int\limits_{T_e}^{T_2'} \rho c \, dT - \int\limits_{T_e}^{T_2'} \rho \Delta H_f \frac{\partial f_s}{\partial t} \frac{\partial T}{\partial t} dT = \int\limits_{T_1}^{T_2} \rho c \, dT - \int\limits_{T_1}^{T_e} \rho c \, dT$$

Finally, the temperature in the volume element at time $t+\Delta t$ is updated to T_2'.

A more accurate approaches can be used by calculating the fraction solid evolution as a function of time from solidification kinetics, thus relaxing the previous assumptions on the fraction solid. They will be discussed in detail in Chapter 7.

6.2   Discretization of governing equations

The continuous variables may be represented by a number of discrete values associated with the volume elements (cells) or volume vertices (nodes) of a computational grid. This process is called discretization. The partial differential equations are approximated by a set of algebraic equations. For a discussion on computational grids the reader is referred to Winterscheidt and Huang (2002)

The discretization equations for the differential equations can be derived in several ways:

- finite differences method (FDM) based on Taylor-series
- variational formulation, based on calculus of variations
- boundary element method (BEM)
- method of weighted residuals - control-volume method (CVM), or finite elements method (FEM)

In the following sections the finite difference method, as well as the control-volume method will be introduced. The application of finite differences to solidification modeling will also be discussed.

6.2.1   The Finite Difference Method - Explicit formulation

The finite difference calculus is a technique for differentiating functions by employing only arithmetic operations. The method is largely known as the Euler method. However, Taylor's series can also be used to derive the relevant equations. While a good understanding of Taylor's series is fundamental to FDM applications, a simpler approach can be used to derive some of the basic FDM equations.

Let us produce a finite difference discretization of the heat conduction equation. If, for the time being, we ignore heat generation, the 1-D format is:

$$\frac{\partial T}{\partial t} = \alpha \frac{\partial^2 T}{\partial x^2} \tag{6.7}$$

The first step is to divide the physical domain over which the calculation will be performed, in a number of volume elements or nodes. Higher the number of nodes, higher is the accuracy of the solution. A graphic representation of the problem, where the computational space has been divided in nodes, is given in Figure 6.2. Let us write the discretization equation. The time derivative can be approximated as:

$$\frac{\partial T}{\partial t} \cong \frac{T_i^{n+1} - T_i^n}{t^{n+1} - t^n} = \frac{T_i^{n+1} - T_i^n}{\Delta t} \tag{6.8}$$

where the superscript n refers to the time level. In terms of Taylor's series this is a *forward difference representation.*

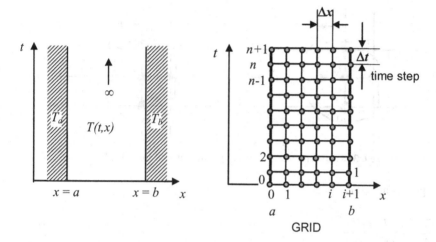

Figure 6.2. Graphic representation of the problem described by Eq. (6.7) and temporal and spatial grid. Boundary conditions are: at $x = a$, $T = T_a$, and at $x = b$, $T = T_b$.

The gradient at time n between nodes i and $i+1$, or between nodes i and $i-1$ is, respectively:

$$\left(\frac{\partial T}{\partial x}\right)_{right} \cong \frac{T_{i+1}^n - T_i^n}{\Delta x} \qquad \left(\frac{\partial T}{\partial x}\right)_{left} \cong \frac{T_i^n - T_{i-1}^n}{\Delta x} \tag{6.9}$$

where the subscript i refers to the spatial level. The second derivative in Eq. (6.7) can now be written as:

$$\frac{\partial}{\partial x}\left(\frac{\partial T}{\partial x}\right) \cong \frac{1}{\Delta x}\left[\left(\frac{\partial T}{\partial x}\right)_{right} - \left(\frac{\partial T}{\partial x}\right)_{left}\right] \quad \text{or} \quad \frac{\partial^2 T}{\partial x^2} = \frac{T_{i+1}^n - 2T_i^n + T_{i-1}^n}{(\Delta x)^2} \tag{6.10}$$

This is a *central difference representation* in terms of Taylor's series. Substituting the two difference representations in Eq. (6.7) and solving for T_i^{n+1} yields:

$$T_i^{n+1} = T_i^n + \frac{\alpha(\Delta t)}{(\Delta x)^2}\left(T_{i+1}^n - 2T_i^n + T_{i-1}^n\right) \tag{6.11}$$

Note that at time step $n + 1$ all values are known for step n. This is known as an *explicit solution*. The temperature in a node i can be found by "marching forward in time".

Let us now apply these principles to a 2-D problem. The graphic representation is given in Figure 6.3. Alternative notations commonly used are i–j and N-S-E-W, as illustrated in the figure. Also for time *old* and *new* instead of n and $n+1$.

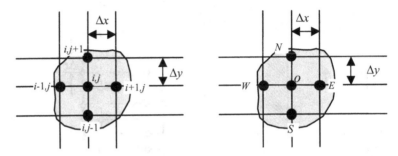

Figure 6.3. Representation of the computational grid for 2D calculations.

The governing equation is:

$$\frac{1}{\alpha}\frac{\partial T}{\partial t} = \frac{\partial^2 T}{\partial x^2} + \frac{\partial^2 T}{\partial y^2} \tag{6.12}$$

Central difference representations are:

$$\frac{\partial^2 T}{\partial x^2} = \frac{T_{i+1,j}^n - 2T_{i,j}^n + T_{i-1,j}^n}{(\Delta x)^2} \quad \text{and} \quad \frac{\partial^2 T}{\partial y^2} = \frac{T_{i,j+1}^n - 2T_{i,j}^n + T_{i,j-1}^n}{(\Delta y)^2} \tag{6.13a}$$

or, alternatively:

$$\frac{\partial^2 T}{\partial x^2} = \frac{T_E^{old} - 2T_o^{old} + T_W^{old}}{(\Delta x)^2} \quad \text{and} \quad \frac{\partial^2 T}{\partial y^2} = \frac{T_N^{old} - 2T_o^{old} + T_S^{old}}{(\Delta y)^2} \tag{6.13b}$$

The forward difference representation for $\partial T/\partial t$ is the same as for 1-D that is Eq. (6.8). Then, taking $\Delta x = \Delta y = \Delta s$, and combining with Eq. (6.13), the 2-D discretization equation is:

$$T_{i,j}^{n+1} = \frac{\alpha(\Delta t)}{(\Delta s)^2}\left(T_{i-1,j}^n + T_{i+1,j}^n + T_{i,j-1}^n + T_{i,j+1}^n\right) + \left[1 - 4\frac{\alpha(\Delta t)}{(\Delta s)^2}\right]T_{i,j}^n \tag{6.14a}$$

or, using the notations *old* for n and *new* for $n+1$:

$$T^{new} = \frac{\alpha(\Delta t)}{(\Delta s)^2}\left(T_W^{old} + T_E^{old} + T_S^{old} + T_N^{old}\right) + \left[1 - 4\frac{\alpha(\Delta t)}{(\Delta s)^2}\right]T^{old} \qquad (6.14b)$$

From this point on we will only use this last type of notations. Defining the Fourier number as $Fo = \alpha(\Delta t)/(\Delta x)^2$, the 1-, 2-, and 3- dimensional explicit discretization equations for uniform and equal spatial increments are respectively:

$$T^{new} = Fo\left(T_W^{old} + T_E^{old}\right) + \left[1 - 2Fo\right]T^{old} \qquad (6.15a)$$

$$T^{new} = Fo\left(T_W^{old} + T_E^{old} + T_S^{old} + T_N^{old}\right) + \left[1 - 4Fo\right]T^{old} \qquad (6.15b)$$

$$T^{new} = Fo\left(T_W^{old} + T_E^{old} + T_S^{old} + T_N^{old} + T_B^{old} + T_T^{old}\right) + \left[1 - 6Fo\right]T^{old} \qquad (6.15c)$$

Here the subscripts T and B in the last equation stand for top and bottom (the z-direction), respectively.

The stability of the numerical solution during iterations for successive time steps requires that the coefficients of T^{old} be non-negative. A negative coefficient would yield fluctuations in the numerical solution. Thus, the stability criteria are:

- for 1D $Fo \leq 1/2$
- for 2D $Fo \leq 1/4$
- for 3D $Fo \leq 1/6$

To meet these criteria Δs and Δt must be appropriately selected.

To solve Eq. (6.15) two boundary conditions (BC) and an initial condition (IC) are necessary. The BCs shown on Figure 6.2, *prescribed temperature* and *constant initial temperature*, are written as:

- BC1 $T(a, t) = T_a$
- BC2 $T(b, t) = T_b$
- IC $T(x, 0) = T_{initial}$

A more useful boundary condition is *prescribed flux at the interface* (Figure 6.4). Heat conservation on the shaded area having unit depth gives:

$$hA\left(T - T_{mold}\right) = -v\rho c\frac{\partial T}{\partial t} + kA\frac{\partial T}{\partial x} \qquad \text{or, in discretized format:}$$

$$\rho c\frac{(\Delta s)^2}{2}\left(\frac{T^{new} - T^{old}}{\Delta t}\right) = k\Delta s\left(\frac{T_E^{old} - T^{old}}{\Delta s}\right) + h\Delta s\left(T_{mold} - T^{old}\right)$$

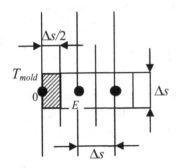

Figure 6.4. Computational grid for derivation of the discretization equation for the prescribed flux boundary condition.

After rearranging, we have:

$$T^{new} = 2Fo\left(T_E^{old} + \frac{h\Delta s}{k}T_{mold}\right) + \left[1 - 2Fo\left(\frac{h\Delta s}{k} + 1\right)\right]T^{old} \tag{6.16}$$

For an exterior nodal point subject to convective boundary conditions in a 2D system, a similar derivation results in:

$$T^{new} = Fo\left(T_S^{old} + 2T_E^{old} + T_N^{old} + 2\frac{h\Delta s}{k}T_{mold}\right) + \left[1 - 2Fo\left(\frac{h\Delta s}{k} + 2\right)\right]T^{old} \tag{6.17}$$

The stability criteria for convective boundary conditions are:

for 1D: $Fo \le \frac{1}{2}\left(\frac{h\Delta s}{k} + 1\right)^{-1}$ for 2D: $Fo \le \frac{1}{2}\left(\frac{h\Delta s}{k} + 2\right)^{-1}$

The most stringent of the stability requirements applicable in a given problem dictates the maximum time step that can be used for calculation.

Let us now consider the 1D case with heat generation. Using the enthalpy method, the governing equation is Eq. (6.4). Using a central difference representation for $\partial^2 T/\partial x^2$, Eq. (6.10), and a forward difference representation for $\partial T/\partial t$, Eq. (6.8), and then rearranging yields the sought off discretization equation:

$$T^{new} = T^{old} + \frac{Fo}{1 - \frac{\Delta H_f}{c}\frac{\partial f_S}{\partial T}}\left(T_E^{old} - 2T^{old} + T_W^{old}\right) \tag{6.18}$$

In order to solve this equation it is necessary to know $f_S(T)$. The thermophysical parameters can be calculated with the mixture theory relationships, *e.g.*, Eq. (3.7).

An example of the implementation of a 1D finite difference method (FDM) scheme for the solidification of a steel casting in conjunction with the enthalpy method is presented in Application 6.1. A 2D FDM problem is solved in Application 6.2.

In the preceding discussion, the FDM method based on forward, backward, and central differences has been introduced and used to derive discretization equations for the energy transport equation. This explicit formulation that only requires information obtained in a preceding time step is also called the *Euler method*. More precise numerical approximation methods can be used, such as the Runge-Kutta method, which is a fourth order (truncation error of the order h^5). These are all one-step methods, in which each step uses only values obtained in the preceding step. There are also multi-step methods (*e.g.*, Adams-Moulton method) in which more than one preceding step is used (Kreyszig, 1988).

6.2.2   The Finite Difference Method - implicit formulation

Eq. (6.1) will be discretized using the following representations:

- central difference representation for $\partial^2 T / \partial x^2$ and $\partial^2 T / \partial y^2$

$$\frac{\partial^2 T}{\partial x^2} = \frac{T_E^{new} - 2T^{new} + T_W^{new}}{(\Delta x)^2} \qquad \text{and} \qquad \frac{\partial^2 T}{\partial y^2} = \frac{T_N^{new} - 2T^{new} + T_S^{new}}{(\Delta y)^2}$$

- backward difference representation for $\partial T / \partial t$:

$$\partial T / \partial t = \left(T^{new} - T^{old}\right) / \Delta t$$

For $\Delta x = \Delta y = \Delta s$, the discretization equation is:

$$(1 + 4Fo)T^{new} = Fo\left(T_W^{new} + T_E^{new} + T_S^{new} + T_N^{new}\right) + T^{old}$$

Note that the future temperature of node n depends on the future temperatures of the adjacent nodes and on its present temperature. Consequently, a set of simultaneously algebraic equations must be solved for each time step. There are as many equations as many nodes. Because all coefficients are positive, there is no stability problem. This is called an *implicit solution* and also the *Crank-Nicholson method*. For details see for example Poirier and Geiger (1994).

6.2.3   The Finite Difference Method - general implicit and explicit formulation

Generalization of the discretization equations previously derived allows easy formulation of boundary conditions for any number of spatial dimensions:

- explicit

$$T_n^{new} = T_n^{old} + \Delta t \left[\sum_m \frac{T_m^{old} - T_n^{old}}{R_{mn} C_n} \right]$$

- implicit

$$T_n^{new} = T_n^{old} + \Delta t \left[\sum_m \frac{T_m^{new} - T_n^{new}}{R_{mn} C_n} \right]$$

where m are the nodes adjacent to node n, $R_{mn} = \Delta s/(k\, A_{k,mn})$ for conduction, $R_{mn} = 1/(h_{mn}\, A_{h,mn})$ for convection, and $C_n = \rho\, c\, v_n$. Here, $A_{k,mn}$ and $A_{h,mn}$ are the areas for conductive and convective heat transfer between nodes m and n, respectively; and v_n is the volume element determined by the value of Δs at node n. Boundary nodes are included by proper formulation of R_{mn} and C_n.

6.2.4   Control-volume formulation

The fundamental concept used in the control-volume method is that the governing partial differential equation is constrained to a finite control-volume over which the specific phase quantity (mass, enthalpy, momentum) must be conserved. As an example, the control-volume statement will be developed for the transient heat conduction equation with heat generation:

$$\rho c \frac{\partial T}{\partial t} = \nabla \cdot (k \nabla T) + \dot{Q}_{gen}$$

Integrating over the control-volume element v, and using the divergence theorem, which states that $\int_v \nabla \cdot (k\nabla T)dv = \int_A (k\nabla T) \cdot \mathbf{n} dA$, we have:

$$v \rho c \frac{\partial \overline{T}}{\partial t} = \int_A (k\nabla T) \cdot \mathbf{n} dA + v \overline{\dot{Q}}_{gen} \quad \text{and, since } \nabla T \cdot \mathbf{n} = \partial T / \partial n:$$

$$v \rho c \frac{\partial \overline{T}}{\partial t} = \int_A k \frac{\partial T}{\partial n} dA + v \overline{\dot{Q}}_{gen} \qquad (6.19)$$

where $\mathbf{n}$ is the unit outward normal vector, $\partial/\partial n$ is its derivative along the outward normal to the surface of the control volume, and $\overline{T}$ and $\overline{\dot{Q}}_{gen}$ are suitable averages of the temperature and the energy generation rate over the control volume. The advantage of this formulation is that the numerical solution of Eq. (6.19) is fully conserving, since it is derived on the basis of global conservation, unlike finite difference equations that are derived based on interface flux conservation. This equation is used to derive discretization equations for various problems. For details of these derivations as well as for the formulation of various terms in the diffusion equations, the reader is referred to the specialized texts by Patankar (1980) and Özisik (1994).

6.3 Solution of the discretized equations

The explicit method introduced in a previous section is very simple computationally. However, it has the disadvantage that the time step is restricted by stability considerations, which may result in unacceptable high computational time.

The implicit formulation has the advantage that it is unconditionally stable. However, a large number of equations may have to be solved simultaneously requiring a large computer memory. Typical methods of solving large sets of algebraic equations include direct methods such as Gauss elimination or the Thomas algorithm, or iterative methods such as Gauss-Seidel iteration. Additional complications may result if the equation to be solved becomes non-linear because of the source term or because of time-dependent thermophysical properties. Linearization methods must then be used. For a detailed discussion of these methods, the reader is referred to Özisik (1994).

For realistic description of the physical phenomenon, all transport equations must be coupled and solved simultaneously. The main difficulty in solving these equations is the unknown pressure field required in the computation of the velocity field. A number of numerical methods have been developed to tackle this problem, e.g., SIMPLE (Semi-Implicit Method for Pressure Linked Equations) (Patankar, 1980), SIMPLEC (Simple Revised Consistent) (Van Doormaal and Raithby, 1984), VOF (Volume of Fluid) (Hirt and Nichols, 1981). However, complications that still linger are not disclosed by producers of commercial codes, as pointed out in the review of fluid flow modeling by Ohnaka (1993).

6.4 Macrosegregation modeling

Several numerical macrosegregation models have been proposed, e.g., Benon and Incropera 1989, Fellicelli et al. 1991, Diao and Tsai 1993, Schneider and Beckermann 1995, Chang and Stefanescu 1996. The Chang-Stefanescu model will be reviewed in some detail in the following paragraphs.

The transport equations previously introduced were simplified in order to describe mathematically phenomena of engineering interest. The following basic assumptions were used:

- only liquid and solid phases are present (i.e., L, S, $L + S$, no pores formation)
- the density of the liquid phase is different than that of the solid
- the properties of the liquid and solid phases are homogeneous and isotropic
- complete diffusion of solute in liquid, no diffusion in solid; since the solid diffusion coefficient is four orders of magnitude smaller than the liquid diffusion coefficient, this is a reasonable assumption

There are two stages during the solidification process:

- Stage I: equiaxed grains move freely with the liquid, *i.e.*, $V = V_S = V_L$, the viscosity of the mixture is described by a relative viscosity; Darcy flow is not important.
- Stage II: a rigid dendritic skeleton is established (dendrite coherency is reached) and $V_S = 0$, the viscosity value returns to liquid viscosity; Darcy flow becomes significant.

Liquid flow is driven by thermal and solutal buoyancy, as well as by solidification contraction. To solve the problem at hand it is now necessary to develop the governing transport equations with the appropriate source terms.

Conservation of mass (continuity) is described by Eq. (3.2). The energy transport equation for a two-phase system was derived based on Eq. (3.3). A detailed discussion of the energy transport equation is provided in section 5.1. Conservation of species and momentum will be discussed in the following paragraphs.

Conservation of species
The governing equation is Eq. (3.4), where the source term is:

$$S_C = \nabla \cdot \left[\rho D\nabla(C_L - C)\right] - \nabla \cdot \left[\rho f_s (\mathbf{V} - \mathbf{V_s})(C_L - C_S)\right]$$

In the first stage, $V = V_S = V_L$, and equation Eq. (3.4) can be simplified to:

$$\frac{\partial}{\partial t}(\rho C) + \nabla \cdot (\rho C\mathbf{V}) = \nabla \cdot (\rho D\nabla C_L) \tag{6.20}$$

In the second stage the relative velocity is $V_r = V_L - V_s = V_L$. Then, Eq. (3.4) becomes:

$$\frac{\partial}{\partial t}(\rho C) + \nabla \cdot (\rho C_L \mathbf{V}) = \nabla \cdot (\rho D\nabla C_L) \tag{6.21}$$

The only difference between these two equations is in the advectiv term. In the first stage, the average of solid and liquid concentrations is used in Eq. (6.20) because solid and liquid move together, whereas liquid concentration is used in Eq. (6.21) because of zero solid velocity. After further manipulations (see Chang/Stefanescu 1996 or Stefanescu 2002) the final equation for the second stage can be written as:

$$\rho f_L \frac{\partial C_L}{\partial t} + \rho\mathbf{V}\nabla \cdot C_L = \nabla \cdot (\rho D\nabla C_L)$$
$$+ \rho C_L (1 - k)\frac{\partial f_s}{\partial t} - (C_L - C)\nabla \cdot (\rho\mathbf{V}) \tag{6.22}$$

This equation indicates that the variation of solute in the liquid phase in a given volume element should be equal to the net loss or gain of solute due to convection, diffusion, interfacial reaction, and solidification contraction. Quite interestingly, this equation can be reduced to the local solute redistribution equation when the assumptions made by Flemings and Nereo (1967) are used (see Stefanescu 2002). In addition, under the assumption of equal solid and liquid density it reduces to the Scheil equation. Thus, it is apparent that Eq. (6.22) can be used as a general equation for macrosegregation calculation without many simplifying assumptions.

Conservation of momentum
The governing equation is Eq. (3.8). Since two-phase flow has to be described, the source term must include an additional term. Assuming flow through the mushy zone to be Darcy type flow, the pressure drop through the mushy zone is calculated from Eq. (4.33) as:

$$\nabla P = -\mu g_L V_L / K \tag{6.23}$$

Then, ignoring again the additional viscous term (interfacial interaction term) the source term is:

in the x direction: $S = -\left(\mu^* / K\right)\left(\rho / \rho_L\right)\left(V^x - V_S^x\right)$

in the y direction: $S = -\dfrac{\mu^*}{K}\dfrac{\rho}{\rho_L}(V^y - V_S^y) + \rho g\left[\beta_T (T - T_o) + \beta_C (C_L - C_{L,o})\right]$

In these equations, the RH1 term is the drag force term (assuming Darcy flow), and the RH2 term is the buoyancy term describing natural convection based on the Boussinesq approximation.

The treatment of the viscosity and of the permeability in the two-phase system requires further discussion. We can divide the computation in three stages. For each of these stages, the viscosity and the permeability must satisfy the conditions summarized in Table 6.1.

Table 6.1. Velocity, viscosity and permeability in the Chang-Stefanescu model

State	f	V	Viscosity	Permeability
liquid	$f_L = 1$	$V = V_L$	μ_L	∞
mushy	$f_L + f_S = 1$	$V_L = V_S$	μ^*	∞
mushy	$f_L + f_S = 1$	$V_S = 0$	μ_L	K

Many models were derived for liquid flowing into a permeable medium, and, in most cases, the local viscosity was considered to be the weighted average of the liquid and solid viscosity. However, when equiaxed solidification is considered, in the early stage of solidification when the solid fraction is small, the Darcy flow is not important. This is because the equiaxed grains will move with the liquid. Darcy flow becomes significant when dendrite coherency is reached, that is when a rigid dendritic skeleton is established. As suggested by Bennon and Incropera (1987), an effective viscosity should be used instead. A hybrid model was developed by Oldenburg and Spera (1992) to account for this behavior. A switching function was defined to control the transition. This switching function should handle different solidification morphologies, such as eutectic and dendritic. In dendritic solidification, the critical switching value will be relatively small compared to eutectic solidification.

The variation of the viscosity of the mixture, μ^*, as a function of solid fraction can be described with equations of the form of Eq. (4.45). In the Chang-Stefanescu model the empirical equation proposed by Metzner (1985) modified to include a *switching function* was used:

$$\mu^* = \mu_L \left[1 - F_\mu \, g_S / A \right]^{-2}$$

(6.24)

where A is a crystal constant, depending on the aspect ratio and surface roughness of the crystal, and F_μ is the switching function for viscosity. The switching function was introduced to smooth the transition from the first to the second stage. The form of the switching function is a modification of that suggested by Oldenburg and Spera (1992), as follows:

$$F_\mu(g_S) = 0.5 - \frac{1}{\pi} \tan^{-1} [s(g_S - g_s^{cr})]$$

(6.25)

where s is a constant (typical value 100), and g_s^{cr} is the critical solid fraction. The critical solid fraction is, for example, 0.27 for the aluminum alloy A201 (Bäckerud, 1990). The value of A is equal to 0.68 for smooth spherical crystals (aspect ratio = 1). For equiaxed dendritic crystals (aspect ratio from 1 to 10), the value of A was calculated to be 0.3. The viscosity of the mixture is employed in the first stage of solidification, and then turned off in the second stage (*i.e.*, the viscosity becomes again that of the liquid).

For equiaxed grains, the permeability can be regarded as isotropic. Thus, a *permeability function* can be obtained based on the theory of flow through a porous medium. The Carman-Kozeny equation (Carman, 1937), similar to the previously discussed Blake-Kozeny equation, was used:

$$K = C_2 \frac{(1 - g_S)^3}{g_S^2 F_K} = \frac{(1 - g_S)^3 d^2}{180 g_S^2 F_K}$$

(6.26)

where F_K is another switching function $F_K(g_S) = 1 - F_\mu(g_S)$, C_2 is a function of grain size, and d is the grain size (*e.g.*, 50 μm). The switching function shuts off permeability in the first stage and turns it on in the second stage.

To obtain a solution of the derived transport equations a numerical model was developed (see Chang and Stefanescu 1996, and Chang 1994 for details). Results of calculation of the concentration along a directionally solidified casting are shown in Figure 6.5a. It is noticed that *positive segregation* is formed near the chill followed by *negative-segregation*. This has been demonstrated theoretically by

Kato and Cahoon (1985). The occurrence of positive segregation can be explained as follows. The liquid ahead of the S/L interface is rich in solute because of solute rejection at the S/L interface. Solidification shrinkage induces back flow of enriched liquid into the solidifying region. This creates a region of positive segregation next to the chill. The back-flowing liquid is replaced by liquid that is depleted in solute. This produces a negative segregation zone. Positive segregation occurs only in the regions where the convection induced liquid velocity is very small, and has little effect on solute redistribution. A comparison of model prediction and experimental data is presented in Figure 6.5b.

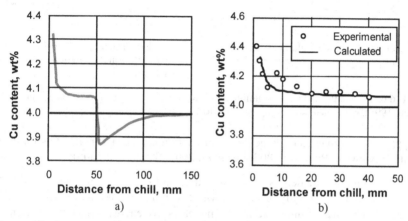

Figure 6.5. Cu concentration in a directionally solidified Al-4% Cu casting: a) Calculated evolution of the average concentration at a given time after the beginning of solidification (Chang and Stefanescu, 1996); b) Comparison of the calculated solute distribution (Chang and Stefanescu, 1996) and experimental data (Kato and Cahoon, 1985).

Figure 6.6a shows the calculated final solute distribution for flow driven only by thermosolutal convection. While the symmetry of the pattern is altered because of the flow pattern, the highest segregation is in the middle of the casting, as expected. The last region to solidify is the region richest in solute.

Figure 6.6b shows the solute redistribution for the case of liquid flow driven by thermosolutal buoyancy and solidification contraction. Solidification contraction imposes a back flow of enriched liquid in the area adjacent to early-solidified regions. This results in the occurrence of isolated, highly segregated regions aligned almost parallel to the right-side wall where solidification starts.

6.5   Macroshrinkage modeling

The prediction of the location and size of shrinkage-related defects is a difficult task. Numerous attempts at answering the problem through complex numerical 3-D models that solve the transport equations are on record. Depending on the assumptions on the physics of the problem and the mathematical apparatus used the differ-

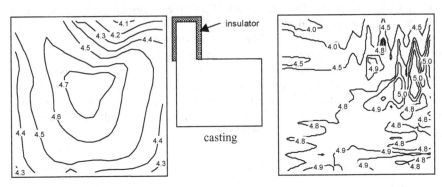

a) Flow driven by thermo-solutal convection.

b) Flow driven by thermosolutal convection and solidification contraction.

Figure 6.6. Calculated macrosegregation maps for an Al-4% Cu alloy (Chang and Stefanescu, 1996). With kind permission of Springer Science and Business Media.

ent approaches to macroshrinkage defect prediction can be summarized as follows (see also reviews in Piwonka 2000, Lee *et al.* 2001, Stefanescu 2005):

- Thermal models: solve energy transport equations to identify the last region to solidify or regions where feeding becomes restricted.
- Thermal/volume calculation models: solve energy transport equations and mass conservation to predict the position of the free surface and of the last region to solidify.
- Thermal/fluid flow models: solve mass and energy transport equations to predict the position of the free surface and of the last region to solidify.
- Transport/stress analysis models.
- Nucleation and growth of gas pores models: compute pore nucleation and growth when the dendrites have formed a coherent network; liquid flow is described as flow through a porous medium.

Many of these models attempt to predict not only macroshrinkage, but also porosity formation, while ignoring the direct contribution of gas evolution. The last type of models is used to predict microshrinkage and will be discussed in detail in Chapter 14.

6.5.1   Thermal models

These models include in the analysis only heat transport and ignore fluid flow and the role of gas. After the pioneering work of Henzel and Keverian (1965) that used a finite difference (FD) mesh to represent the irregular shape of casting and adapted a general purpose transient heat transfer program to solve for the temperature field, it became possible to identify the last region to solidify (hot spots) in a casting by mapping the isotherms. Assuming further that shrinkage cavities are located in the last region to solidify, prediction of position but not of the size of shrinkage cavities could be mapped. This technique has been widely used and its simplicity made it possible to implement it even on Excel sheets (see Application 6.2). However, the assumption that the mold is full at the beginning of solidification and that the liquid has uniform temperature, and ignoring the role of gravity on thermo-solutal convection are serious sources of errors.

Simpler thermal models based on the derivation of some analytical equation used as a criterion function were also proposed. The best known is the Chvorinov (1940) rule, which recognizes that for adequate feeding the riser must solidify after the casting, and consists in comparing the final solidification time, t_f, of the riser and the casting. It relates t_f to the modulus of the casting (volume to cooling surface area ratio, v/A), based on 1-D heat transport across the mold-metal interface Eq. 5.19).

In complex castings, where modulus calculations are cumbersome, numerical discretization schemes have been used to calculate casting and riser moduli. Incorporation of such calculations in casting simulation models has allowed a first approximate prediction of macroshrinkage defects (Upadhya and Paul 1992, Suri and Yu 2002).

The observation that occurrence of porosity can be avoided or at least minimized by maintaining a minimum thermal gradient, G_T, in the casting is at the basis of the Pellini (1953) criterion. It states that a shrinkage defect may occur in a region where the thermal gradient is smaller than a critical value, $G_T < G_T^{cr}$, at the end of solidification (when the fraction solid is $f_S = 1$, or when a critical fraction solid is reached ($f_S < f_S^{cr}$). This critical fraction solid may be that when flow is interrupted or dramatically decreased (for example when dendrites reach coherency).

Sigworth and Wang (1993) have proposed a "geometric" model in which the critical thermal gradient required to avoid microshrinkage is a function of the angle of the inner feeding channel inside the casting, θ. Microshrinkage will not occur if:

$$G_T^{cr} \geq \left(1 + \frac{2\Delta T_{SL}\, k_S}{\rho_S\, \Delta H_f}\frac{t_c}{l^2}\right) \cdot \frac{\rho_S\, \Delta H_f\, l}{4 k_S\, t_c} \cdot \tan\theta$$

where ΔT_{SL} is the solidification interval, t_c is the solidification time for the center of the plate, k_S is the thermal conductivity of the solid, ρ_S is the solid density, l is half the thickness of the plate, and ΔH_f is the heat of fusion. The model seems to work fine for narrow freezing range alloys, including steel. They also point out that surface energy effects may prevent feeding of the casting once dendrite coherency is reached in the riser.

Lee et al. (1990) developed a criterion function for wide freezing range alloys. The critical value under which feeding becomes difficult and porosity occurs was derived to be $G_T \cdot t_f^{2/3} / V_S$, where V_S is the solidification velocity.

6.5.2   Thermal/volume calculation models

These models solve the heat transport problem and attempt to predict defect locations through simple change in volume calculations based on mass conservation,
thus avoiding rigorous flow analysis of the molten metal during solidification.
Imafuku and Chijiiwa (1983) were the first to propose such a model for prediction
of the shape of macroshrinkage in steel sand castings. The main assumptions of the
model are: i) gravity feeding occurs instantly (liquid metal moves only under the
effect of gravity, solidification velocity is much smaller than flow velocity); ii)
liquid metal free surface is flat and normal to the gravity vector; iii) the volume of
shrinkage cavity is equal to the volume contraction of the metal; iv) macroscopic
fluid flow exists as long as the fraction solid, f_S, is less than a critical fraction of
0.67. The net change in volume because of shrinkage is calculated with
$\Delta v = \beta \cdot v \cdot \Delta f_L$, where v is the initial volume, $\beta = (v_S - v_L)/v_L = = (\rho_S - \rho_L)/\rho_L$
is the shrinkage ratio, ρ is the density and the subscripts L and S stand for liquid
and solid, respectively. If the volume loss is on the surface, *e.g.* on top of the riser,
it is compensated by lowering the level of the liquid metal. If it is inside the casting
it is compensated by introducing a void. In either case the shape of the void is
governed by the solidification sequence and by the value chosen for the critical
fraction solid. This model can predict the position and size of pipe shrinkage and of
macroporosity. By introducing a shrinkage ratio that is function of temperature (see
for example Hummer, 1988), Suri and Paul (1993) extended the previous model to
ductile iron.

A similar approach was used more recently by Beech *et al.* (1998). Heat conduction in the casting and mold was coupled with calculation of the volume change
because of solidification contraction for each isolated liquid region of the casting at
any given time. The volume change was calculated as $\Delta v = \beta v_o g_s F$, where v_o is
the initial volume of the element, g_s is the volume fraction of solid in the volume
element, and F is the fluid fraction. The fluid fraction has been originally defined
by Hirt and Nichols (1981), when they developed the *Volume of Fluid* (VOF)
method. If the volume element is empty $F = 0$, while $F = 1$ if the volume element is
full. A partially occupied volume has a value of F between 0 and 1. This volume is
subtracted from the top of the liquid region. The top of the liquid region is defined
by the direction of the gravity vector. A feeding criterion based on a drag force
coefficient, K, is introduced to describe the feeding process. K is a function of the
local solid fraction, as follows:

$$K = \begin{cases} 0 & if \quad f_S \le f_S^{coh} \\ \infty & if \quad f_S < f_S^{coh} \end{cases}$$

For pure metals it was assumed that the solid fraction for coherency is unity. When
the drag force coefficient K is infinite no feeding occurs; otherwise the shrinkage is
fully fed. The method was validated by simulating the solidification of T-shaped
castings (Figure 6.7).

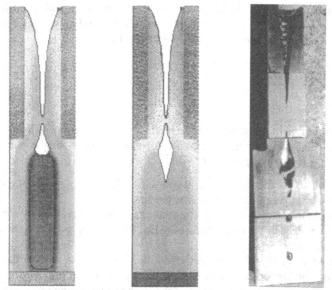

a) during solidification   b) end of solidification   c) end of solidification

Figure 6.7. Simulation (a and b) and experimental (c) results for cylindrical castings (Beech *et al.*, 1998).

Jiarong *et al.* (1995) developed a 3-D model for calculation of pipe shrinkage and macroporosity (closed shrinkage) in hypereutectic ductile iron based on the variation of the total volume of the casting over a time step:

$$\Delta v = \sum \Delta v_L + \sum \Delta v_{Gp} + \sum \Delta v_{Ge} + \sum \Delta v_{\gamma} + \Delta v_{mold}$$

where Δv_L is the liquid contraction, Δv_{Gp} is the primary graphite expansion, Δv_{Ge} is the eutectic graphite expansion, Δv_{γ} is the eutectic austenite contraction, and Δv_{mold} is the volume change due to mold wall movement. A similar equation can be written for hypoeutectic iron. While the equations for the volume changes in the metal are fully given no explanation is provided on how mold expansion was treated. No details were given on the implementation of this equation in the model.

6.5.3   Thermal/fluid flow models

These models tackle more or less comprehensively the whole transport problem but ignore the contribution of gas rejected by the solidifying melt to porosity formation. The physics of pore formation can be described through the following pressure balance equation: $P_{shr} > P_{amb} + P_{st}$.

An early model by Walther *et al.* (1956) assumed that void pores form because the section of the channel along which feed metal travels continuously narrows during solidification until the pressure drop ultimately ruptures the liquid in the channel. Further assuming shrinkage driven mass flow and conduction heat transport, an expression for the pressure drop along the channel, P_{shr}, was derived. It is given here in the format modified by Piwonka and Flemings (1966):

$$P_{st} - P_{shr} = \frac{32 \mu \beta \varsigma^2 L^2}{(1-\beta) r^4} \qquad (6.27)$$

where μ is the dynamic viscosity of the melt, L is the length of liquid zone in the casting, r is the radius of liquid channel (central cylinder) in the partially solid casting, and $\varsigma = k_{mold} (T_f - T_o) / \rho_S \Delta H_f \sqrt{\pi \alpha_{mold}}$ where k_{mold} is the thermal conductivity of the mold, T_f is the melting point of the metal, and α_{mold} is the thermal diffusivity of the mold. Assuming further that the first pore forms when $P_{shr} = 1$ atm and that once formed, the pore occupies the space previously occupied by the melt, the radius of the pore can be calculated from Eq. (6.27). Reasonable agreement was obtained with experiments.

Niyama et al. (1982) further elaborated on Pellini's idea in their development of a criterion for low-carbon steel castings. They used Darcy's law in cylindrical coordinates and expressed the pressure drop in the mushy zone as an inverse function of the ratio $G_T / \sqrt{dT/dt}$. Shrinkage defects form in the region where the ratio is smaller than a critical value, to be determined experimentally. While this criterion works well for low-carbon steel, its application by many non-ferrous foundries is questionable (see for example Spittle et al., 1995).

Attempting to improve this model, Huang et al. (1993) and then Suri et al. (1994) performed a 1-D analysis of the conservation of mass and momentum in the two-phase interdendritic region. It was found that the only significant term in the momentum equation responsible for loss of liquid pressure is that of the friction drag. Assuming flow along channels in the mushy zone, the nondimensional frictional drag on the feeding fluid can be expressed as: $F_{drag} = C_o V_L'$. Here, V_L' is the nondimensional liquid velocity and C_o is the "feeding resistance number" that controls the feeding in the mushy region expressed as:

$$C_o = \frac{N \mu \Delta T_{SL}}{\rho_L V_S \beta G_T d^2}$$

where N is a numerical constant (16π for columnar dendrites and 216 for equiaxed dendrites), and d is the characteristic length scale of the solid phase, i.e. either the primary dendrite arm spacing or the equiaxed grain diameter. C_o can be used as a criterion for microporosity formation. A high value indicates a high resistance to feeding and thus higher potential for pore formation. Indeed, experimental verification of this criterion showed excellent correlation between C_o and the percent porosity in an equiaxed A356 alloy.

Another proposed criterion is that by Hansen and Sahm (1988):

$$G_T / \left[(dT/dt)^{1/4} \cdot \left(V_L^{mush} \right) \right]^{1/2}$$

where V_L^{mush} is the flow velocity through the fixed dendrite skeleton. As opposed to previously described criteria, this criterion is scale and shape independent.

Bounds et al. (2000) developed a 3-D code that predicts pipe shrinkage, macro-porosity, and misruns in shaped castings. They reduced the multi-phase (S, L, G) problem to single-phase by modifying the standard transport equations as proposed by the mixture-theory model. The conservation of mass was rewritten as $\nabla \cdot V = S_{met} + S_G$, where V is the mixture velocity, and S_{met} and S_G are the source terms corresponding to the metal shrinkage and gas evolution respectively. S_{met} was evaluated numerically with the iterative scheme:

$$S_{met}^{new} = S_{met}^{old} - \frac{1}{\rho_{met}} \left(\frac{\partial f_{met}\, \rho_{met}}{\partial t} + \nabla \cdot \left(\rho_{met}\, V_{met} \right) \right)$$

where V_{met} is the metal component of the mixture velocity, and the other symbols are as before.

The mixture theory relationships were used to evaluate the density, fraction and velocity of the metal. The equivalent fluid component of velocity is given by $\rho V = \rho_{met}\, V_G + \rho_G\, V_G$ where no subscript indicates mixture quantities, and the subscript G stands for gas. The mixture density is given by $\rho = f_{met}\, \rho_{met} + (1 - f_{met})\, \rho_G$, where the metal fraction is tracked using a scalar advection equation $\partial f_{met}/\partial t + \nabla \cdot (f_{met}\, V_{met}) = S_{met}$. The momentum conservation equation was modified to account for the presence of fluid and solid phases by the introduction of an effective viscosity, μ_{eff}. A Darcy source term, S_D, was added as a momentum sink to describe flow through the fixed dendritic network after coherency:

$$\frac{\partial}{\partial t}(\rho V) + \nabla \cdot (\rho V \cdot V) = \nabla \cdot \left(\mu_{eff} \cdot \nabla V \right) - \nabla P + \rho g + S_D$$

where ρg is the body force. The fraction solid was obtained from the energy conservation equation.

The flow model used to calculate the pressure drop was based on the Chang-Stefanescu model (1996) (see section 6.4). By extending the Chang-Stefanescu model to dendritic-eutectic and columnar-equiaxed morphologies Bounds et al. were able to show that the pressure drop increases by five orders of magnitude as the flow changes from semisolid feeding to interdendritic feeding.

The model does not deal with the complexities of pore nucleation and growth but rather assumes that below a specific pressure, arbitrarily taken as a value below atmospheric, the volume vacated by metal shrinkage is filled by gas. In this way macroporosity is predicted based only on the pressure drop in the liquid when the feed path is obstructed.

Surface-connected macroporosity occurs under the same circumstances as internal porosity with the additional requirement that the pressure drop in the liquid must be sufficient to draw air into the casting through the permeable mold. This condition is modeled through the boundary condition $Q_{air} = K^{-1}(1 - f_{met})(P_{amb} - P)$, where Q_{air} is the flow rate and K is the permeability. For $P \geq P_{amb}$, f_{met} takes the value calculated for the metal and for $P < P_{amb}$, it is taken as zero. This prevents liquid metal permeating the mold when $P \geq P_{amb}$.

Numerical predictions with this model are presented in Figure 6.8 for the case of short and long freezing range alloys. For short freezing range alloys a solid skin forms at the mold-metal interface. Once the feeding path becomes obstructed, the pressure drop in the solidifying region triggers macroporosity formation. For long freezing range alloys, the mushy zone extends to the surface of the casting. The air can be drawn into the casting through the semi-solid surface, and shrinkage may form at the hot spots on the surface.

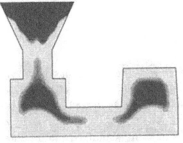

a) pipe shrinkage and closed shrinkage in short freezing range alloy

b) surface shrinkage in long freezing range alloy

Figure 6.8. Predicted macroshrinkage in different type alloys (Bounds *et al.* 2000). With kind permission of Springer Science and Business Media.

6.6   Applications of macro-modeling of solidification

The use of solidification models to help production of better castings is today accepted in the foundry industry. Foundries that are not actively using one of the many commercial models are aware of them, and most accept that soon they will find it necessary to use them to remain competitive. The development of solidification simulation models has even become a matter of national pride: each major casting producing country in the West has developed at least one (and often more than one) model for its foundrymen to use. With rapid prototyping and concurrent engineering applications growing, modeling is clearly an established tool for metalcasters. But models of the casting process can and must include more than merely the solidification event.

Modeling implementation into the foundry industry has been led by casting solidification models. The first models did not incorporate fluid flow, and it was simply assumed that once the mold is full, the temperature is uniform across the casting. However, as the technology matured it was soon noticed that this assumption is a source of significant error, in particular for casting with relatively large variations in section size. Subsequently, mold-filling models were developed and implemented. Today, any casting solidification package that claims state of the art level includes both mold filling and solidification modeling.

More ambitious goals have also been set for casting models. By combining mold filling and solidification models with knowledge-based systems for gating and risering, a completely computerized solution for casting design becomes possible (Upadhya and Paul, 1994). The architecture of such a package is summarized in Figure 6.9.

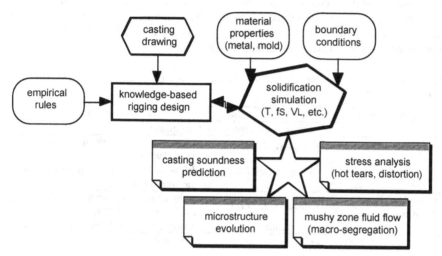

Figure 6.9. Architecture of a comprehensive casting modeling system (Stefanescu and Piwonka, 1996). Copyright 1966 American Foundry Soc., used with permission.

The objectives of a complete computer based casting design are to design and evaluate the process, and to evaluate process output, which includes casting integrity, surface quality, microstructure, and mechanical properties. It can also be used as a process control tool. Knowledge-based systems attempt to design the process (design of casting rigging), while mold filling and solidification models strive to do the rest. Process modeling has penetrated the metal casting industry and casting markets, and has become integral part of advanced manufacturing processes.

It must be made clear that what the foundryman needs and what the models deliver is not the same thing. As far as the foundryman is concerned, the objective of the modeling is to provide information on casting quality. However, after intricate calculations the models can only deliver some physical quantities such as temperature, solidification time, composition, pressure, fluid velocity, etc. A summary of "promises and realities" deriving from objectives is given in Table 6.2. Matching the foundryman's objectives with models deliverables is still very much an area of research and continuous development.

Some of the quantities calculated by the models can be used directly in the form of property maps (e.g. temperature, solidification time, fraction of solid, composition, velocity etc.). Composition mapping can be used to predict such features as micro- and macro-segregation. Examples of macro-segregation mapping have been given in Figure 6.6.

As is probably clear to the reader, many features of interest to the foundryman, such as microshrinkage, surface quality, mechanical properties, etc. cannot or are very difficult to obtain through direct calculation. The mathematical complexity of the numerical models for shrinkage prediction and the lack of reliable database have led a number of investigators to develop simpler analytical equations termed "criterion functions" to predict when and where there is a high probability of defect formation in a casting. Criterion functions are simple empirical rules that relate the local conditions (e.g., cooling rate, solidification velocity, thermal gradient, etc.) to the shrinkage defect susceptibility. Some of the criteria functions used to predict casting soundness have been given earlier in this section. Others and microstructure prediction criteria are summarized in Table 6.2.

Table 6.2. Summary of foundry requirements and mold filling and solidification macro-models deliverables

Foundry objectives	Model deliverables	Calculations based on
casting soundness, surface quality		
macro-shrinkage	$T, t_f, f_S, G, \dot{T}$	criteria functions (Ohnaka 1986, Sahm 1991)
	volume deficit (Δv)	diffusive energy and convective mass transport (Jiarong et al. 1995)
micro-shrinkage/ porosity	$T, t_f, f_S, G, \dot{T}, V_L, P$	criteria functions (Niyama *et al.* 1982, Lee *et al.* 1990)
	pressure map	pressure balance (Bounds *et al.* 2000)
missruns, cold shuts	$G, \dot{T}, T_S > T_{min}$	criteria functions
	filling time	convective - diffusive, energy and mass transport
casting appearance - surface quality	penetration index	pressure balance at metal /mold interface and interfacial reactions (Stefanescu *et al.* 1996)
casting composition		
macro-segregation	composition map	convective - diffusive energy + mass transport (Mehrabian *et al.* 1970, Poirier *et al.* 1991, Schneider/Becker-mann 1995, Chang/ Stefanescu 1996)
casting dimensions - distortions	stress map	residual stress (Thomas 1993)
casting microstructure and mechanical properties		
fraction phase	$C^{macro}, T, t_f, f_S, G, \dot{T}$	criteria functions
SL interface stability	$G/V \geq \Delta T_{SL}/D$	criteria functions
phase transition	$\dot{T}$	criteria functions
columnar-to-equiaxed transition	$G_L < G_{min}$	criteria functions
dendrite arm spacing	$\lambda_I = ct. \cdot \dot{T}^n$, $\lambda_{II} = ct. \cdot t_f^{1/3}$ t_f: local solidif. time	criteria functions
gray-to-white transition in cast iron	$V < V_{max}$ or $\dot{T} < \dot{T}_{cr}$	criteria functions
mechanical properties of casting	$C^{macro}, \dot{T}$	criteria fct. based on composition and cooling rate
	$C^{micro}, f_\alpha, N, \lambda$	criteria fct. based on microstructure

Nomenclature

C^{macro}: macro-scale composition	P: pressure
C^{micro}: micro-scale composition	t_f: final solidification time
f_α: fraction of phase	ΔT_{SL}: solidification interval
G: temperature gradient	
N: number of grains per unit volume	Δv: total volume variation
	λ: phase spacing

Most of the criteria discussed here are size (scale) and shape dependent. Hansen and Sahm (1993) have developed a number of scaling relationships that can be used to develop nondimensional criteria or criteria that are scale and shape independent.

Most criteria are shape dependent. That is, if the shape of the casting is changed the constants involved are changed. The classic example is that of the Chvorinov criterion $t_f = ct. \cdot M^2$ where $M = v/A$, where M is the casting modulus. This criterion is size independent but shape dependent.

The Niyama criterion can be transformed into a dimensionless criterion if written as $G_S \cdot v^{-1/2} \left(\dot{T}_S \cdot \Delta T_{SL} \right)^{-1/2}$, where v is the kinematic viscosity (in m^2 s) and ΔT_{SL} is the solidification interval.

6.7   Applications

Application 6.1

Consider a cubic casting, having a volume of 0.001 m^3, poured in a sand mold. The casting material is eutectic cast iron and the pouring temperature is 1350 °C. Assume resistance at the interface on two of the opposite faces of the cube and adiabatic conditions on the other four. Consider also that the solidification interval is 10 °C. Calculate the solidification time of this casting using a 1D FDM scheme. Plot the corresponding cooling curves at different positions throughout the casting. Compare with calculation with the Chvorinov equation. Required data are found in Appendix B.

Answer:
A 1D FDM explicit formulation will be used. To decrease the computational time, since the casting is symmetric, the calculation will be performed only for half of the casting having the length $l = 0.5 \, (v)^{1/3}$. A number of i nodes will be used for the casting, as shown in Figure 6.10. The node $i + 1$ is used for the second boundary condition at the center of the casting.

The governing equation is Eq. (6.4) in its discretized form Eq. (6.18). Using the same procedure as in Application 5.2 to discretize the fraction of solid, the time evolution of the fraction solid is $\partial f_S / \partial T = -1/\Delta T_o$, where ΔT_o is the solidification interval. Thus, the discretized governing equation is:

$$T^{new} = T^{old} + \frac{Fo}{a}\left(T_E^{old} - 2T^{old} + T_W^{old}\right) \quad \text{where} \quad a = 1 + \frac{\Delta H_f}{c \cdot \Delta T_o} \tag{6.28}$$

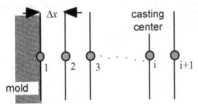

Figure 6.10. Problem discretization.

Note that ΔH_f is expressed in J/kg. To solve this equation we need an initial condition and two boundary conditions. The initial condition is $T_{init} = 1350$ °C. The first BC in node 1 is a convective boundary condition. Its basic discretized form is given by Eq. (6.16). However, this equation must be modified to include heat generation:

$$h A\left(T - T_{mold}\right) = v \, \rho \Delta H_f \frac{\partial f_s}{\partial t} - v \rho c \frac{\partial T}{\partial t} + k A \frac{\partial T}{\partial x}$$

or, in discretized format:

$$T^{new} = \frac{2Fo}{a}\left(T_E^{old} + b \cdot T_{mold}\right) + \left[1 - \frac{2Fo}{a}(b+1)\right]T^{old} \quad \text{where} \quad b = \frac{h\,\Delta x}{k} \tag{6.29}$$

The second BC in node i is a zero flux condition. That is: $\partial T/\partial x = 0$. An additional node, $i + 1$, past the symmetry line is added. In discretized form BC2 is:

$$T_{i+1} = T_i \tag{6.30}$$

The time step is obtained from the stability criterion, $\Delta t \le \left(\Delta x^2 / 2\alpha\right)\left(h\Delta x/k + 1\right)^{-1}$. The grid size is $\Delta x = l/(i-1)$.

The spreadsheet is organized as shown in Table 6.3. This is a temporal (columns) and spatial (rows) grid. The times in column A result from the summation of the time step. The first row shows the nodes. Node 1 is next to the mold and node 10 is in the center of the casting. Node 11 is the additional node past the symmetry line. The numbers in columns B through C are temperature. Row 1 is for the initial condition, $T_{init} = 1350$ °C. Columns B and L are for the boundary conditions. Eq. (6.28) is used in all the interior nodes.

Table 6.3. Spreadsheet structure for the 1D heat transport problem.

	A	B	C	D		J	K	L
	Time	i1	i2	i3		i9	i10	i11
	Eq.	(6.29)	(6.28)	(6.28)		(6.28)	(6.28)	(6.30)
1	0.00	1350	1350	1350		1350	1350	1350
2	2.60	1287	1350	1350		1350	1350	1350
3	5.20	1274	1326	1350		1350	1350	1350
n	704.6	797	844	890		1114	1144	1144

Since the constant a becomes 1 before and after solidification, a in Eq. (6.28) is substituted with the following IF statement: IF(OR($T > T_{eut}$, $T < (T_{eut} - \Delta T)$), 1, a). Eqs. (6.29) and (6.30) are used for the boundary cells, as shown in the table.

The computational results are shown in Figure 6.11. Several bumps are seen on the cooling curves of nodes 1, 4 and 7. This is because these nodes "feel" the influence of eutectic solidification of adjacent nodes. A finer mesh or an implicit method should be used to avoid this problem.

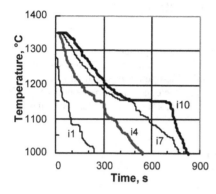

Figure 6.11. Cooling curves at different positions in the eutectic iron casting.

Another concern is that the same eutectic arrest temperature is shown for all nodes. This is unrealistic, since higher cooling rate should result in lower eutectic arrest. This is a consequence of the simplifying assumption that the fraction solid is only function of temperature.

As seen from the table, the solidification time (the time at which the center of the casting is solid) is of 704 s. When Chvorinov's equation (Eq. 5.19) is used for calculation with $\Delta T_{super} = 1350 - 1154 = 196°C$, the solidification time is 1626 s. This is considerably larger than that obtained from the 1D FDM calculation. This means that the choice of heat transfer coefficient is not correct. If h is decreased, the solidification time is increased.

Application 6.2

Consider an L-shaped mold cavity with the dimensions given in Fig. (a). A 0.6 % C steel is poured in this mold at an initial temperature of 1550 °C. Calculate the solidification time of this casting using a 2D FDM scheme and the enthalpy method.

Answer:
A 2-D grid is superimposed on the casting, as shown in Figure 6.12. Four different types of nodes are observed. Discretized equations must be written for each node, as follows:

- inside node: from Eq. (6.4) written in 2D format, and using the enthalpy method as in previous examples we obtain:

$$T^{new} = \frac{Fo}{a}\left(T_E^{old} + T_W^{old} + T_N^{old} + T_S^{old}\right) + \left(1 - \frac{4Fo}{a}\right)T^{old} \quad \text{where} \quad a = 1 + \frac{\Delta H_f}{c \cdot \Delta T}$$

- side node (left side):

$$T^{new} = \frac{Fo}{a}\left(T_N^{old} + T_S^{old} + 2T_E^{old} + 2bT_{mold}\right) + \left[1 - \frac{2Fo}{a}(b+2)\right]T^{old} \quad \text{where} \quad b = \frac{h \cdot \Delta s}{k}$$

- outside corner node (top left):

$$T^{new} = \frac{2Fo}{a}\left(T_S^{old} + T_E^{old} + 2bT_{mold}\right) + \left[1 - \frac{4Fo}{a}(b+1)\right]T^{old}$$

• inside corner node

$$T^{new} = \frac{2}{3}\frac{Fo}{a}\left(T_N^{old} + T_E^{old} + 2T_W^{old} + 2T_S^{old} + 2bT_{mold}\right) + \left[1 - \frac{4}{3}\frac{Fo}{a}(b+3)\right]T^{old}$$

The time step is selected based on the stability criterion presented in Application 6.1. Then, a number of cells are selected on the Excel sheet such that each cell corresponds to a node in Figure 6.12. We will call this grid 1. The initial temperature is written in each cell. A second group of cells is selected in a similar way and the equations corresponding to each node is written in the cells. We will call this grid 2. Then, a macro is written using the following sequence:

• COPY grid 2 (cells with equations)
• PASTE SPECIAL, VALUE in grid 1

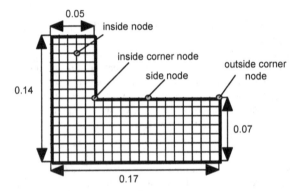

Figure 6.12. Casting dimensions and grid.

In this way an iteration in time has been created. The macro is saved, a short cut key is assigned to it and the calculation can be started. By holding the short cut key down, continuous iterations are obtained. Before starting the iterations, grid 2 will look as shown in Table 6.4. A similar iteration can be created between two other cells to record the time corresponding to the temperature in the cells. The data can be output using Excel surface chart type. An example of the graphic output after 116 s is given in Figure 6.13. The central area of the casting is still in a mushy state.

Table 6.4. Iteration grid.

1386	1468	1468	1468	1386											
1468	1550	1550	1550	1468											
1468	1550	1550	1550	1468											
1468	1550	1550	1550	1468											
1468	1550	1550	1550	1468											
1468	1550	1550	1550	1468											
1468	1550	1550	1550	1495	1468	1468	1468	1468	1468		1468	1468	1468	1468	1386
1468	1550	1550	1550	1550	1550	1550	1550	1550	1550		1550	1550	1550	1550	1468
1468	1550	1550	1550	1550	1550	1550	1550	1550	1550		1550	1550	1550	1550	1468
1468	1550	1550	1550	1550	1550	1550	1550	1550	1550		1550	1550	1550	1550	1468

1468	1550	1550	1550	1550	1550	1550	1550	1550	1550		1550	1550	1550	1550	1468
1468	1550	1550	1550	1550	1550	1550	1550	1550	1550		1550	1550	1550	1550	1468
1468	1550	1550	1550	1550	1550	1550	1550	1550	1550		1550	1550	1550	1550	1468
1386	1468	1468	1468	1468	1468	1468	1468	1468	1468		1468	1468	1468	1468	1386

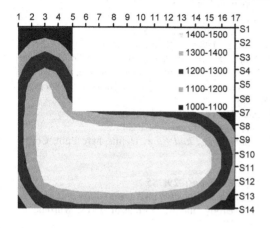

1 2 3 4 5 6 7 8 9 10 11 12 13 14 15 16 17

1400-1500 S1
S2
1300-1400 S3
1200-1300 S4
1100-1200 S5
S6
1000-1100 S7
S8
S9
S10
S11
S12
S13
S14

Figure 6.13. Isotherms after 116 s. Initial temperature was 1550 °C.

References

Bäckerud L., Chai G., and Tamminen J., 1990, *Solidification Characteristics of Aluminum Alloys: Volume 2, Foundry Alloys*, AFS/Skanaluminimu, Des Plaines, Illinois

Beech J., Barkhudarov M., Chang K., Chin S.B., 1998 in: *Modeling of Casting Welding and Advanced Solidification Processes VIII*, B.G. Thomas and C. Beckermann, eds., The Minerals, Metals and Materials Soc., Warrendale. PA, p.1071

Bennon W.D., Incropera F.P., 1987, *Int. J. Heat Mass Transfer* **30**:2161,2171

Bounds S., Moran G., Pericleous K., Cross M. and Croft T.N., 2000, *Metall. Mater. Trans. B* **31B**:515

Carman P.C., 1937, *Trans. Inst. Chem. Eng.* **15**:150

Chang S., 1994, Numerical Modeling of Micro- and Macro-Segregation in Casting Alloys, *PhD Disseratation*, The Univ. of Alabama, Tuscaloosa

Chang S., Stefanescu D.M., 1996, *Metall. Mater. Trans.* **27A**:2708

Chvorinov N., 1940, *Giesserei* **27**:201

Diao Q.Z., Tsai H.L., 1993, *Metall. Trans.* **24A**:963

Fellicelli S.D., Heinrich J.C., and Poirier D.R., 1991, *Metall. Trans.* **22B**:847

Flemings M.C. and Nereo G.E., 1967, *Trans. AIME* **239**:1449

Hansen P. N., Sahm P.R., and Flender E., 1993, *Trans. AFS* **101**:443

Hansen P.N. and Sahm P.R., 1998, in: *Modeling of Casting and Welding Processes IV*, eds. A.F. Giamei and G.J. Abbaschian, , TMS, Warrendale, PA, p.33

Henzel J.G. and Keverian J., 1965, *J. of Metals* **17**:561

Himemiya T. and Umeda T., 1998, *ISIJ Intern.* **38**:730

Hirt C.W. and Nichols B.D., 1981, *J. Computational Physics* **39**:201

Huang H., Suri V.K., EL-Kaddah N. and Berry J.T., 1993, in *Modeling of Casting, Welding and Advanced Solidification Processes VI*, eds. T.S. Piwonka, V. Voller and L. Katgerman, TMS, Warrendale, Pa, p.219

Hummer R., 1988, *Cast Metals* **1**:62

Imafuku I. and Chijiiwa K., 1983, *AFS Trans.* **91**:527

Jiarong L.I., Liu B., Xiang H., Tong H., and Xie Y., 1995, in: *Proceedings of the 61st World Foundry Congress*, International Academic Publishers, Beijing China, p.41

Kato H. and Cahoon J.R., 1985, *Metall. Trans.* **16A**:579

Kreyszig E., 1988, *Advanced Engineering Mathematics*, John Wiley & Son, New York

Mehrabian R., Keane M., and Flemings M.C., 1970, *Metall. Trans.* **1**:1209

Metzner A.B., 1985, *Rheology of suspensions in polymeric liquids*, **29**:739

Lee P.D., Chirazi A., and See D., 2001, *J. Light Metals*, **1**:15

Lee Y.W., Chang E. and Chieu C.F., 1990, *Met. Trans. B* **21B**:715

Niyama E., Uchida T., Morikawa M. and Saito S., 1982, *AFS Cast Metals Research J.* **7**:52

Ohnaka I., 1986, in *State of the Art of Computer Simulation of Casting and Solidification Processes*, H. Fredriksson ed., Les Editions de Physique, Les Ulis, France, p.211

Ohnaka I., 1993, in *Modeling of Casting, Welding and Advanced Solidification Processes VI*, Eds. T. S. Piwonka et al., TMS, p.337

Oldenburg C.M. and Spera F.J., 1992, *Numer. Heat Transfer B* **21**:217

Özisic M.N., 1994, *Finite Difference Methods in Heat Transfer*, CRC Press

Patankar, S.V., 1980, *Numerical Heat Transfer and Fluid Flow*, Hemisphere Publ. Corp., New York

Pellini W.S., 1953, *Trans. AFS* **61**:61

Pham Q.T., 1986, *International J. of Heat & Mass Transf.* **29**:285

Piwonka T.S., 2000, in: *Proc. Merton C. Flemings Symposium on Solidification and Materials Processing*, eds. R. Abbaschian, H. Brody, and A. Mortensen, TMS, Warrendale Pa., p.363

Piwonka T.S. and Flemings M.C., 1966, *Trans. AIME*, **236**:1157

Poirier D.R., Nandapurkar P.J., and Ganesan S., 1991, *Metall. Trans.* **22B**:1129

Poirier D.R., and Geiger G.H., 1994, *Transport Phenomena in Materials Processing*, TMS Minerals Metals Materials, Warrendale Pa. pp. 571-598

Sahm P.R., 1991, in *Numerical Simulation of Casting Solidification in Automotive Applications*, C. Kim and C.W. Kim eds., TMS, p.45

Schneider M.C. and Beckermann C., 1995, *Metall. Trans.* **26A**:2373

Sigworth G.K. and Wang C., 1993, *Met. Trans. B*, **24B**:365

Spittle J.A., Almeshhedani M. and Brown S.G.R., 1995, *Cast Metals* **7**:51

Suri V.K. and Paul A.J., 1993, *Trans. AFS* **144**:949

Suri V.K., Paul A.J., EL-Kaddah N. and Berry J.T., 1994, *Trans. AFS* **138**:861

Suri V.K. and Yu K.O., 2002, in *Modeling for Casting and Solidification Processing*, K.O. Yu editor, Marcel Dekker, New York, p.95

Stefanescu D.M., Giese S.R., Piwonka T. S. and Lane A., 1996, *AFS Trans.* **104**:1233

Stefanescu D.M. and T.S. Piwonka, 1996, in *Applications of Computers, Robotics and Automation to the Foundry Industry*, Proceedings of the Technical Forum, 62nd World Foundry Congress, Philadelphia, PA, CIATF, American Foundrymen's Soc., Inc., p.62

Stefanescu D.M., 2002, *Science and Engineering of Casting Solidification*, Kluwer Academic/Plenum Publishers, New York, 342p

Stefanescu D.M., 2005, *Int. J. Cast Metals Res.* **18**(3):129-143

Thomas D.G., 1965, *J. of Colloid Science* **20**:267

Upadhya G.K. and Paul A.J., 1992, *Trans. AFS* **100**:925

Van Doormaal J.P. and G.D. Raithby, 1984, *Numer. Heat Transfer*, **7**:147

Walther W.D., Adams C.M., and Taylor H.F., 1956, *Trans. AFS* **64**:658

Winterscheidt D.L. and Huang G.X., 2002, in *Modeling for Casting and Solidification Processing*, K.O Yu editor, pp17-54

7

MICRO-SCALE PHENOMENA AND INTERFACE DYNAMICS

Interface dynamics deals with phenomena occurring at scales smaller than the macro-scale used so far in our analysis of casting and solidification, but larger than the atomic scale. This scale is usually called micro-scale (micrometer size). Phenomena occurring at this scale determine the microscopic shape of the interface.

For solidification to occur, energy must be transported out of the part (or volume element), as discussed in Chapter 5. The equations describing this transport for the particular case of thermal resistance in the mold, or resistance at the mold/metal interface are equations (5.32) and (5.24), respectively. Discretized in time-stepping format they are (see also Eq. 5.35):

$$T^{new} = T^{old} - \frac{A}{v \rho c} \sqrt{\frac{k_m \rho_m c_m}{\pi t^{n+1}}} \left(T^{new} - T_m\right)\Delta t + \frac{\Delta H_f}{c} \Delta f_S \qquad (7.1)$$

$$T^{new} = T^{old} + \frac{\rho \Delta H_f v \Delta f_S - hA\left(T^n - T_0\right)\Delta t}{\rho c v} \qquad (7.2)$$

These equations can be solved to obtain the current temperature in the volume of interest based on the temperature of the previous time step, assuming the change in fraction solid, Δf_S, is known. In macro-transport analysis (MT) Δf_S is assumed some function of temperature. However, if information is available on the fraction solid evolution at the micro-scale level, then Δf_S can be calculated directly, without additional assumptions.

The fraction solid is the sum of the volume of the solid grains divided by the total volume, that is $f_S = N \cdot v_S$, where N is the number of grains per unit volume and v_S is the sum of the volumes of the solid grains. The change in time of the fraction solid is then:

$$\frac{\partial f_S}{\partial t} = \frac{\partial}{\partial t}\left(\frac{\partial N}{\partial t} v + \frac{\partial v}{\partial t} N\right)$$

where $\partial N/\partial t$ is the grain nucleation and $\partial v/\partial t$ is their growth. It is thus clear that the two fundamental phenomena that determine the fraction solid evolution are grain nucleation and grain growth. The same phenomena are also responsible for the solidification interface morphology. They will be discussed in some detail in the following sections.

7.1   Nucleation

To grasp a good understanding of nucleation, a treatment of the problem at the nano-scale level is required. However, for the present discussion, it will be sufficient to consider nuclei as some solid particles of micron-scale size that can serve as substrates for growing grains, without necessarily explaining their nature. The problem is then to establish their population distribution throughout the melt. A more detailed analysis of nucleation will be conducted in the chapter on nano-scale phenomena.

Once the solid has nucleated and started growing, the shape of the liquid/solid interface is the result of competition between the effects of various transport phenomena. To describe interface dynamics, and thus formulate the growth velocity of the grain, it is necessary to discuss in depth the role of diffusive and convective energy and mass transport on interface stability, as well as the resulting interface shape (grain morphology).

Let us start by discussing the formulation of the nucleation rate, $\dot{N}$. Nucleation is of extreme importance to microstructure evolution. It affects solidification undercooling, heat evolution during solidification, and thus cooling rate, as well as the final number of grains.

It is convenient to classify the types of nuclei available in the melt as resulting from:

- homogeneous nucleation
- heterogeneous nucleation
- dynamic nucleation

Homogeneous nucleation, which implies that growth is initiated on substrates having the same chemistry as the solid, is not common in casting alloys.

Heterogeneous nucleation is based on the assumption that the development of the grain structure occurs upon a family of substrates of chemistry different than that of the solid. These substrates have variable potencies and population densities. Such a heterogeneous model has some obvious justification when substrate particles are deliberately introduced into a melt to promote equiaxed grain formation. This is common practice in liquid processing of metallic alloys. The formal heterogeneous nucleation theory is based on the assumption that nucleation is only a function of the temperature and potency of an existing nucleant.

However, in many cases deliberate additions are not made to the melt, and yet equiaxed grain formation takes place in bulk liquid at small undercoolings. Experiments show that dynamic conditions in the liquid may influence nucleation. At least two mechanisms have been proposed for *dynamic nucleation*, the big bang mechanism and the crystal fragmentation mechanism.

Before discussing the details of these models let us analyze the behavior of the liquid metal in the proximity of the S/L interface. When liquid metal is poured into the mold, liquid motion is induced from the pouring momentum as well as from thermo-solutal convection. Thermo-solutal convection is generated by the difference in density within the molten metal. The metal close to the interface is colder, and thus in most cases denser, than the metal in the middle of the casting. Consequently, flow in the direction of the gravity vector will develop close to the interface. Ascending flow will develop toward the middle of the casting (Figure 7.1). This is thermal convection. Similarly, the differences in composition next to the S/L interface and in the bulk liquid will induce differences in density, which in turn will produce flow parallel to the interface. The direction of the flow will depend on the density difference between the solute and the solvent. Because of these combined effects, the liquid at the S/L interface is in motion.

The *big bang mechanism* (Chalmers, 1962) assumes that grains can grow from the pre-dendritic nuclei formed during pouring by the initial chilling action of the mold. These grains are then carried into the bulk by fluid flow and survive until superheat has been removed. This model relies on the action of convective currents within the melt.

Compelling experimental evidence for this mechanism has been provided for example by Davies (1973) and by Ohno (1987). Figure 7.2 shows the results of experiments with a steel net inserted in the middle of a crucible where an Al-2% Cu alloy was allowed to solidify. It is that in the lower part only columnar grains are formed. In the upper part, a large number of small equiaxed grains are seen in the vicinity of the sieve. It is argued that these grains have originated at the mold wall, then have been carried by convection currents in the middle and the ingot, and have been prevented from sinking by the sieve.

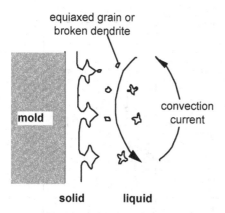
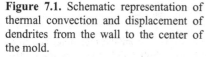

Figure 7.1. Schematic representation of thermal convection and displacement of dendrites from the wall to the center of the mold.

Figure 7.2. Equiaxed grain accumulation on a steel sieve inserted in a solidifying Al-2% Cu alloy (Ohno, 1987). With kind permission of Springer Science and Business Media.

Another line of thinking invokes the argument of fragmentation of existing crystals through ripening and local remelting of columnar dendrites (*e.g.*, Jackson *et al.*, 1966). Indeed, for single-phase alloys, a dendrite detachment mechanism has been shown to operate on transparent organic alloys (Figure 7.3). Nuclei for the equiaxed zone in the middle of an ingot originate from the detached dendrite arms that are carried to the center of the mold by convection currents. If the center of the mold is still above the liquidus, the crystals swept into the center of the mold remelt (Figure 7.3a). If the center of the mold is undercooled, these crystals act as nuclei for equiaxed grains (Figure 7.3b). This is often referred to as *big-bang nucleation*. In the case of eutectics, low gravity experiments have also shown that for regions solidified under low gravity, where convection currents are dramatically reduced, the number of eutectic grains is smaller than for the regions solidified under high gravity (Figure 7.4).

a) 1 min after pouring   b) 2 min after pouring   c) 2.5 min after pouring

Figure 7.3. Broken dendrite branches transported in the center of the ingot by liquid convection in an NH_4Cl-H_2O system ($T_L = 50°C$) poured at $75°C$ (Jackson *et al.*, 1966).

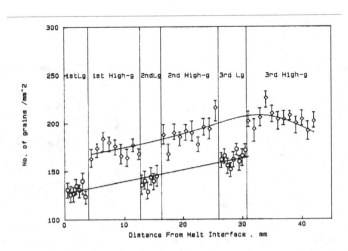

Figure 7.4. Variation of eutectic grain density with distance from the melt interface in a directional solidification experiment conducted during parabolic flight (Tian *et al.*, 1990). With kind permission of Springer Science and Business Media.

Fragmentation can also be induced through increased convection, ultrasonic vibrations, or a pressure pulse. In the last two cases, nucleation follows because of the change in equilibrium temperature caused by the pressure changes during the collapse of cavitation bubbles. A more detailed discussion of dynamic nucleation can be found in the treatment by Flood and Hunt (1988).

Based on the preceding discussion, it is reasonable to assume that, even in the presence of deliberate grain refining additions, there do exist, at all times, other inherent identifiable "nuclei". It is therefore not surprising that estimation of the volumetric density of nucleation sites before and during solidification of casting alloys is not a trivial problem.

Evaluation of nucleation laws required to calculate the volumetric density of growing grains, is the weak link in the computer simulation of microstructure evolution. A summary of the models proposed for the quantitative description of nucleation is given in the following sections.

7.1.1   Heterogeneous nucleation models

Without thermosolutal convection to transport dendrite fragments from the mushy zone to the bulk liquid, it is reasonable to assume that nucleation of equiaxed grains is based on heterogeneous nucleation mechanisms. While at least two significant methods, based on the heterogeneous nucleation theory have been developed, they are empirical in essence, and rely heavily on metal- and process-specific experimental data. The *continuous nucleation model* (Figure 7.5a) assumes a continuous dependency of N on temperature. Some mathematical relationship is then provided to correlate nucleation velocity, $\partial N/\partial t$, with undercooling, ΔT, cooling rate, or temperature. A summation procedure is carried on to determine the final number of nuclei.

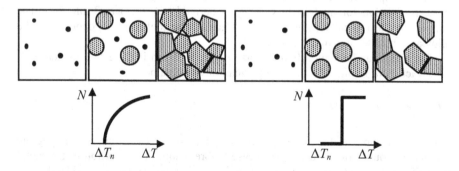

a) continuous nucleation   b) instantaneous nucleation (site saturation)

Figure 7.5. Schematic comparison between assumptions of instantaneous and continuous nucleation models (Stefanescu, 1995).

The *instantaneous nucleation model* assumes site saturation that is all nuclei are generated at the nucleation temperature, T_N (Figure 7.5b). Again, an empirical relationship must be provided to correlate the final number of nuclei (grains) in a volume element with ΔT or $\dot{T}$.

A summary of the basic equations and of the parameters that must be assumed or experimentally evaluated is given in Table 7.1. It is seen that all models require either two or three fitting parameters.

Table 7.1. Summary of nucleation models (Stefanescu, 1995).

Model	Type	Basic equation	Fitting parameters	Eq.
Oldfield (1966)	continuous	$\dfrac{dN}{dt} = -n\mu_1 (\Delta T)^{n-1} \dfrac{dT}{dt}$	n, μ_1	(7.3)
Maxwell / Hellawell (1975)	continuous	$\dfrac{dN}{dt} = (N_s - N_i)\mu_2 \exp\left[-\dfrac{f(\theta)}{\Delta T^2 (T_p - \Delta T)}\right]$	N_S, θ	(7.4)
Thévoz et al. (1989)	continuous (statistical)	$\dfrac{\partial N}{\partial(\Delta T)} = \dfrac{N_s}{\sqrt{2\pi}\,\Delta T_\sigma} \exp\left[\dfrac{(\Delta T - \Delta T_N)^2}{2(\Delta T_\sigma)^2}\right]$	$N_S, \Delta T_N, \Delta T_\sigma$	(7.5)
Goettsch / Dantzig (1994)	continuous (statistical)	$N(r) = \dfrac{3N_s}{(R_{max} - R_{min})^3}(R_{max} - r)^2$	N_S, R_{max}, R_{min}	(7.6)
Stefanescu et al. (1990)	instantaneous	$N = a + b \cdot \dot{T}$	a, b	(7.7)

In Oldfield's (1966) continuous model, a power law function, $N = \mu_1(\Delta T)^n$, fitting the experimental data on cast iron, was used to evaluate the final number of grains. A typical experimental and fitted N-ΔT dependency is given in Figure 7.6. From this, an equation for nucleation velocity is derived (Eq. (7.3) in Table 7.1). The exponent n has typically values between 1 and 2. The coefficient μ_1 depends on the alloy and includes inoculation effects.

Eq. (7.3) was later modified (Lacaze et al., 1989) to include the residual volume fraction of liquid. The following equation was obtained:

$$dN = -n\mu_1 (\Delta T)^{n-1} f_L dT \tag{7.8}$$

However, it appears that for equiaxed solidification this correction is negligible because the nucleation process will cease before a significant portion of the metal has solidified. Calculations for the peritectic Al-Ti system (Maxwell and Hellawell, 1975) showed that when nucleation was completed the fraction of solid was a mere 10^{-4}. Nevertheless, it is claimed that such an equation gives a better description of nucleation during directional solidification (Lesoult, 1991).

In the more fundamental model developed by Maxwell and Hellawell (1975), Eq. (7.4) in Table 7.1, N_s is the number of heterogeneous substrates, N_i is number of particles that have nucleated at time i, T_p is the peritectic temperature, and $f(\theta)$ is

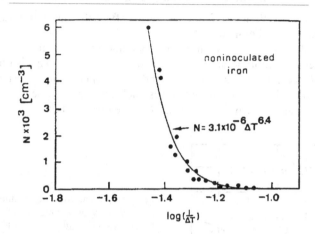

Figure 7.6. Correlation between undercooling and volumetric nucleation density of eutectic grains in eutectic cast iron (Mampey, 1988).

the classic function of contact angle. The derivation of Eq. (7.4) will be discussed later, at the nano-scale level.

Other continuous nucleation models introduced some statistical functions to help describe the rather large size distribution of grains sometimes encountered in castings. A Gaussian (normal) distribution of number of nuclei with undercooling was introduced by Thévoz et al. (1989), Eq. (7.5) in Table 7.1. In this equation, ΔT_N is the average nucleation undercooling and ΔT_σ is the standard deviation. The same distribution was used by Mampey (1991) to model spheroidal graphite iron solidification. However, rather than applying this distribution to the number of nuclei, he applied it to the size of nuclei. To avoid the complication of having to specify θ for heterogeneous nucleation, the width of the substrate was used as a function of undercooling ($K_2/\Delta T$).

Goettsch and Dantzig (1994) assumed a quadratic distribution of the number of grains as a function of their size, $N = a_o + a_1 r + a_2 r^2$. This allows calculation of the number of nuclei of a given radius r, $N(r)$, as a function of the total number of substrates, the maximum grain size, R_{max}, and the minimum grain size, R_{min} (see Eq. (7.6) in Table 7.1).

The main assumption used in the instantaneous nucleation model (Stefanescu et al., 1990) is that all nuclei are generated at the nucleation temperature. The fundamentals of the model are based on Hunt's (1984) equation for heterogeneous nucleation:

$$\frac{dN}{dt} = (N_s - N_i)\mu_3 \exp\left(-\frac{\mu_4}{\Delta T^2}\right) \tag{7.9}$$

Calculations for eutectic cast iron showed that all substrates became nuclei over a very short time. Therefore, this equation can be substituted by

$\partial N / \partial t = N_s \delta(T - T_N)$, where δ is the Dirac delta function. Integration gives a total number of nuclei of N_S at T_N. While μ_3 and μ_4 affect the nucleation rate, only N_S will determine the final grain density.

Thus, the dependency between cooling rate and grain density reflects a direct correlation between cooling rate and the number of substrates. This is illustrated by Eq. (7.7) in Table 7.1. In this equation a and b are experimental constants. This is the most common form of the instantaneous nucleation law. Some typical numbers for cast iron of various carbon equivalents, CE, are given in Figure 7.7.

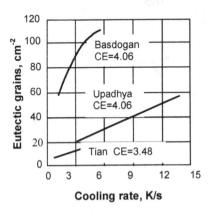

Figure 7.7. Variation of volumetric eutectic grain density as a function of cooling rate for cast iron (Upadhya *et al.* 1990, Basdogan *et al.* 1982, Tian and Stefanescu 1993).

The question is now which nucleation model works best? In principle they all work, since they are based on fitting experimental data. Thus, the issue is which one fits experimental data better. This is debatable. The main difference between the Oldfield and the Thévoz et al. models is the use of second or third order polynomial, respectively, to fit the data. In other words, they are using two and three adjustable parameters, respectively. Two adjustable parameters seem to be sufficient in most cases.

In general, for alloys that solidify with narrow solidification interval the instantaneous nucleation model is recommended, since it saves computational time. The use of the continuous nucleation model runs into computation complications, related to the definition of the dimensions of the micro-volume element (diffusion distance), when applied to equiaxed dendritic solidification, unless complete solute diffusion is assumed in the liquid.

Experimental evaluation of heterogeneous nucleation laws has been traditionally oversimplified. Typically, the final number of grains at the end of solidification is used to compute a nucleation law. However, as demonstrated through liquid quenching experiments (Tian and Stefanescu, 1993), the evaluation of a nucleation law from the final grain density may result in inaccurate data, since grain coalescence plays a significant role as seen in Figure 7.8. Indeed, the final eutectic grain density in cast iron was found to be smaller by up to 27% than the maximum number of grains developed during solidification.

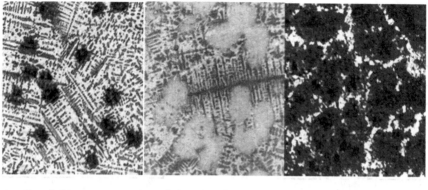

a) early solidification   b) late solidification   c) after solidification (room
                                                                 temperature)

Figure 7.8. Nucleation and coalescence of eutectic grains in cast iron (Tian, 1992).

7.1.2   Dynamic nucleation models

Because of natural convection in front of the columnar mushy zone, thermosolutal convection will disperse dendrite fragments into the bulk liquid. Nucleation of the equiaxed grains will thus depend not only on the potency of heterogeneous substrates, but also on the crystal fragments resulting from dynamic nucleation. As suggested by Steube and Hellawell (1992), a quantitative representation of nucleation should include the following steps the kinetics of fragment formation or "crystal multiplication", the transport of fragments from the liquid or mushy zone into the bulk liquid because of natural convection or forced stirring, the survival time of crystal fragments in the bulk liquid above the liquidus temperature, and the growth and sedimentation rates of fragments which survive long enough to enter a region below the liquidus temperature. This approach is physically more correct than that strictly based on heterogeneous nucleation. However, formulation of these steps with sufficient confidence is difficult at best.

7.2   Micro-solute redistribution in alloys and microsegregation

A comprehensive theoretical treatment of interface morphology requires accurate tracking of the solutal field during solidification that is a microsegregation model. Assessment of microsegregation occurring in solidifying alloys is also important because it influences mechanical properties.

The concept of solute redistribution, discussed previously at the macro-scale level, can be extended straightforward to the micro-scale. Consider for example a micro-volume element extending from the middle of a dendrite to the middle of the interdendritic liquid region (Figure 7.9 left). Similarly, the case of equiaxed grains can be considered. Further, assume that the curvature of the S/L interface is infinite, *i.e.*, the interface is planar. Then, the concepts used to develop equations for

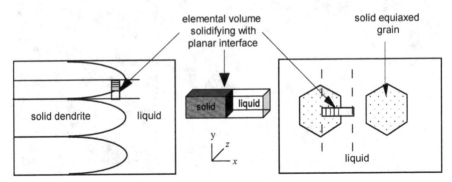

Figure 7.9. Schematic drawing for the calculation of micro-solute redistribution during solidification.

macroscopic solute redistribution can also be used for the evaluation of microscopic solute redistribution, *i.e.*, micro-segregation.

Since the dendrite geometry is rather complicated, all models start by assuming some simplified volume element over which calculations are performed (Figure 7.10). A summary of the major assumptions used in some analytical microsegregation models is given in Table 7.2. Quantitative evaluation of the extent of microsegregation can be done using some of the equations previously derived for the macro-scale. The basic equations for microsegregation models are given in Table 7.3. Many of these equations have been previously discussed and are repeated here for convenience.

The Scheil equation (7.11) can be used to calculate microsegregation when solid diffusivity is very small. However, the diffusion of solute into the solid phase can affect microsegregation significantly, especially toward the end of solidification. For example, calculations by Brooks *et al.* (1991) showed that little solid-state diffusion occurs during the solidification and cooling of primary austenite solidified welds of Fe-Ni-Cr ternary alloys, whereas structures that solidify as ferrite may become almost completely homogenized because of diffusion.

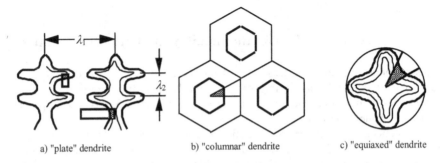

a) "plate" dendrite   b) "columnar" dendrite   c) "equiaxed" dendrite

Figure 7.10. Schematic representation of models for microsegregation. A "plate dendrite" is essentially a 1D dendrite, a "columnar dendrite" is a 2D dendrite, an "equiaxed dendrite is a 3D dendrite.

Table 7.2. Major assumptions used in analytical microsegregation models.

Model	Geometry	Solid diffusion	Liquid diffusion	Partition coefficient	Growth	Coarsening
Lever rule	no restriction	complete	complete	variable	no restriction	No
Scheil (1942)	no restriction	no	complete	constant	no restriction	No
Brody-Flemings (1966)	no restriction	incomplete	complete	constant	no restriction	No
Clyne-Kurz (1981)	no restriction	spline fit	complete	constant	no restriction	No
Ohnaka (1986)	linear, columnar	quadratic equation	complete	constant	linear parabolic	No
Sarreal-Abbaschian (1986)	no restriction	limited	complete	constant	no restriction	No
Kobayashi (1988)	columnar	limited	complete	constant	linear	No
Nastac-Stefanescu (1993)	plate, columnar, equiaxed	limited	limited	variable	no restriction	Yes

Table 7.3. Equations for models in Table 7.2.

Model	Equation	Eq. no
Lever rule	$C_S = kC_o\big/\big[(1-f_S)+kf_S\big]$	(7.10)
Scheil (1942)	$C_S = kC_o(1-f_S)^{k-1}$	(7.11)
Brody-Flemings (1966)	$C_S = kC_o\big[1-(1-2\alpha k)f_S\big]^{(k-1)/(1-2\alpha k)}$ with $\alpha = 4D_S t_f/\lambda^2$	(7.12)
Clyne-Kurz (1981)	$C_S = kC_o\big[1-(1-2\Omega k)f_S\big]^{(k-1)/(1-2\Omega k)}$ with $\Omega = \alpha\big[1-\exp(-1/\alpha)\big]-0.5\exp(-1/2\alpha)$	(7.13)
Ohnaka (1986)	$C_S = kC_o\big[1-(1-2\beta k)f_S\big]^{(k-1)/(1-2\beta k)}$ with $\beta = 2\gamma/(1+2\gamma)$ $\gamma = 8D_S t_f/\lambda_I^2$	(7.14)
Kobayashi (1988)	$C_S = kC_o\,\xi^{(k-1)/(1-\beta k)}\big\{1+\Gamma\big[0.5(\xi^{-2}-1)-2(\xi^{-1}-1)-\ln\xi\big]\big\}$ with $\xi = 1-(1-\beta k)f_S$ $\Gamma = \beta^3 k(k-1)\big[(1+\beta)k-2\big](4\gamma)^{-1}(1-\beta k)^{-3}$	(7.15)
Nastac-Stefanescu (1993)	$C_S^* = kC_o\left[1-\dfrac{(1-k)f_S}{1-(m+1)\big(kI_S^{(m+1)}+I_L^{(m+1)}\big)}\right]^{-1}$ $f_S = \big(r^*/r_f\big)^{m+1}$ see text for I_S and I_L	(7.16)

Note than for $f_S = 1$, the Scheil equation predicts $C_S = \infty$, which is impossible. When examining for example Figure 4.4 it is seen that when the composition of the solid reaches the maximum solubility in the solid, C_{SM}, the composition of the liquid reaches the eutectic composition. Consequently, the remaining liquid will solidify as eutectic, and thus, the Scheil equation is not valid anymore. However, the *Scheil* equation can be used to calculate the amount of eutectic that will form at the end of single-phase alloys solidification, as $f_E = 1 - f_{SM}$, where f_E is the fraction of eutectic and f_{SM} is the fraction of primary phase formed when the composition of the solid reached C_{SM}.

Clyne and Kurz (CK) used the Brody-Flemings (BF) model and added a spline fit (the term Ω in Eq. (7.13)) forcing predictions of the BF model to match predictions by Scheil equation and the equilibrium equation for infinitesimal and infinite diffusion coefficient, respectively. This relation has no physical basis. All these models are 1D Cartesian and describe "plate" dendrite solidification.

The BF and CK analyses were used to explain microsegregation in Al-Cu and Al-Si alloys at cooling rate up to 200 K/s (Sarreal and Abbaschian, 1986). For higher cooling rate a new equation based on the BF model that includes the effects of dendrite tip undercooling and eutectic temperature depression was developed.

Ohnaka (1986) proposed a model for a "columnar" dendrite (Figure 7.10b, Eq. (7.14)). Complete mixing in the liquid and parabolic growth was assumed. Based on an assumed profile ($C_s = A + Bx + Cx^2$), an equation for solute redistribution in the solid, that includes Clyne-Kurz equation, was derived. Note that for $D_S = 0$ this equation reduces to the Scheil equation, and for $D_S \to \infty$ it becomes the equilibrium equation. However, the diffusion equation was not directly solved. Prior knowledge of the final solidification time is required.

Nastac and Stefanescu (1993) (NS) have proposed a complete analytical and a numerical model for 'Fickian' diffusion with time-independent diffusion coefficients and zero-flux boundary condition in systems solidifying with equiaxed morphology (Figure 7.10c, Eq. (7.16)). The model can be used for "plate" or "equiaxed" dendrites. It takes into account solute transport in the solid and liquid phases and includes overall solute balance. The model allows a comprehensive treatment of dendritic solidification through calculation of the fraction of solid with an MT-TK model.

The main features of the model are as follows:

- solute transport in the solid and liquid phases is by diffusion with diffusion coefficients independent of concentration; diffusion depends only on the radial coordinate, r; Fick's second law must be satisfied in each phase:

$$\frac{\partial C_S}{\partial t} = \frac{1}{r^m} \frac{\partial}{\partial r} \left(r^m D_S \frac{\partial C_S}{\partial r} \right) \quad \text{and} \quad \frac{\partial C_L}{\partial t} = \frac{1}{r^m} \frac{\partial}{\partial r} \left(r^m D_L \frac{\partial C_L}{\partial r} \right)$$

where $m = 1, 2$ or 3, for plate, cylindrical and spherical geometry, respectively;

- closed system is assumed (no solute flow into or out of the volume element); the overall solute balance can then be written in integral form as:

$$\frac{1}{\rho v} \int_v \rho C(r,t) dv = C_o$$

- the boundary conditions are as follows:
  at the interface $C_S^* = k\, C_L^*$
  at $r = 0$ $D_S \cdot \partial C_S / \partial r = 0$
  at $r = R_f$ $D_L \cdot \partial C_L / \partial r = 0$, where R_f is the final radius of the domain

The final exact analytical solution obtained through the method of separation of variables is Eq. (7.16) in Table 7.3. The values of the coefficients I_S and I_L are as follows:

- for spherical geometry:

$$I_S^{(3)} = \frac{2 f_S}{\pi^2} \sum_{n=1}^{\infty} \frac{1}{n^2} \exp\left[-\left(\frac{n\pi}{f_S^{1/3}} \right)^2 \frac{D_S t}{r_f^2} \right] \quad \text{and}$$

$$I_L^{(3)} = 2 f_S^{2/3} \left(1 - f_S^{1/3} \right) \sum_{n=1}^{\infty} \frac{1}{\alpha_n^2} \exp\left[-\left(\frac{\alpha_n}{1 - f_S^{1/3}} \right)^2 \frac{D_L t}{r_f^2} \right]$$

where α_n is the n-th root of the equation $\alpha_n / \tan(\alpha_n) = 1 - f_S^{1/3}$;

- for plate geometry:

$$I_S^{(1)} = \frac{2 f_S}{\pi^2} \sum_{n=1}^{\infty} \frac{1}{(n-0.5)^2} \exp\left[-\left(\frac{(n-0.5)\pi}{f_S} \right)^2 \frac{D_S t}{r_f^2} \right] \quad \text{and}$$

$$I_L^{(1)} = \frac{2}{\pi^2} \left(1 - f_S \right) \sum_{n=1}^{\infty} \frac{1}{(n-0.5)^2} \exp\left[-\left(\frac{(n-0.5)\pi}{1 - f_S} \right)^2 \frac{D_L t}{r_f^2} \right]$$

Note that for D_S, $D_L \to \infty$, Eq. (7.16) becomes the equation for equilibrium solidification. Also, for $D_S = 0$ and $D_L \to \infty$ at $f_S = 0$ it yields $C_S^* = k\, C_o$, and at $f_S = 1$ it predicts $C_S^* \to \infty$.

However, for this last set of conditions the Scheil equation is not obtained. The reason is that the Fickian diffusion equation exhibits a singularity at $D_S = 0$. Thus, Eq. (7.16) should not be used for the particular case of $D_S = 0$.

Another limitation of the model comes from the use of the method of separation of variables. It can be demonstrated that the method holds only when the solidification velocity is much smaller than the diffusion velocity (or the diffusion time is much smaller than the solidification time). This amounts to assuming a solid-state back-diffusion coefficient ($\alpha = D_S t_f / l^2$) larger than one. Conversely, the condition can be expressed as a Péclet number smaller than one. While this is typically the case at the micro-scale level, the condition is not necessarily met for the macro-scale.

A comparison of predictions of niobium redistribution in Inconel 718 by various models is presented in Figure 7.11. Less than 2% by volume Laves phases was measured by Thompson et al. (1991). Further discussion of the applicability of various models is offered through Application 7.1.

As pointed out by Battle (1992) in his excellent review of the modeling of solute segregation, when using analytical models to evaluate microsegregation, it must be assumed that all physical properties are constant. The solid-state concentration can only be calculated at the interface, and cannot be modified by subsequent solid diffusion. In other words, only the trace of the solid-state concentration can be plotted. To obtain the average composition in the liquid and solid, respectively, the overall mass balance equations for liquid and solid must be used. Thus, a numerical scheme is required.

Many numerical micro-segregation models have also been proposed. However, the use of numerical segregation models in macro-micro solidification codes is impractical because the computational time is significantly increased.

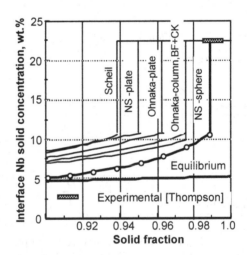

Figure 7.11. Comparison of various models for Nb redistribution in Inconel 718 solidified with equiaxed morphology. Initial Nb content was 5.25 wt.% (Nastac and Stefanescu, 1993).

Ogilvy and Kirkwood (1987) further developed the BF model to allow for dendrite arm coarsening in binary and multicomponent alloys. For binary systems the basic equation is:

$$C_L^*(1-k)\frac{dX}{dt} = D\frac{\partial C}{\partial x} + \frac{dC_L}{dt}\left(\frac{\lambda}{2} - X\right) + \frac{C_L - C_o}{2}\frac{d\lambda}{dt} \tag{7.17}$$

Here, X is the distance solidified. Thus, $f_S = 2X/\lambda$. The end term represents the increase in the size of the element due to arm coarsening, which brings in liquid of average composition that requires to be raised to the composition of the existing liquid. This equation was solved numerically under the additional assumptions of constant cooling rate and liquidus slope. Also, a correction factor for fast diffusing species was added.

Matsumiya et al. (1984) developed a 1D multi-component numerical model in which both diffusions in liquid and solid were considered. Toward the end of solidification, especially for small partition coefficients, a lower liquid concentration than predicted by the analytical models was obtained.

Yeum et al. (1989) proposed a finite difference method to describe microsegregation in a "plate" dendrite that allowed the use of variable k, D, and growth velocity. However, complete mixing in liquid was assumed.

Battle and Pehlke (1990) developed a 1D numerical model for "plate" dendrites that can be used either for the primary or for the secondary arm spacing. Diffusion was calculated in both liquid and solid, and dendrite arm coarsening was considered.

An integral profile method was used by Himemyia and Umeda (1998) to develop a numerical model that can calculate microsegregation assuming finite diffusion in both the liquid and the solid. The ordinary differential equations can be easily solved, for example, through the Runge-Kutta method.

Further complications arise when multi-component systems are considered. Chen and Chang (1992) have proposed a numerical model for the geometrical description of the solid phases formed along the liquidus valley of a ternary system for plate dendrites. Constant growth velocity, variable partition coefficients and the BF model for diffusion were used.

In many solidification models, the microsegregation problem is simplified by assuming infinite diffusivity in the liquid for all elements, and no diffusivity in the solid for substitutionally dissolved elements. Nevertheless, experiments conducted by Hillert et al. (1999) on Fe-Cr-C alloys demonstrate that while indeed C and Cr have infinite diffusivity in the liquid, and C can be assumed as having infinite diffusivity in the fcc solid, the back-diffusion of chromium cannot be ignored. In addition, back-diffusion during cooling after the end of solidification should not be neglected either.

To decrease computational time in complex solidification simulation packages it is preferably to use analytical models for microsegregation in conjunction with numerical schemes for energy and mass transport.

Let us now analyze the implications of microsegregation on microstructure. Consider for example the solidification of a spherical grain. Solidification starts with composition $k\,C_o$ and ends with composition C_E. When $k < 1$, $C_o < C_E$. In other words, if solid diffusivity is relatively small, the center (core) of the grain that solidifies first is poorer in solute than the outside shell, (Figure 7.12). If conversely $k > 1$ (right hand corner of a phase diagram), it is also possible to have the core of the grain richer in solute than its shell. This phenomenon is called *coring*. Rapid solid diffusion, or extended exposure to high temperature will decrease the extent of microsegregation, since the solid will become increasingly chemically homogeneous.

In industrial applications microsegregation is evaluated by the *microsegregation ratio* (C_{Smax}/C_{Smin}), and by the amount of non-equilibrium second phase in the case of alloys that form eutectic compounds.

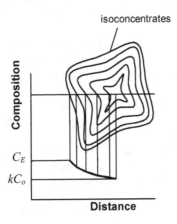

Figure 7.12. Composition variation across a grain resulting from micro-segregation.

The intensity of microsegregation depends on the value of the partition coefficient. The farther away this value is from unity, the larger the segregation. Some typical values for the partition coefficient of various elements in binary iron alloys are given in Appendix B.

7.3   Interface stability

In the preceding discussion we have not concerned ourselves with the morphology of the S/L interface. For calculation purposes it has been considered to be reasonably smooth. However, this is seldom the case in solidification of castings. It will be demonstrated that the thermal and compositional field ahead of the solidifying interface determines its morphology. If such influences are not considered, there is no reason for the interface to loose its planar morphology and become unstable. To evaluate the evolution of interface morphology, interface stability arguments are used. A perturbation is assumed to form at the interface. Then, if the perturbation is damped out in time, the interface is considered to be stable. If the perturbation is amplified in time, the interface is unstable.

7.3.1   Thermal instability

Since in pure substances there is no constitutional undercooling, only the instabilities resulting from the thermal field must be considered. In the case shown in Figure 7.13a, the temperature decreases continuously from the liquid to the solid. It is said that a positive thermal gradient exists.

Solidification will start at the mold/liquid interface, on some nuclei on the mold wall, and proceed toward the liquid. If the temperature at the S/L interface is equal to the solidification temperature, the interface is at equilibrium and cannot move. A small kinetic undercooling is required to drive the process. If a thermal instability (a local perturbation) should form and grow at the interface, it will find itself in an environment where the temperature is higher than its melting point. Consequently, this perturbation could not grow, will disappear and the interface will remain stable. Its morphology will be flat (planar). Note that on Figure 7.13a, several grains are shown in the solid region. The number of grains that will form in the solid is a function of the *nucleation potential* at the mold/liquid interface.

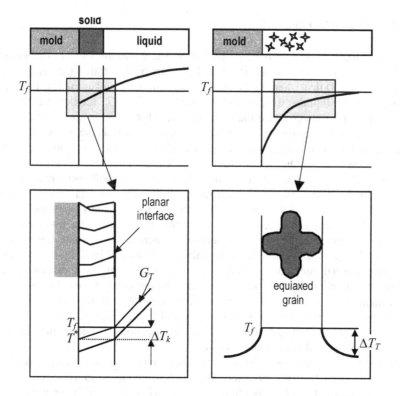

a) low thermal undercooling, positive gradient, planar interface

b) high thermal undercooling, negative gradient, equiaxed solidification

Figure 7.13. Formation of various types of structures because of instabilities generated by the thermal field.

If the melt becomes highly undercooled because of lack of nucleation sites at the mold wall, solidification can start on nuclei forming in the bulk liquid, away from the interface (Figure 7.13b). Since growth conditions in the liquid are isotropic, the new crystals will have a spherical shape at the beginning of solidification. The crystals growing in the liquid from these spherical nuclei are called *equiaxed crystals*. Because the latent heat of fusion is evolved at the grain surface, the temperature will be higher at the L/S interface than in the bulk liquid. It is said that a negative thermal gradient has occurred. The crystal surface will find itself in an undercooled environment, and will continue to grow. Local surface instabilities will also grow at the interface, and the final shape of the equiaxed crystal will not be spherical but dendritic.

Thus, in a pure metal where only thermal instabilities can occur, there are two types of possible structures:

- planar (the interface is stable) when the thermal gradient is positive
- dendritic equiaxed (the interface is unstable) when the thermal gradient is negative.

7.3.2   Solutal instability

Interface instability can also be promoted by the evolution of the compositional field ahead of the growing interface. It has been demonstrated (see for example Figs. 2.7 and 2.8) that when the thermal gradient in the liquid at the S/L interface is smaller than the liquidus temperature gradient, i.e., $G_T < G_L$, the liquid at the interface is at a lower temperature than its liquidus. This liquid is constitutionally undercooled. Instabilities growing in this region will become stable, because they will find themselves at a temperature lower than their equilibrium temperature. They will continue to grow. On the contrary, if $G_T > G_L$, the interface will remain planar.

For small constitutional undercooling, the instabilities will only grow in the solidification direction, and a cellular interface will result. This is shown schematically in Figure 7.14, and supported with pictures resulting from experimental work in Figure 7.15 b and c. The planar-to-cellular transition occurs at a gradient $G_{p/c}$.

However, as the constitutional undercooling increases because of lower thermal gradient, the spacing between the cells increases (this will be demonstrated later), and constitutional undercooling may also occur in the y-direction, perpendicular to the growth direction. Instabilities will develop on the sides of the cells, resulting in the formation of dendrites. This is the cellular-to-dendrite transition. It takes place at a temperature gradient $G_{c/d}$ (Figure 7.14 and Figure 7.15d). Both cellular and dendritic growth occurring from the wall in the direction opposite to the heat transport can be described as *columnar growth*.

If constitutional undercooling is even higher, equiaxed grains can be nucleated in the liquid away from the interface. The dendritic-to-equiaxed transition occurs at $G_{d/e}$. If the thermal gradient is almost flat, i.e., $G_T = 0$, the driving force for the columnar front will be extremely small. A complete equiaxed structure is expected.

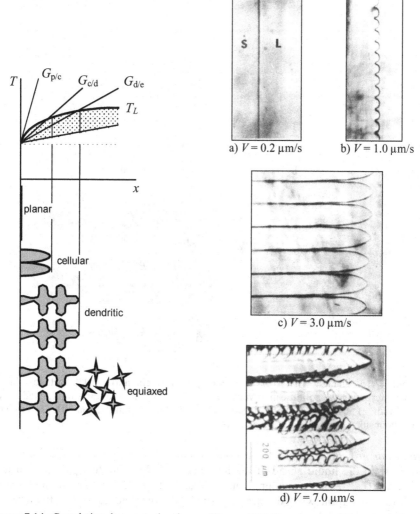

a) $V = 0.2\ \mu m/s$ b) $V = 1.0\ \mu m/s$

c) $V = 3.0\ \mu m/s$

d) $V = 7.0\ \mu m/s$

Figure 7.14. Correlation between the thermal gradient at the interface and the interface morphology.

Figure 7.15 The change of the morphology of the liquid/solid interface as a function of growth velocity in a transparent organic system (pivalic acid-0.076% ethanol) directionally solidified under a thermal gradient of 2.98K/mm (Trivedi and Kurz, 1988).

Additional information on the transition from planar to cellular and then to dendritic solidification can be obtained from Figure 7.16. Grain growth competition and the survival of the dendrites oriented preferentially in the direction of heat extraction are demonstrated. Both cellular and dendritic growth occurring from the wall in the direction opposite to the heat transport can be described as

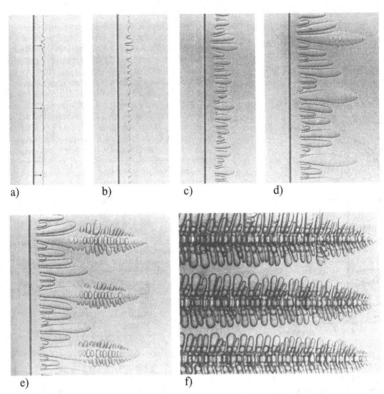

Figure 7.16. Change in interface morphology of a succinonitrile - 4% acetone solution when increasing the solidification velocity from 0 to 3.4μm/s at a temperature gradient of 6.7K/mm: a) 50s; b) 55s; c) 65s; d) 80s; e) 135s; f) 740s. Magnification 30X (Trivedi and Somboonsuk, 1984). Reprinted with permission from Elsevier.

columnar growth. If constitutional undercooling is even higher, equiaxed grain can nucleate in the liquid ahead of the interface. The dendritic-to-equiaxed transition occurs at $G_{d/e}$. If the thermal gradient is almost flat, *i.e.* $G_T = 0$, the driving force for the columnar front will be extremely small. A complete equiaxed structure will result.

Let us now have a closer look at the diffusion field at the tip of the perturbation. The driving force for the growth of the solutal perturbation is the composition gradient at the tip, $C_L^* - C_o$. Using the notations in Figure 7.17, this driving force can be expressed in a nondimensional form as:

$$\Omega_c = \frac{C_L^* - C_o}{C_L^* - C_S^*} = \frac{C_L^* - C_o}{C_L^* (1 - k)} \tag{7.18}$$

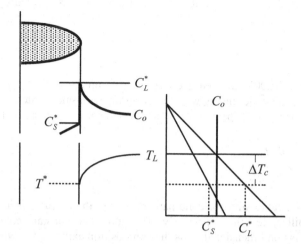

Figure 7.17. Definition of solutal supersaturation and of constitutional undercooling.

where Ω_c is the solutal supersaturation. Note that the supersaturation can vary between 0 (no transformation) and 1 (maximum transformation rate). ΔT_c is the constitutional undercooling. It can vary between 0 and ΔT_o.

Let us now try to quantify the conditions under which the planar-to-columnar transition occurs. The criterion for instability formation is $G_T < G_L$. The temperature gradient along the liquidus line at the interface is:

$$\left(G_L\right)_{x=0} = \left(\frac{dT_L}{dx}\right)_{x=0} = m_L\left(\frac{dC_L}{dx}\right)_{x=0}$$

The concentration gradient at the interface can be evaluated from the composition of the boundary layer, Eq. 4.5, to yield:

- for steady state

$$\left(\frac{dC_L}{dx}\right)_{x=0} = -C_o\frac{1-k}{k}\frac{V}{D_L}$$

- for non-steady state

$$\left(\frac{dC_L}{dx}\right)_{x=0} = -C_L^*\left(1-k\right)\frac{V}{D_L}$$

Substituting these equations in the criterion for instability formation one obtains the *criterion for constitutional undercooling*:

- for steady state

$$\frac{G_T}{V} < -\frac{m_L C_o\left(1-k\right)}{kD_L} = \frac{\Delta T_o}{D_L} \qquad (7.19a)$$

- for non-steady state:

$$\frac{G_T}{V} < -\frac{m_L C_L^*\left(1-k\right)}{D_L} \qquad (7.19b)$$

The criterion for the cellular-to-dendritic transition was derived by Laxmanan (1987) as:

$$G_T/V < \Delta T_o/(2D_L)$$

Trivedi *et al.* (2003) derived a different criterion for the cellular-to-dendritic transition based on experiments with the succinonitrile-salol system. They argued that there is a critical cell spacing, λ_{cd}, above which cells transform to dendrites:

$$\lambda_{cd} = \left(\frac{10.8}{C_o}\right)\left(\frac{\Delta T_o}{G_T}\frac{D}{V}\frac{\Gamma}{\Delta T_o}\right)$$

Note that the ratios in the second parenthesis are the characteristic thermal, solutal, and capillary length, respectively. The numerical constant contains system constants and is nondimensional, as is the composition expressed in weight %.

This simple analysis of interface stability explains the four possible S/L interfaces found in the experimental solidification of metal and alloys, under low and moderate growth velocities, as summarized in Figure 7.18. Note that by changing either the thermal gradient or the solidification velocity, the interface morphology can change from planar to equiaxed, or vice-versa, for a fixed composition C_o.

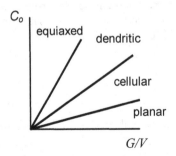

Figure 7.18. Influence of composition, thermal gradient and growth velocity on interface morphology.

In commercial alloys, cast in sand or even some metal molds, interface instability will occur because the thermal gradient is typically very small (see Application 7.2).

The constitutional undercooling criterion in Eq. (7.19) ignores the effect of interfacial energy which should inhibit the formation of perturbations, since an additional energy is required if the interface area is increased. In addition, the influence of the temperature gradient in the solid has also been ignored. A more complete analysis will now be introduced.

7.3.3   Thermal, solutal, and surface energy driven morphological instability

When the combined effects of the thermal and solutal field are considered (Figure 7.19), two types of growth can be defined for the perturbation:

- constrained: the growth rate is controlled by the temperature gradient ahead of the interface (there is no thermal undercooling) and by constitutional undercooling; heat flows from the melt, to the perturbation, to the mold, that is the melt is the hottest; if the perturbation is stable, columnar structure results; no recalescence will be seen on the cooling curve;
- unconstrained: the growth rate is controlled by the thermal and constitutional undercooling; heat flows from the perturbation (grain), to the melt, to the mold, that is the grain is the hottest; if the liquid is undercooled, the grain grows freely in the liquid; equiaxed structure results; recalescence will be seen on the cooling curve.

Thus, in constrained growth the driving force is the constitutional undercooling, ΔT_c, which is determined by the G_L - G_T difference. Unconstrained growth is controlled by both solutal and thermal undercooling, $\Delta T_c + \Delta T_T$. In addition, in both cases the curvature of the perturbation will have to be considered through the change in local surface energy. An additional undercooling, ΔTr, will be added. The stability of a perturbation must be evaluated as a function of the thermal and solutal field, and of the local surface energy (curvature).

Following the approach introduced by Mullins and Sekerka (1964) and modi-

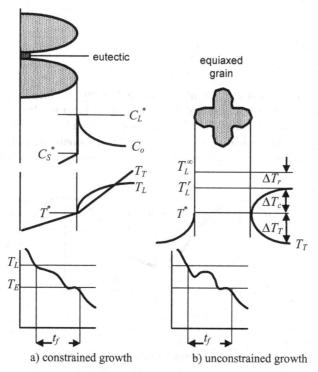

a) constrained growth   b) unconstrained growth

Figure 7.19. The combined effects of the temperature and solutal field on interface stability.

fied by Kurz and Fisher (1989), first a planar, unperturbed interface is considered. The general form of the governing equation to solve is:

$$\frac{\partial^2 \phi}{\partial y^2} + \frac{V}{\Gamma}\frac{\partial \phi}{\partial y} = 0 \tag{7.20}$$

where ϕ is the phase quantity and Γ is the general diffusion coefficient. This equation must be solved for both solute and heat diffusion. Ignoring solute diffusion in the solid because it is very small, the variables in the above equation are $\phi = C_L, T_L, T_S$, and the diffusivities are $\Gamma = D_L, \alpha_L, \alpha_S$. Note that this equation written for solute diffusion (Eq. 4.4) has been solved before to obtain the shape of the diffusion boundary layer (Eq. 4.5). The solutions for the planar interface are:

$$C_L = C_o - \frac{D_L G_L}{mV}\exp\left(-\frac{Vy}{D_L}\right) \tag{7.21a}$$

$$T_L = T_o + \frac{\alpha_L G_T^L}{V}\left[1 - \exp\left(-\frac{Vy}{\alpha_L}\right)\right] \tag{7.21b}$$

$$T_S = T_o + \frac{\alpha_S G_T^S}{V}\left[1 - \exp\left(-\frac{Vy}{\alpha_S}\right)\right] \tag{7.21c}$$

where C_o is the concentration at infinity, while T_o is the interface temperature.

If the interface looses planarity, it can be assumed that small sinusoidal perturbations are formed (Figure 7.20). The interface can then be described as a sinusoidal function, for example $y = \varepsilon \sin(\omega x)$, where ε is the amplitude and $\omega = 2\pi/\lambda$ is the wave number. The problem is now to evaluate whether the perturbation is stable or not.

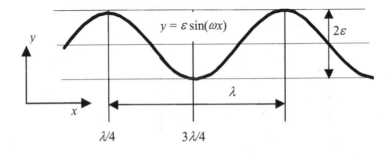

Figure 7.20. Elements of sinusoidal perturbations.

Using the perturbation technique, a term having the same form as the perturbation in Figure 7.20 is added to the three equations that represent the solution for the stable interface:

$$C_L = C_o - \frac{D_L G_L}{mV}\exp\left(-\frac{Vy}{D_L}\right) + C_1 \varepsilon\sin(\omega x)\exp(-b_c y) \qquad (7.22a)$$

$$T_L = T_o + \frac{\alpha_L G_T^L}{V}\left[1 - \exp\left(-\frac{Vy}{\alpha_L}\right)\right] + C_2 \varepsilon\sin(\omega x)\exp(-b_L y) \qquad (7.22b)$$

$$T_S = T_o + \frac{\alpha_S G_T^S}{V}\left[1 - \exp\left(-\frac{Vy}{\alpha_S}\right)\right] + C_3 \varepsilon\sin(\omega x)\exp(-b_S y) \qquad (7.22c)$$

where we must have $b = V/2\alpha + \left[(V/2\alpha)^2 + \omega^2\right]^{1/2}$ to satisfy Eq. (7.20), and C_1, C_2, C_3, b_c, b_L, and b_S must be determined. Assuming that the velocity of the perturbed interface is $V = \dot{\varepsilon}\sin(\omega x)$, after further manipulations and further simplifications, we obtain:

$$\frac{\dot{\varepsilon}}{\varepsilon} = -\Gamma\omega^2 - G_T \xi_T + G_L \xi_c \qquad (7.23)$$

Here, G_T is the conductivity weighted temperature gradient, given by:

$$G_T = \frac{k_S G_T^S + k_L G_T^L}{k_S + k_L} \qquad (7.24)$$

The quantities ξ_T and ξ_c are:

$$\xi_T = 1 - \left(1 + 4\pi^2 P_T^{-2}\right)^{-1/2} \qquad \xi_c = 1 - 2k_V \left[\left(1 + 4\pi^2 P_c^{-2}\right)^{1/2} - 1 + 2k_V\right]^{-1} \qquad (7.25)$$

In this last equation, the quantities $P_T = V\lambda/2\alpha$ and $P_c = V\lambda/2D$ are called the thermal and solutal *Péclet numbers*, respectively. k_V is given by Eq. (2.31).

Equation (7.23) is plotted in Figure 7.21. For $\dot{\varepsilon}/\varepsilon < 0$, since $d\varepsilon/dt$ must be negative, *i.e.* the perturbation amplitude decreases, the perturbation is unstable. For the smaller gradient, G_1, it is seen that there is a range where $\dot{\varepsilon}/\varepsilon > 0$, and thus the perturbation is stable. At $\lambda \geq \lambda_i$ the perturbation becomes stable and continues to grow. Let us try to evaluate λ_i.

Under the assumption of small Péclet numbers, $\omega >> V/D$ or $>> V/\alpha$, the parameters ξ_T and ξ_c are taken as unity. This is the case for typical casting processes. Then, for $\dot{\varepsilon}/\varepsilon = 0$, Eq. (7.23) reduces to $-\Gamma\omega^2 - G_T + G_L = 0$, or:

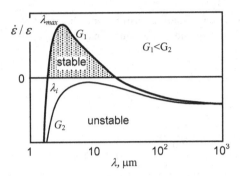

Figure 7.21. Relative growth velocity of instability as a function of interspace.

$$\lambda_i = 2\pi \sqrt{\frac{\Gamma}{G_L - G_T}} \qquad (7.26)$$

When $G_L = G_T$, the perturbation becomes $\lambda_i = \infty$, which means that the interface is planar, and thus stable. This is the previously discussed criterion for constitutional undercooling. If $G_T \ll G_L = V\Delta T_o / D_L$ then:

$$\lambda_i = 2\pi \sqrt{\frac{D}{V} \frac{\Gamma}{\Delta T_o}} \qquad (7.27)$$

where $\delta_c = D/V$ is the diffusion length and $d_o = \Gamma/\Delta T_o$ is the capillary length.

At low solidification velocity the solute gradient destabilizes the interface (the diffusion length is high). When, because of increased solidification velocity, the diffusion length becomes of the size of the capillary length, i.e., $\delta_c = d_o$, the interface is again stable. This is because the boundary layer has approached atomic dimensions and the surface energy has a stabilizing effect. The *absolute limit of stability* has been reached. Thus, the condition for absolute stability is $D/V = \Gamma/\Delta T_o$. In the formal derivation of the absolute stability criterion (Huntley and Davis, 1993), the partition coefficient appears in the equation because of non-equilibrium effects. The solidification velocity required to attain absolute stability, V_{as}, can then be calculated as:

$$V_{as} = \frac{D\Delta T_o(V)}{\Gamma k_V} \qquad (7.28)$$

ΔT_o is a function of velocity. However, note that for rapid solidification $k_V \to 1$. When $V > V_{as}$ difusionless solidification occurs.

The correlation between the various parameters influencing interface stability is shown graphically in Figure 7.22. It is seen that increasing C_o increases the region of instability. A higher G increases the chances for planar interface solidification, but does not change the absolute stability line. Finally, a higher solidification velocity is conducive first to interface destabilization. However, further increase in velocity beyond the limit of absolute stability stabilizes the interface.

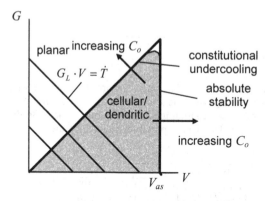

Figure 7.22. Influence of composition, thermal gradient and solidification velocity on interface stability.

7.3.4   Influence of convection on interface stability

In the analysis of interface stability presented so far, the role of thermosolutal convection has been ignored. The results of numerical analysis of coupled thermosolutal convection and morphological stability for a lead-tin alloy, solidified in the vertical-stabilized Bridgman configuration, are summarized in Figure 7.23.

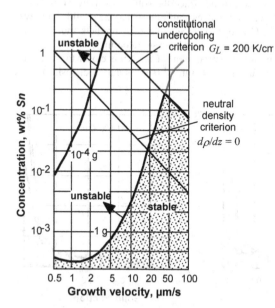

Figure 7.23. Concentration-growth velocity stability diagram for a Pb-Sn alloy (Coriell *et al.*, 1980).

While the constitutional undercooling criterion predicts stability in the entire field under the constitutional line, when thermosolutal convection is taken into account, the stability field is decreased to the shaded region. In low gravity (10^{-4} g), where convection is lower, calculation predicts higher stability for constant boundary conditions (the instability curve moves to the left).

7.4   Applications

Application 7.1

Consider a micro-volume element in a casting. Calculate and compare the compositional profiles in the solid after solidification assuming two different materials for the casting, Fe – 0.13% C and Al - 5% Cu, for the following segregation models, lever rule, Scheil, Brody-Flemings (BF), and Nastac-Stefanescu (NS). The data required for calculation are given in the following table (SI units).

Alloy	k	D_S	D_L	t_f	R_f
Fe - 0.13% C	0.17	$1 \cdot 10^{-9}$	$2 \cdot 10^{-8}$	180	$5 \cdot 10^{-5}$
Al - 5% Cu	0.145	$5.5 \cdot 10^{-13}$	$3 \cdot 10^{-9}$	93	$2.3 \cdot 10^{-5}$

Answer:
Linear solidification and plate geometry will be assumed. Then, $t = f_S \cdot t_f$. Alternatively, $t = f_S^2 \cdot t_f$ could be used for parabolic solidification. The Excel spreadsheet was used for calculation. It can be seen that five terms are sufficient for I_S to converge. I_L seems to have little influence for the diffusivity values used here. Calculation results are presented in the Figure 7.24. It is seen that for the Fe-0.13% C alloy the lever rule and the NS model predict the same amount of segregation. This is not surprising, since the solid diffusivity of carbon is very high (interstitial diffusion), for practical purposes - infinite. Thus, the equilibrium lever rule is satisfactory when modeling Fe-C alloys. The Scheil model predicts much higher segregation, while the BF model much lower. They both have shortcomings.

On the contrary, for the Al-5% Cu alloy the prediction of the NS is closer to the Scheil model because of the very low solid diffusivity of Cu in Al (solid solution diffusion). This

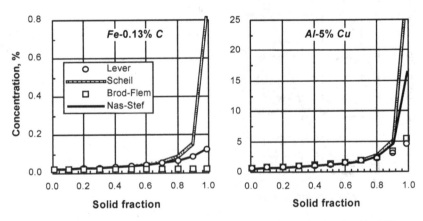

Figure 7.24. Comparison of the interface solid concentration calculated with different models.

infers that the Scheil model is a reasonable approximation for slow diffusing substitutional elements. However, the NS model is more accurate. The BF model predicts slightly higher segregation than the equilibrium equation.

Application 7.2

Calculate the critical thermal gradient for the planar to cellular transition for a 0.6% carbon steel cast in a cube having the volume of $0.001 m^3$, poured into a silica sand mold.

Answer:

The database required for this calculation is taken from Appendix B as follows: $C_o = 0.6$, m = -65, $k = 0.35$, $D_L = 2 \cdot 10^{-8}$ m/s. If the solidification velocity calculated in Application 5.1 is used ($2.7 \cdot 10^{-5}$ m/s), from Eq. (7.19a) it is calculated that $G_L = 9.78 \cdot 10^5$ K/m. This is a very large number, much higher than the thermal gradient typically existing in the casting. Compare for example with the gradient $G_T = 1.86 \ 10^{-3}$ K/m calculated in Application 5.3. Thus, clearly $G_T < G_L$.

References

Basdogan M.F., Kondic V. and Bennett G.H.J., 1982, *Trans. AFS* **90**:263
Battle T.P. and. Pehlke R.D, 1990, *Metall. Trans.* **21B**:357
Battle T.P., 1992, *International Materials Reviews* **37**,6:249
Brody H.D. and Flemings M.C., 1966, *Trans. Met. Soc. AIME* **236**:615
Brooks J.A.. Baskes M.I. and Greulich F.A., 1991, *Metall. Trans.* **22A**:915
Chalmers B., 1962, *J. Aust. Inst. Met.* **8**:225
Chen S.W. and Chang Y.A., 1992, *Metall. Trans.* **23 A**:965
Clyne T.W. and Kurz W., 1981, *Metall. Trans.* **12A**:965
Coriell S.R., Cordes M.R., Boetinger W.J., and Sekerka R.F., 1980, *J. Crystal Growth* **49**:22
Davies G.J., 1973, in: *Solidification and Casting*, Chapter 6, Applied Science Publishing Co., London
Flood S.C. and Hunt J.D., 1988, in: *Metals Handbook Ninth Edition*, vol. 15, D.M. Stefanescu ed., ASM International, Metals Park, Ohio p.130
Goettsch D.D. and Dantzig J.A., 1994, *Metall. and Mat. Trans.* **25A**:1063
Himemyia T. and Umeda T., 1998, *ISIJ International* **38**:730
Hillert M., Höglund L. and Schlan M., 1999, *Metall. and Mater. Trans.* **30A**:1653
Hunt J.D., 1984, *Mat. Sci. and Eng.* **65**:75
Huntley D.A. and Davis S.H., 1993, *Acta metal. mater.* **41**:2025
Jackson K.A., Hunt J.D., Uhlman D., and Seward III T.P., 1966, *Trans AIME* **236**:149
Johnson W.A. and Mehl R.F., 1939, *Trans. AIME* **135**:416
Kanetkar C.S., Chen I.G., Stefanescu D.M., El-Kaddah N., 1988, *Trans. Iron and Steel Inst. of Japan* **28**:860
Kobayashi S., 1988, *Trans. Iron Steel Inst. Jpn.* **28**:728
Kurz W. and Fisher D.J., 1989, *Fundamentals of Solidification*, 3rd ed., Trans Tech Publications, Switzerland
Lacaze J., Castro M. and Lesoult G., 1989, in: *Advanced Materials and Processes, vol. 1*, H. E. Exner and V. Schumacher eds., Informationsgesellschaft Verlag p.147
Laxmanan V., 1987, *J. Crystal Growth* **83**:391
Lesoult G., 1991, in: *Modeling of Casting, Welding and Advanced Solidification Processes-VI*, M. Rappaz, M. R. Ozgu and K. W. Mahin editors, TMS, Warrendale Pa. P.363
Mampey F., 1988, in: *55 th International Foundry Congress*, CIATF, Moscow paper 2I

Mampey F., 1991, in: *Modeling of Casting, Welding and Advanced Solidification Processes-VI*, M. Rappaz, M. R. Ozgu editors, TMS, Warrendale Pa. p.403

Matsumiya T., Kajioka H., Mizoguchi S., Ueshima Y. and Esaka H., 1984, *Trans. Iron and Steel Inst. of Japan* **24**:873

Maxwell I. and Hellawell A., 1975, *Acta Metall.* **23**:229

Mullins W.W. and Sekerka R.F., 1964, *J. Appl. Phys.* **35**:444

Nastac L. and Stefanescu D.M., 1993, *Metall. Trans.* **24A**:2107

Ohnaka I., 1986, *Trans. Iron Steel Inst. Jpn.* **26**:1045

Ogilvy A.J.W. and Kirkwood D.H., 1987, *Applied Scientific Research* **44**:43

Ohno A., 1987, *Solidification. The Separation Theory and its Practical Applications*, Springer-Verlag, Berlin

Oldfield W., 1966, *ASM Trans.* **59**:945

Sarreal J.A. and Abbaschian G.J., 1986, *Metall. Trans.* **17A**:2863

Scheil E., 1942, *Z. Metallk.* **34**:70

Stefanescu D.M., Upadhya G. and Bandyopadhyay D., 1990, *Metall. Trans.* **21A**:997

Stefanescu D.M., 1995, *ISIJ International* **35**:637

Steube R.S. and Hellawell A., 1992, in: *Micro/Macro Scale Phenomena in Solidification*, C. Beckermann et al. editors, Am. Soc. Mech. Eng., New York, HTD-vol. 218, AMD-vol. 139 p.73

Thévoz P., Desbioles J.L., and Rappaz M., 1989, *Metall. Trans.* **20A**:311

Thompson R.G., Mayo D.E., and Radhakrishnan B., 1991, *Metall. Trans.* **22A**:557

Tian H., Stefanescu D.M. and Curreri P., 1990, *Metall. Trans.* **21A**:241

Tian H., 1992, *Ph. D. Dissertation*, University of Alabama, Tuscaloosa

Tian H. and Stefanescu D.M., 1993, in: *Modeling of Casting, Welding and Advanced Solidification Processes-VI*, T. S. Piwonka, V. Voller and L. Katgerman editors, TMS, Warrendale Pa. p.639

Trivedi R. and K. Somboonsuk, 1984, *Mat. Sci. and Eng.* **65**:65-74

Trivedi R. and Kurz W., 1988, in: *Metals Handbook Ninth Edition*, vol. 15, D.M. Stefanescu ed., ASM International, Metals Park, Ohio p.114

Trivedi R., Shen Y.X., and Liu S., 2003, *Metall. Mater. Trans.* **34A**:395

Upadhya G., Banerjee D.K.,. Stefanescu D.M and Hill J.L., 1990, *Trans. AFS* **98**:699

Yeum K.S., Laxmanan V. and Poirier D.R., 1989, *Metall Trans.* **20A**:2847

8

CELLULAR AND DENDRITIC GROWTH

Many of the alloys used in practice, such as steel, aluminum-copper alloys, nickel-base and copper-base alloys, are single phase alloys, which means that the final product of solidification is a solid solution. Depending on the thermal and compositional field, cellular or, in most practical cases, dendritic morphology will occur. In other cases, even when the room temperature microstructure is mostly eutectic some primary phases solidify before the eutectic. They can be solid solutions, carbides, intermetallic phases, inclusions, etc. Their morphology affects mechanical properties, and thus, understanding how this morphology can be controlled is a mater of significant practical importance. A detailed discussion of primary phase growth, and in particular of dendrite growth, will be provided in the following sections.

8.1 Morphology of primary phases

The interface morphology of primary phases can be classified in faceted and non-faceted. Whether a phase grows faceted or non-faceted depends mostly on its entropy of fusion. A discussion of the criterion for faceting during growth will be provided in Chapter 15, *Atomic Scale Phenomena*. At this point it is enough to state that, in general, if $\Delta S_f/R < 2$, where R is the gas constant, non-faceted growth is expected. This is mostly the case for metals. If $\Delta S_f/R > 2$, faceted growth will occur, which is common for non-metals. Some typical examples of faceted growth in metal/non-metal systems are given in Figure 8.1. The faceting behavior is also common in some transparent organic materials such as salol (Figure 8.2).

Regardless of morphology, the driving force for growth is the undercooling. Uneven undercooling on the growing surfaces of the crystal will determine dramatic changes in its morphology. A schematic sequence of the shape change of a faceted primary phase growing in the liquid from a cubic crystal to a dendrite is presented in Figure 8.3. At the corners of the cube divergent transport occurs and the thermal as well as the solutal undercooling are larger than on the facets. Consequently, the corners will grow faster, resulting in the degeneration of the cube into a dendrite. Thus, for a faceted phase the {111} planes are the preferred growth direction.

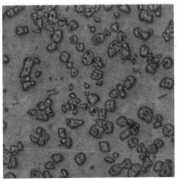

a) vanadium carbides in a Fe-C-V alloy

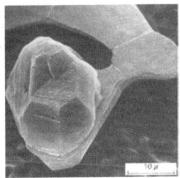

b) graphite crystal in Ni-C alloy (Lux *et al.*, 1975)

c) primary Si crystal with (111) facets in Al-Si alloy (Elliot, 1983)

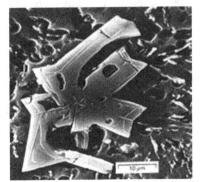

d) star-like primary Si crystal in Al-Si alloy (Elliot, 1983)

Figure 8.1. Faceted growth in metal /non-metal systems.

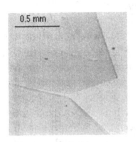

Figure 8.2. Faceted cells in salol (Hunt and Jackson, 1966).

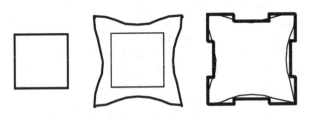

Figure 8.3. Schematic representation of the growth of a faceted dendrite.

The more interesting problem is that of the morphology of primary non-faceted phases as encountered in commercial alloys such as steel, cast iron, aluminum alloys and superalloys. At the onset of constitutional undercooling, instabilities appear on the interface as segregations associated with depressions (nodes) (Figure 8.4a). As the undercooling increases, these nodes become interconnected by interface depressions, forming first elongated cells (Figure 8.4b), and eventually a hexagonal cellular substructure (Figure 8.4c).

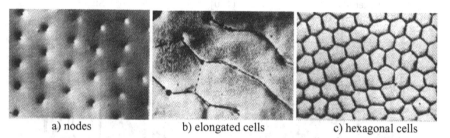

| a) nodes | b) elongated cells | c) hexagonal cells |

Figure 8.4. Evolution of segregation substructure as a function of constitutional undercooling; cross-section view (Biloni and Boettinger, 1970).

According to the theory of constitutional undercooling, as the undercooling increases, the cells should gradually change into dendrites. The question is how does this transition occur? Regular cells grow in the direction of heat extraction, which is typically perpendicular to the S/L interface. When the solidification velocity is increased, because of higher requirements for atomic transport, the main growth direction becomes the preferred crystallographic growth direction of the crystal (Figure 8.5a). The preferred crystallographic growth directions for some typical crystals are given in Table 8.1. At the same time with the change in growth direction, the cross section of the cell deviates from a circle to a Maltese cross, and eventually secondary arms are formed (Figure 8.5d).

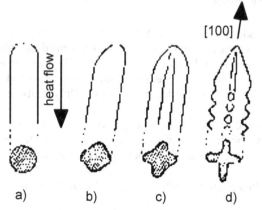

[100]

a)   b)   c)   d)

Figure 8.5. Sequential change of interface morphology as the solidification velocity increases: a) cell growing in the direction of heat extraction; b) cell growing in the [100] direction; c) cell /dendrite; d) dendrite (Morris and Winegard, 1969). Reprinted with permission form Elsevier.

The preferred crystallographic growth directions for some typical crystals are given in Table 8.1. The orientation of the growing dendrite with respect to the direction of the heat flow can affect significantly the dendrite morphology, as exemplified in Figure 8.6.

Table 8.1. Preferred crystallographic growth directions.

Crystal structure	Growth direction	Example
fcc	[100]	Al
bcc	[100]	δFe
bc tetragonal	[110]	Sn
hcp	$[10\bar{1}0]$	ice flakes, graphite
	[0001]	$Co_{17}Sm_2$

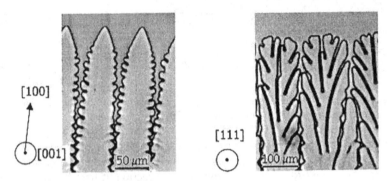

Figure 8.6. Effect of crystalline anisotropy on the morphology of directionally solidified dendrites; growth velocity 35μm/s, heat extraction upward; thin films of a CBr$_4$–8mol% C$_2$Cl$_6$ alloy (Akamatsu *et al.*, 1995). Reprinted with permission from Phys. Rev. Copyright 1995 by the American Physical Soc. www.aps.org

8.2   Analytical tip velocity models

Cells are a relatively simple periodic pattern of the S/L interface. Dendrites that evolve during solidification of metals are complex patterns characterized by side branches (primary, secondary, tertiary, etc.). Describing mathematically the temporal evolution of such patterns is a challenging endeavor. Both analytical and numerical models have been proposed to describe dendritic growth. Only analytical models will be discussed in this chapter.

The analytical models are limited in scope, attempting to describe solely dendrite tip kinetics, as determined by the thermal and solutal field, and by capillarity. Dendritic array models include also an analysis of transport from the root to the tip.

8.2.1   Solute diffusion controlled growth (isothermal growth) of the dendrite tip

Consider a needle-like crystal growing in the liquid. Assume diffusion-controlled growth, which means that the only driving force for growth is the concentration gradient (curvature and thermal undercooling ignored).

In a first approximation, let us assume that the tip of the crystal is a hemi-spherical cap (hemi-spherical approximation), as shown in Figure 8.7. Flux balance at the interface gives:

$$\pi r^2 V(C_L - C_S) = -2\pi r^2 D\left(\frac{dC}{dr}\right)_{tip} \quad \text{or} \quad VC_L(1-k) = -2D\left(\frac{dC}{dr}\right)_{r=r_o} \tag{8.1}$$

The solution of this equation is (see inset for derivation):

$$P_c = \Omega_c \tag{8.2}$$

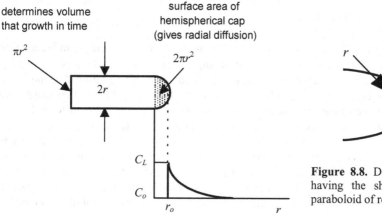

determines volume
that growth in time

πr^2

surface area of
hemispherical cap
(gives radial diffusion)

$2\pi r^2$

$2r$

C_L

C_o

r_o

r

r

Figure 8.8. Dendrite tip having the shape of a paraboloid of revolution.

Figure 8.7. Diffusion field ahead of a hemispherical needle.

where the solutal Péclet number as defined earlier is:

$$P_c = V r/2D \tag{8.3}$$

and the solutal supersaturation, Ω_c, is given by Eq. (7.18). Substituting in Eq. (8.2) we obtain:

$$V = 2D_L \Omega_c / r \tag{8.4}$$

Derivation of the growth velocity of the hemispherical needle, Eq. (8.2).
To find the composition gradient at the tip of the crystal it is necessary to solve the steady state diffusion equation in radial coordinates with no tangential diffusion:

$$\frac{d^2C}{dr^2} + \frac{2}{r}\frac{dC}{dr} = 0 \quad \text{or} \quad r^2\frac{d^2C}{dr^2} + 2r\frac{dC}{dr} = 0 \quad \text{or} \quad \frac{d}{dr}\left(r^2\frac{dC}{dr}\right) = 0$$

The general solution of this equation is $C = C_1 + C_2/r$, where C_1 and C_2 are constants. The following boundary conditions are used:

at $r \rightarrow \infty$	$C = C_o$	thus	$C_1 = Co$
at $r = r_o$	$C = C_L$	thus	$C_2 = r_o (C_L - C_o)$

Then:

$$\left(\frac{dC}{dr}\right)_{r=r_o} = \left[-\frac{r_o}{r^2}(C_L - C_o)\right] = -\frac{C_L - C_o}{r_o}$$

Substituting in Eq. (8.1) we obtain: $\dfrac{V r_o}{2D} = \dfrac{C_L - C_o}{C_L(1-k)}$, and since the solutal supersatura-

tion is $\Omega_c = (C_L - C_o)/[C_L(1-k)]$ and the solutal Péclet number is $P_c = V r_o/(2D)$ we obtain the final solution Eq. (8.2).

Eq. (8.4) gives the growth velocity of the hemispherical needle. It indicates that velocity depends on tip radius, r, and on supersaturation, Ω_c, which is the driving force. However, velocity is not uniquely defined since this equation does not have a unique solution for V, but rather pairs of solutions for V and r. In other words, the solution of the diffusion equation does not specify whether a dendrite will grow fast or slow, but only relates the tip curvature to the dendrite rate of propagation.

The other problem with this solution is that the shape defined by this velocity is not self-preserving. In other words, the hemispherical cap does not grow only in the x-direction, but also in all r-directions, meaning that the needle thickens as it grows. Experimental work on dendrite growth has demonstrated that the dendrite tip preserves its shape. Consequently, another solution must be found for the diffusion problem.

If it is assumed that the dendrite tip has the shape of a paraboloid of revolution (Figure 8.8), which is self-preserving, the solution to the steady state diffusion equation given by Ivantsov (1947) is:

$$I(P_c) = \Omega_c \quad \text{where} \quad I(P_c) == P_c \exp(P_c) E_1(P_c) = P_c \exp(P_c)\int_{P}^{\infty}\frac{\exp(-x)}{x}dx \quad (8.5)$$

Here, $E_1(P_c)$ is the exponential integral function. This solution is valid for both the solutal diffusion (P_c and Ω_c) and the thermal diffusion (P_T and Ω_T).

There are several approximations of the Ivantsov number, I(P), that can be used in numerical or analytical calculations (see inset). Since I(P) is a function of both V and r, the problem of evaluating an unique velocity is still to be solved.

Approximation of the Ivantsov number.

$$I(P) = \cfrac{P}{P + \cfrac{1}{P + \cfrac{1}{1 + \cfrac{1}{P + \cfrac{2}{1 + \cfrac{2}{P + ...}}}}}}$$

The continued fraction approximation is:

Note that the zero-th approximation of the continued fraction approximation of the Ivantsov function is $I_0(P) = P$, that is the hemispherical approximation.

Typically, for casting solidification $P < 1$. For this case the following approximation can be used (Kurz and Fisher, 1989):

$$I(P) = P \cdot \exp(P) \cdot [a_0 + a_1 \cdot P + a_2 \cdot P^2 + a_3 \cdot P^3 + a_4 \cdot P^4 + a_5 \cdot P^5 - \ln(P)]$$

where $a_0 = -0.57721566$, $a_1 = 0.99999193$, $a_2 = -0.24991055$, $a_3 = 0.05519968$, $a_4 = -0.00976004$, $a_5 = 0.00107857$.

For limiting values of the Péclet number, the Ivantsov function for a paraboloid of revolution can be approximated as (Trivedi and Kurz, 1994):

for $P \ll 1$: $I(P) \approx -P \ln P - 0.5772\, P$
for $P \gg 1$: $I(P) \approx 1 - 1/P + 2/P^2$

8.2.2   Thermal diffusion controlled growth

During solidification, a thermal gradient is imposed over the system. Thermal diffusion will drive the process. In pure metals, this will be the only driving force for growth. If it is assumed that the driving force for perturbation growth is only the thermal gradient (thermal dendrite), similar equations to those obtained for the diffusion-controlled growth can be derived:

$$P_T = \Omega_T \quad \text{with} \quad P_T = \frac{Vr}{2\alpha} \quad \text{and} \quad \Omega_T = \frac{\Delta T_T}{\Delta H_f / c} \tag{8.6a}$$

and alternatively:

$$I(P_T) = \Omega_T \tag{8.6b}$$

where P_T is the thermal Péclet number and Ω_T is the thermal supersaturation. For the hemispherical approximation a derivation of the particular case of Eq. (8.6b) is given in the inset.

Derivation of the correlation between the thermal Péclet number and thermal supersaturation.

Temperature flux balance at the interface gives: $\pi r^2 V \dfrac{\Delta H_f}{c} = -2\pi r^2 \alpha_L \left(\dfrac{dT}{dr} \right)_{r=r_o}$

To calculate the thermal gradient at the interface we need the temperature of the tip. The solution of the steady state diffusion equation in radial coordinates is $T = C_1 + C_2/r$. Applying the boundary conditions: i) at $r \to \infty$ - $T = T_{bulk}$ and ii) at $r = r_o$ - $T = T_f$, the solution becomes $T = T_{bulk} + \Delta T \cdot r_o / r$.

Thus, the temperature gradient at the tip is $(dT/dt)_{r=r_o} = -\Delta T_T / r_o$. Substituting in the flux balance equation and rearranging:

$$\frac{Vr}{2\alpha} = \frac{\Delta T_T}{\Delta H_f / c} \quad \text{or} \quad P_T = \Omega_T$$

All the diffusion models discussed above conclude that at steady state, the tip of the dendrite will advance in the liquid following the simple law $V \cdot r = \text{const}$. This means that there is no unique solution, since multiple pairs of V and r satisfy this relationship. However, experimental work has demonstrated that for each under-cooling a unique value of tip velocity and radius is obtained. The problem is then to find the additional constrains that impose a unique dendrite tip radius from the multiple solutions offered by the diffusion models.

8.2.3   Solutal, thermal, and capillary controlled growth

To obtain a unique solution it is necessary to find additional criteria that define the tip radius. Several models have been proposed:

- the *extremum* criterion
- the marginal stability criterion
- the microsolvability theory

The *extremum* criterion
As discussed earlier, at high solidification velocities, when the diffusion length becomes of the size of the solute capillary length $(D/V = \Gamma / \Delta T_o)$, the interface becomes planar. Thus, as shown in Figure 8.9, the maximum velocity of a dendrite tip is limited by the absolute stability. The *extremum* criterion implies that the perturbation will grow at the maximum possible velocity and the minimum possi-ble undercooling. These conditions are satisfied by the velocity corresponding to the radius tip of the dendrite r_e.

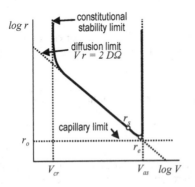

Figure 8.9. Growth velocity - tip radius correlation; the full line is the V-r correlation.

An expression for V can be obtained for example for a perturbation driven only by the solutal and curvature undercooling (solutal perturbation), starting from:

$$\Delta T = \Delta T_c + \Delta T_r$$

where

$$\Delta T_c = -m(C_L - C_o) = -m(C_L - C_S)\Omega_c = -m(1-k)C_L\Omega_c$$
$$= m(k-1)C_L P_c = -m(1-k)C_L \frac{Vr}{2D}$$

and $\Delta T_r = \frac{2\Gamma}{r}$ (8.7)

Substituting these last two equations in the total undercooling equation we have:

$$\Delta T = -m(1-k)C_L^* \frac{Vr}{2D} + \frac{2\Gamma}{r} \tag{8.8}$$

According to the *extremum* criterion, it will be assumed that growth proceeds at the minimum undercooling (the maximum of the curve) which can be obtained from $\partial \Delta T / \partial r = 0$, that is:

$$r_e = \left[\frac{4D\Gamma}{m(k-1)C_L^*} \right]^{1/2} V^{-1/2} = \sqrt{\frac{4\Gamma}{G_L}} \tag{8.9}$$

Substituting in Eq. (8.8) an equation for tip velocity is obtained:

$$V = \mu \Delta T^2 \quad \text{with} \quad \mu = \frac{D_L}{4\Gamma m(k-1)C_L^*} \tag{8.10}$$

Note that for steady state solidification, $C_L^* = C_o / k$.

However, experimental evidence (Nash and Glicksman, 1974) demonstrates that such velocities are considerable higher than the measured ones. This is shown schematically in Figure 8.9, where r_s is the position of the experimental point and r_e is the value calculated from the *extremum* criterion.

The marginal stability criterion
Langer and Müller-Krumbhaar (1978) performed a linear stability analysis for an Ivantsov parabola dendrite tip region in a pure undercooled melt. A small departure from the parabolic shape, caused by interface energy, was introduced in the system. It was concluded that dendrite tip radii are not stable at values smaller than predicted by the extremum criterion, or larger than a certain critical value. At such large radii tip splitting will occur to decrease the radius. They proposed that this largest radius is selected by the dendrite during its growth (*marginal stability* criterion). A number of models discussed in the following paragraphs use this criterion to obtain a unique dendrite tip radius.

To formulate the growth velocity of the dendrite tip for the most general case it is assumed that the tip is a growth instability driven by the kinetic, solutal, thermal, and capillary undercooling:

$$\Delta T = \Delta T_k + \Delta T_c + \Delta T_T + \Delta T_r \qquad (8.11)$$

A comprehensive treatment of this problem applicable to a wide range of Péclet numbers was given by Boettinger *et al.* (1988). ΔT_k was formulated through Eq. 2.26, ΔT_T through Eq. (8.6a), and ΔT_r as $2\Gamma/r$. The constitutional undercooling was written as:

$$\Delta T_c = -m_L(V)(C_L - C_o) = m_L C_o \left[1 - \frac{m_L(V)/m_L}{1 - (1 - k_e)\Omega_c} \right] \qquad (8.12)$$

Here, $m_L(V)$ is the velocity dependent liquidus slope given in Eq. 2.32. When all these equations are introduced in Eq. (8.11), the values of dendrite tip velocity and tip radius for a given ΔT can be calculated. A dependency as shown in Figure 8.9 is obtained. To obtain a unique value for dendrite tip velocity it was assumed that growth occurs at the limit of stability (marginal stability criterion). In other words, the perturbation will grow with the shortest stable wavelength, *i.e.*, $r_s = \lambda_i$. This implies that if the tip radius of the perturbation is smaller than λ_i, the radius will tend to increase, while if it is larger than λ_i, additional instabilities will form and the radius will decrease. Then, for $\dot{\varepsilon}/\varepsilon = 0$, Eq. (7.23) gives:

$$r^2 = \frac{\Gamma/\sigma^*}{G_L \xi_c - G_T \xi_T} \quad \text{with} \quad \sigma^* = (4\pi^2)^{-1} \qquad (8.13)$$

Note that this equation can be used to derive dendrite growth velocity equations for both slow solidification rates typical for castings ($P<1$) and rapid solidification rates ($P>>1$). The modified Eq. (8.11) and Eq. (8.13) can now be concomitantly solved numerically to give a unique solution for the dendrite tip velocity.

In their model, Lipton *et al.* (1984) also started from Eq. (8.11), ignored the kinetic undercooling, and used the formulations for ΔT_c, ΔT_T, and ΔT_r as before:

$$\Delta T = \Delta T_T + \Delta T_c + \Delta T_r$$
$$= \frac{\Delta H_f}{c_p} I(P_T) + m C_o \left[1 - \frac{1}{1 - (1 - k)I(P_c)} \right] + \frac{2\Gamma}{r} \qquad (8.14)$$

The dendrite tip radius is derived form the marginal stability theory as:

$$r = \frac{\Gamma}{\sigma^*} \left[\frac{\Delta H_f}{c_p} P_T - P_c \frac{m C_o (1 - k)}{1 - (1 - k)I(P_c)} \right]^{-1} \qquad (8.15)$$

From the solution of the diffusion field the tip velocity is formulated as:

$$V = 2\alpha P_T/r \qquad (8.16)$$

Finally, the solutal and thermal Péclet numbers are correlated by:

$$P_c = P_T (\alpha/D) \tag{8.17}$$

Here, σ^* is the dendrite tip selection parameter $\approx 1/(4\pi^2)$. This parameter will be discussed later in more detail. Since velocity is introduced through the thermal and solutal supersaturations, either the hemispherical ($P = \Omega$) or the paraboloid ($I(P) = \Omega$) approximations can be used.

The dendrite tip velocity for equiaxed dendrites growing at small undercooling can be calculated from the preceding four equations which are solved by numerical iterations.

A more complete solution was derived by Trivedi and Kurz (1994). They started with Eq. (8.11) and obtained an equation similar to Eq. (8.14), but with a different formulation for the solutal undercooling (second RH term) as follows:

$$\Delta T_c = \frac{k \Delta T_o I(P_c)}{1 - (1-k) I(P_c)} \tag{8.18}$$

The capillary term in Eq. (8.14) is generally negligible for metals at low undercooling (the case of shaped castings), but is significant under rapid solidification conditions. Substituting the values for the thermal and constitutional gradients in (8.13), the general dendrite tip radius selection criterion in undercooled alloys is obtained:

$$V r^2 \left(\frac{k_V \Delta T_o(V)}{\Gamma D} \right) \left(\frac{1}{1 - (1-k_V) I(P_c)} \right) \xi_c + V r^2 \left(\frac{\Delta H_f / c_L}{2 \Gamma \alpha_L \beta} \right) \xi_L = \frac{1}{\sigma^*} \tag{8.19}$$

where $\Delta T_o(V)$ is the velocity dependent solidification interval and $\beta = 0.5[1 + (k_S/k_L)]$. This equation is valid for slow as well as rapid cooling unconstrained growth. The modified Eq. (8.14) and Eq. (8.19) completely describe the dendrite growth problem and can be solved numerically to obtain the growth velocity.

A simple analytical solution can be obtained for a solutal dendrite under the assumption of small Péclet number ($P_c = 0$, $\xi_c = 1$):

$$V r^2 = \frac{4\pi^2 \Gamma D_L}{m(k-1) C_o} \tag{8.20}$$

For constrained growth (directional solidification) they proposed the equation:

$$V \left(\frac{k \Delta T_o}{D} \right) \left(\frac{C_L^*}{C_o} \right) \xi_c - G_T = \frac{\Gamma}{\sigma^* r^2} \tag{8.21}$$

Purely analytical solutions can be obtained with further simplifying assumptions. Following the derivation proposed by Nastac and Stefanescu (NS) (1993) for unconstrained growth (equiaxed dendrites), ignoring kinetic undercooling and assuming that the effect of surface energy (curvature) is introduced through the limit of stability criterion, only the solutal and thermal undercooling must be considered. The dendrite tip velocity equation is (see inset for derivation):

$$V = \mu_{eq} \Delta T^2 \quad \text{with} \quad \mu_{eq} = \left[2\pi^2 \Gamma \left(\frac{m(k-1)C_L^*}{D_L} + \frac{\Delta H_f}{c\alpha_L} \right) \right]^{-1} \tag{8.22}$$

The growth coefficient can also be written as:

$$\mu_{eq} = \left(\mu_c^{-1} + \mu_T^{-1} \right)^{-1} \quad \text{with} \quad \mu_c = \frac{D_L}{2\pi^2 \Gamma m(k-1)C_L^*} \quad \text{and} \quad \mu_T = \frac{c\alpha_L}{2\pi^2 \Gamma \Delta H_f}$$

For steady state, C_L^* will be substituted with C_o/k. The growth coefficient is a constant only for steady state, since C_L^* is constant only for steady state. An equation similar to Eq. (8.22) can be obtained if a paraboloid of revolution shaped tip is assumed. Then, the $I(P) = \Omega$ relationships must be used in the derivation.

This equation describes unconstrained growth (equiaxed dendrites) since it includes thermal undercooling ahead of the interface. For some metallic alloys, it can be calculated that the contribution of thermal undercooling is negligible as compared to that of solutal undercooling (see Application 8.1). Then, the velocity equation can be simplified. Since $\Delta T_T = 0$ and thus $G_T = 0$, tip velocity is given by Eq. (8.22) with the growth coefficient given by the equation for μ_c.

Derivation of the Nastac-Stefanescu equation.
When using the assumptions used by the model the total undercooling is:

$$\Delta T = \Delta T_c + \Delta T_T \tag{8.23}$$

where ΔT_c is given by Eq. (8.7) and ΔT_T is calculated from Eq. (8.6a). If it is further assumed that the tip of the instability is of hemispherical shape, and substituting the value of P_c and $P_T = \Omega_T$ it is obtained that:

$$\Delta T = \frac{Vr}{2} \left(\frac{m(k-1)C_L^*}{D_L} + \frac{\Delta H_f}{c\alpha_L} \right) \tag{8.24}$$

The liquidus and thermal gradients are:

$$G_L = -m \frac{\partial C_L}{\partial r} = \frac{V}{2D_L} m C_L^*(k-1) = \frac{P_c \Delta T_c}{r \, \Omega_c} = \frac{\Delta T_c}{r}$$

$$G_T = -\frac{\partial T_T}{\partial r} = -\frac{V}{2\alpha_L}\frac{\Delta H_f}{c} = -\frac{P_T}{r}\frac{\Delta T_T}{\Omega_T} = -\frac{\Delta T_T}{r}$$

The first of these two equations was obtained from Eq. (4.5). A factor of ½ was introduced to describe the flux at the hemispherical tip. The last equation is valid for a negative gradient, which occurs during equiaxed solidification. Then, assuming that growth occurs at the limit of stability $(r = \lambda_i)$, Eq. (7.26) gives $r = 2\pi\sqrt{\Gamma/(G_L - G_T)}$. Substituting the expressions for G_L and G_T, and using Eq. (8.23) the tip radius is:

$$r = 4\pi^2\Gamma/\Delta T$$

(8.25)

Substituting this expression for r in Eq. (8.24) the equation for the hemispherical tip velocity, Eq. (8.22), is obtained.

For the case of columnar dendrites (constrained growth), there is no thermal undercooling. Thus, ignoring the kinetic undercooling, the basic undercooling equation simplifies to $\Delta T = \Delta T_c + \Delta T_r$.

Using similar formulations for the undercooling as above, and the hemispherical approximation, Kurz, Giovanola, and Trivedi (KGT) (1986) derived the following equation for columnar dendrites:

$$V = \frac{m(k-1)C_o D_L \Omega_c^2}{\pi^2\Gamma} = \frac{D_L}{\pi^2\Gamma m(k-1)C_o}\Delta T_c^2$$

(8.26)

Both the KGT and NS models are only valid for small Péclet numbers, since this assumption was used to derive Eq. (7.27), which is adopted for the limit of stability criterion.

Let us now evaluate to what degree the simplifications introduced in the analytical Nastac-Stefanescu (NS) model produce deviations from the more accurate semi-analytical Trivedi-Kurz (TK) model. To this effect, a comparison of calculated V-r correlations with the two models and experimental data is shown in Figure 8.10 (for details of the calculation see Application 8.3). Both models are very close to the experimental data on the linear part of the log-log graph. As the growth velocity decreases and a dendritic-to-cellular transition occurs, the radius increases very fast (it should tend to infinity when planar solidification occurs). Only the TK model for columnar growth (simplified Eq. (8.20)) follows well the experimental data in the velocity range smaller than 1μm/s. However, as typical velocities in casting solidification are in the range of 0.01 to 0.5m/s, it is apparent that both models can be used.

A columnar dendrite operates in a constrained environment, since the temperature gradient ahead of the interface is always positive. Its growth is driven by the temperatue gradient. The interface temperature of the columnar dendrite is between T_L and T_S, and the undercooling is mostly constitutional, ΔT_c. The assumption of negligible thermal undercooling is reasonable for most metallic alloys, so that its growth velocity can be calculated with $V_c = \mu_c \Delta T_c^2$. At steady state the columnar

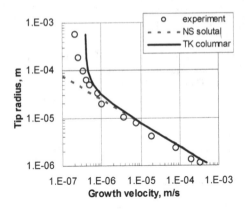

Figure 8.10. Measured and calculated tip radii of cells and dendrites in a Fe-3.08%C-2.01%Si alloys solidified under a thermal gradient of 5000K/m at various velocities. Experimental data are from Tian and Stefanescu (1992).

dendrite can grow at a maximum velocity corresponding to the maximum under-cooling $\Delta T_c = \Delta T_o$. However, steady state can be reached also at an undercooling smaller than ΔT_o, when the solutal velocity, V_c, is equal to the thermal velocity, V_T, calculated from macro-transport considerations. As long as $V_c < V_T$, the tip velocity is simply V_c. If, on the contrary, $V_c > V_T$, dendritic growth is constrained at V_T.

An equiaxed dendrite operates in an unconstrained environment, since the temperature gradient ahead of the interface is negative. Its velocity is again mostly dictated by the constitutional undercooling, and can be calculated with Eq. (8.22), as long as $\Delta T_{bulk} < \Delta T_o$ (see example in Application 8.5). If $\Delta T_{bulk} > \Delta T_o$ ($T^* < T_S$), then, interface equilibrium does not apply anymore, and the partition coefficient is velocity dependent.

Koseki and Flemings (1995) developed a model that includes the combined effects of the undercooled melt and heat extraction through the solid, applicable to chill-casting. It is a hybrid of the models for constrained and unconstrained growth.

Experimental work by Nash and Glicksman (1974) has demonstrated that, indeed, the operating point of the dendrite is close to the value calculated from the limit of stability criterion (r_s on Figure 8.9). In their experiment, they measured concomitantly r, V, and ΔT during the growth of a succinonitrile dendrite. It was also proven that the tip of the dendrite fits a parabolic curve (dotted line on Figure 8.11).

The microsolvability theory
While, as discussed, the marginal stability criterion gives an excellent agreement with most experimental results, there is no physical reason to accept the marginally stable state over the other stable states.

Kessler and Levine (1986) and Bensimon et al. (1987) have found a unique, self-consistent solution to the steady-state dendrite problem (the interface shape obtained from the thermal and solutal field equations with the boundary conditions that includes the effect of surface energy satisfies the shape preserving condition) taking into account the anisotropy of the interface energy around the dendrite tip. This unique solution, known as the microsolvability condition, gives a unique value for the dendrite tip radius.

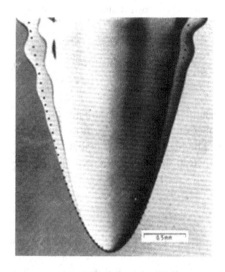

Figure 8.11. Tip of a growing succinonitrile dendrite (Huang and Glicksman, 1981). Reprinted with permission from Elsevier.

The concept that the capillary effect is a singular perturbation which destroys Ivantsov's continuous family of solutions has been confirmed by analytical and numerical studies. While the solvability theory has achieved notable theoretical successes, its quantitative relevance to the interpretation of experimental data has not been established (Barbieri and Langer, 1989).

8.2.4   Interface anisotropy and the dendrite tip selection parameter σ^*

In their analysis, Langer and Müller-Krumbhaar (1978) introduced the dendrite tip selection parameter, σ^* through the relationship:

$$r = \left(\delta_c \, d_o / \sigma_c^* \right)^{1/2} \tag{8.27}$$

where $\delta_c = 2D_L/V$ is the diffusion length, $d_o = \Gamma/\Delta T_o$ is the capillary length, and σ_c^* is the parameter for the solutal case (with $D_S \ll D_L$). Substituting these relationships this equation can be rewritten as:

$$V r^2 = 2\Gamma D / \left(\sigma_c^* \, \Delta T_o \right) \tag{8.28}$$

Note that this equation is identical with Eq. (7.29) when assuming $\lambda_i = r$, and using the notation $\sigma^* = 1/4\pi^2$. In the marginal stability theory this parameter is considered to be a constant equal to 0.02533.

Similarly, a parameter for the purely thermal case (for $\alpha_S = \alpha_L$) can be derived, using for example Eqs. (8.6a) and (8.25), to obtain:

$$\frac{1}{\sigma_T^*} = V r^2 \left(\frac{\Delta H_f / c_L}{2\Gamma \alpha_L} \right) \tag{8.29}$$

In order to analyze the dendrite tip selection parameter for undercooled alloys, Trivedi and Kurz (1994) substituted Eqs. (8.28) and (8.29) in Eq. (8.19) to obtain:

$$\left(\frac{2}{\sigma_c^*}\right)\left(\frac{C_L^*}{C_o}\right)\xi_c + \left(\frac{1}{\sigma_T^*}\right)\left(\frac{\xi_L}{\beta}\right) = \frac{1}{\sigma^*} \tag{8.30}$$

They stated that for an alloy system, σ^* is constant, but σ_c^* and σ_T^* are not.

The microsolvability theory produces equations similar in form as those derived through the marginal stability criterion. The main difference is that σ^* is not a numerical constant any more, but rather a function of the interface energy anisotropy parameter, ε, as follows:

$$\sigma^* = \sigma_o\, \varepsilon^{1.75} \tag{8.31}$$

where σ_o is of the order of unity and the definition of ε is discussed in the following paragraphs.

Crystalline materials are characterized by anisotropic S/L interface energy. For cubic crystals, the variation of the interfacial energy γ with orientation θ can be expanded about the dendrite tip orientation ($\theta = 0$), which has four-fold symmetry. This expansion up to the first order term is (Trivedi and Kurz, 1994) $\gamma(\theta) = \gamma_o(1 + \delta \cos 4\theta)$, where δ is the interface anisotropy parameter.

Using the Gibbs-Thomson equation for anisotropic materials, the capillary undercooling is:

$$\Delta T_r = \left(\gamma + d^2\gamma/d\theta^2\right)K/\Delta S_f = \gamma_o\left(1 - \varepsilon \cos 4\theta\right)K/\Delta S_f \tag{8.32}$$

where $\varepsilon = 15\delta$ and is known as the anisotropy coefficient.

The anisotropy in the S/L interface energy strongly affects the tip radius. Higher anisotropy reduces the tip radius (Lu and Liu, 2007).

8.2.5   Effect of fluid flow on dendrite tip velocity

The rather intricate picture of dendritic growth presented so far is made even more complicated by buoyancy-driven convection, which is unavoidable during solidification under terrestrial conditions. It was calculated (Miyata, 1995) that the growth velocity of the dendrite increases with the forced melt flow at a given undercooling. The effect of melt flow becomes particularly significant in the low velocity regime of both dendrite growth and forced flow.

For the case of uniform fluid velocity, u_∞, directed opposite to the crystal growth direction, Ananth and Gill (1991) have proposed a simplified approximation for the effect of fluid flow on the velocity-undercooling relation. Their solution of the three-dimensional Navier-Stokes equation under the approximation of low Reynolds number, Re, resulted in the following expression for the thermal melt undercooling:

$$\Delta T_t = \frac{u_\infty \, r \, \Delta H_f}{2\alpha c_L} \int_1^\infty \exp\left[-\int_1^z f(\eta)\,d\eta \right] dz$$

with the function $f(\eta)$ given by:

$$f(\eta) = \frac{r}{2\alpha}(u_\infty + V) + \frac{1}{\eta} - \frac{V\,r\left(\exp\left[-Re^*/2\right] - \exp\left[-Re^*\eta/2\right]\right)}{Re^*\eta\alpha\,E_1\left(Re^{*/2}\right)} - \frac{V\,r\,E_1\left(Re^*\eta/2\right)}{2\alpha\,E_1\left(Re^*/2\right)}$$

Re^* is defined by the relation $Re^* = V\,r/v + Re$, with the Reynolds number $Re = u_\infty\,r/v$. E_1 denotes the exponential integral function and v is the kinematic viscosity.

Glicksman et al. (1995) measured the dendritic growth velocities and tip radii of succinonitrile in micro-gravity environment experiments (space shuttle Columbia) during isothermal solidification (Figure 8.12). It was observed that convective effects under terrestrial conditions increase the growth velocity by a factor of two at lower undercoolings (< 0.5K). In the undercooling range of 0.47 to 1.7K, the data remained virtually free of convective effects. A diffusion solution to the dendrite problem was not consistent with the experimental data.

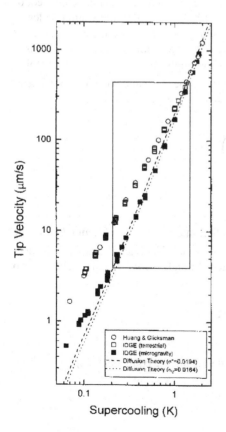

Figure 8.12. Dendrite tip velocity as a function of undercooling for terrestrial and microgravity conditions (Glicksman et al., 1995).

8.2.6   Multicomponent alloys

Eq. (8.22) can be extended to dilute multicomponent systems, by writing the growth constant as (Nastac and Stefanescu, 1993):

$$\mu = \frac{1}{2\pi^2}\left[\sum_{i=1}^{n}\left[\Gamma_i\left(\frac{m_i\left(k_i-1\right)C_{Li}^{*\,2}\rho_i}{D_{Li}\sum\limits_{j=1}^{n}C_{Lj}^{*}\rho_j}+\frac{\Delta H_f}{c\alpha_L}\right)\right]\right]^{-1} \tag{8.33}$$

where i is the component, and n is the number of components in the alloy that have a significant contribution upon the tip growth velocity.

The solidification parameters of interest for modeling for calculation of dendrite tip velocity of multicomponent alloys, i.e., the equilibrium liquidus temperature, T_L, the liquidus slope, m_L, and the partition coefficient, k, can be obtained by using thermodynamic calculation, as described by Boettinger et al. (1988). However, when these parameters are used one needs to solve the mass transport equation for each species.

To avoid excessive computational time a combined multicomponent/pseudobinary approach can be used (Nastac et al., 1999). First, thermodynamic calculations are used to obtain the slope of the liquidus line and the partition coefficient for each element at successive temperatures. The algorithm for such calculations is now incorporated in some solidification simulation software (e.g., the SLOPE subroutine in ProCAST). Then, an equivalent slope, $\overline{m}_L$, and partition coefficient, $\overline{k}$, are calculated for each temperature using the equations:

$$\overline{m}_L = \frac{\sum\limits_{i=1}^{n}\left(m_L^i C_L^i\right)}{\overline{C}_L} \quad \text{and} \quad \overline{k} = \frac{\sum\limits_{i=1}^{n}\left(m_L^i C_L^i k^i\right)}{\sum\limits_{i=1}^{n}\left(m_L^i C_L^i\right)}$$

where $\overline{C}_L$ is the sum of all the elements in the liquid, and m_L^i, C_L^i, and k_L^i are the slope, liquid composition, and partition coefficient of individual elements, respectively. Then regression equations were fitted through the m_L-T and k_L-T curves to obtain the temperature dependence of m_L and of k_L:

$$\overline{m}_L = a+b\cdot T+c\cdot T^2 \quad \text{and} \quad \overline{k} = a'+b'\cdot T+c'\cdot T^2$$

where a, b, etc., are known coefficients.

The liquidus temperature can be calculated using the equivalent slope:

$$T_L = T_f + \overline{m}_L \cdot \overline{C}_L$$

or from the weighted average of the slope and liquid composition of each element:

$$T_L = T_f + \sum_i m_L^i C_L^i$$

where T_f is the melting temperature of the pure solvent.

8.3   Dendritic array models

In the preceding discussion only events happening around the tip have been considered. However, most practical interest resides in structures that are the result of the growth of arrays of instabilities. In this case lateral diffusion and diffusion from the root to the tip of the instability must also be included.

Bowers *et al.* (1966) proposed a model in which the contribution of liquid diffusion from root to tip was included (Figure 8.13). Solute balance in the volume element is:

$$D_L \frac{\partial}{\partial x}\left(f_L \frac{\partial C_L}{\partial x} \right) = \frac{\partial \overline{C}}{\partial t}$$

where $\overline{C}$ is the average composition of the volume element at time t. The variation of composition in time is:

$$\frac{\partial \overline{C}}{\partial t} = C_L (1-k)\frac{\partial f_L}{\partial t} + f_L \frac{\partial C_L}{\partial t}$$

The first term on the right hand side is the solute rejected from the growth of the perturbation (growth in x and y-directions). The second term is the solute entering the volume element due to liquid phase diffusion from root to tip (x-direction). The governing equation becomes:

$$\frac{\partial}{\partial x}\left(D_L f_L \frac{\partial C_L}{\partial x} \right) = C_L (1-k)\frac{\partial f_L}{\partial t} + f_L \frac{\partial C_L}{\partial t} \tag{8.34}$$

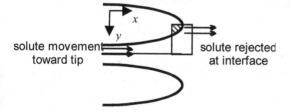

solute movement toward tip

solute rejected at interface

Figure 8.13. Array of instabilities.

For steady-state solidification, solute flux balance at the tip gives $V(C_t - C_o)_{tip}$ $= -D_L(\partial C/\partial x)_{tip}$. For constant concentration gradient along x between perturbations from root to tip the concentration gradient is $\partial C/\partial x = = -m^{-1} dT/dx = G/m$. Substituting this gradient in the previous equation, the tip composition is obtained as:

$$C_t = (1 - a) C_o \quad \text{with} \quad a = D_L G_L / (m V C_o) \tag{8.35}$$

The composition gradient at the interface can then be expressed as:

$$\frac{\partial C}{\partial x} = \frac{V}{D}(C_L - C_o) = \frac{V}{D} a C_o \tag{8.36}$$

Also:

$$\frac{\partial f_L}{\partial x} = \frac{\partial f_L}{\partial t} \frac{\partial t}{\partial x} = -\frac{1}{V} \frac{\partial f_L}{\partial t} \tag{8.37}$$

Substituting (8.36) and (8.37) in the solute balance equation:

$$\frac{\partial f_L}{f_L} = -\frac{\partial C_L}{C_L(1-k) + a C_o}$$

Upon integration between C_o and C_L, we obtain the local solute redistribution equation for dendritic solidification:

$$C_S = k C_o \left[\frac{a}{k-1} + \left(1 - \frac{ak}{k-1}\right)(1 - f_S)^{k-1} \right] \tag{8.38}$$

If $f_S(T)$ is known, the shape of the columnar dendrite can be calculated. The limits $a = 0$ and $a = -(1 - k)/k$ correspond to the condition of Scheil and equilibrium solidification, respectively. For an Al-4% Cu alloy a maximum of 9% eutectic is predicted to form in the interdendritic solidification for $a = 0$, and less for $a < 0$ (see Application 8.2).

The hemispherical approximation analysis of Burden and Hunt (1974), slightly modified by Laxmanan (1985) resulted in the following equation for tip undercooling:

$$\Delta T = \frac{D_L G_T}{V} - \frac{V m C_o (1-k) r}{D_L} - k r G_T + \frac{2 \gamma_{SL} T_L}{\rho_S \Delta H_f r} \tag{8.39}$$

Assuming further that growth proceeds at the *extremum*, that is $\partial(\Delta T)/\partial r = 0$, an equation for the tip radius of an array of instabilities as a function of the solidification velocity can be derived:

$$r = \sqrt{\frac{2\Gamma}{mC_o(k-1)V/D - kG_T}} = \sqrt{\frac{2\Gamma}{G_L - kG_T}}$$

When substituting this equation in Eq. (8.39) a $V(\Delta T)$ equation is obtained. The tip composition was calculated to be:

$$C_t = C_o\left(1 - a - \left(\frac{2\Gamma V(1-k)}{mD_L C_o}\right)^{1/2}\right)$$

8.4   Dendritic arm spacing and coarsening

The dendritic arm spacing (DAS) is a morphological parameter directly related to the mechanical properties of the alloy. In general, the finer is the arm spacing, the higher the mechanical properties. In columnar solidification, both primary and secondary arm spacing can be measured through metallographic analysis. In equi-axed solidification, only the secondary arm spacing is an issue. In the characterization of the fineness of the microstructure, the primary arm spacing is replaced by the number of grains.

8.4.1   Primary spacing

The relationship that allows calculation of the primary spacing, λ_1, is a complicated dependency of solidification velocity and temperature gradient. Two simpler relationships for the primary DAS will be discussed here.

A first relationship can be obtained based on *Flemings* array (Figure 8.13). Ignoring solute diffusion in the *x*-direction, material balance dictates:

$$D_L \frac{\partial^2 C_L}{\partial y^2} = \frac{\partial C_L}{\partial t} \quad \text{where} \quad \frac{dC_L}{dt} = \frac{dC}{dx}\frac{dx}{dt} = -\frac{V_y G_L}{m}$$

Upon substitution in the governing equation we have $(d^2C_L/dy^2) = -(V/D_L)(G/m)$. Integrating between 0 and dC/dy, and between 0 and y yields $dC/dy = -VGy/Dm$. Integrating again between C_o/k and C_L^{max} (in the interdendritic spacing), and between 0 and $\lambda/2$ gives:

$$\lambda^2 = -\frac{8mD}{VG}\left(C_L^{max} - \frac{C_o}{k}\right) \quad \text{or}$$

$$\lambda_1 = ct.(GV)^{-1/2} \quad \text{or, more general,} \quad \lambda_1 = ct.\left(\dot{T}\right)^{-n} \tag{8.40}$$

Examples of cooling rates and dendrite arm spacing are given in Table 8.2.

Kurz and Fisher (1989) have derived a more complex relationship. Assume that the dendrites are half of ellipsoids of revolution (Figure 8.14). Then, the dendritic tip radius is $r = b^2/a$. For a hexagonal arrangement of dendrites $b = 0.58\lambda_1$. From the phase diagram it can be approximated that:

$$a = \frac{\Delta T'}{G} = \frac{T^* - T_E}{G} \approx \frac{\Delta T_o}{G}$$

Table 8.2. Range of cooling rates in solidification processes (Cohen and Flemings, 1985).

Cooling rate, K/s	Production processes	Dendrite arm spacing, μm
10^{-4} to 10^{-2}	large castings	5000 to 200
10^{-2} to 10^3	small castings, continuous castings, die castings, strip castings, coarse powder atomization	200 to 5
10^3 to 10^9	fine powder atomization, melt spinning, spray deposition, electron beam or laser surface melting	5 to 0.05

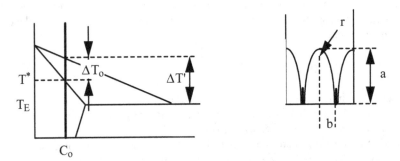

Figure 8.14. Assumptions for calculation of primary DAS (Kurz and Fisher, 1989).

This approximation is increasingly valid, as the composition of the alloy is closer to the eutectic. Then: $(0.58\lambda_1)^2 = r\,\Delta T_o/G$, or $\lambda_1 = \sqrt{3\Delta T_o r/G}$. Since $r = 2\pi\sqrt{D\Gamma/(V\,\Delta T_o)}$:

$$\lambda_1 = \mu_{\lambda 1} \cdot V^{-1/4} \cdot G^{-1/2} \quad \text{where} \quad \mu_{\lambda 1} = 4.3\left[\Delta T_o\, D_L\, \Gamma\right]^{1/4} \tag{8.41}$$

The constant refers to a single-phase alloy.

Earlier, Hunt (1979) has derived a similar equation for primary spacing different only through the numerical constant, which was 2.83 rather than 4.3. Note that all models introduced here demonstrate that the primary spacing is a function of G and V.

Bouchard and Kirkaldy (1997) tested these equations against experimental data for steady-state solidification of cells (28 alloys) and dendrites (21 alloys) in binary alloys. The experimental data summarized by the following equations agree reasonable well with theoretical predictions:

for cells: $\quad \lambda_1 = ct. \cdot \dot{T}^{-0.36 \pm 0.05}$

for dendrites: $\quad \lambda_1 = ct. \cdot V^{-0.28 \pm 0.04} \cdot G^{-0.42 \pm 0.04} \quad$ and $\quad \lambda_1 = ct. \cdot \dot{T}^{-0.3 \pm 0.03}$

However, for unsteady-state flow all equations failed to perform adequately.

Once the primary spacing is established, it will remain constant throughout steady - state solidification, and during cooling in solid state. If non-steady state solidification occurs, the primary spacing will change. Two typical mechanisms for adjustment of primary spacing are presented in Figure 8.15. Engulfing results in the increase of DAS while branching decreases DAS.

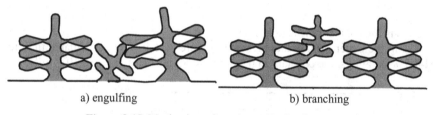

a) engulfing   b) branching

Figure 8.15. Mechanisms for primary DAS adjustment.

8.4.2   Secondary arm spacing

In the early understanding of dendrite growth, it was assumed that the secondary dendrite arm spacing is formed at the beginning of solidification. Then, arms thicken and grow as solidification proceeds. Thus, the final arm spacing, λ_f, was thought to be the same as the initial spacing, λ_o.

Later it was realized that as solidification proceeds, only the larger arms grow. The smaller arms remelt (dissolve) and eventually disappear. Consequently, throughout solidification the secondary arm spacing (SDAS) increases and $\lambda_f > \lambda_o$. This is the dynamic coarsening of dendrites. The effect of coarsening on the SDAS of a transparent organic material is shown in Figure 7.16f. It is seen that the secondary DAS increases with the distance behind the tip.

Many mathematical models have been developed for dendrite coarsening based on the concept that dendrite coarsening is diffusion controlled, the diffusive species under consideration being the solvent. Assuming isothermal coarsening, the growth rate of the distance, λ, between two spherical particles must be proportional to the compositional gradient:

$$d\lambda/dt = ct \cdot \left(\Delta C_L / \lambda\right)$$

(8.42)

The liquid temperature and composition in equilibrium with a solid surface depends on the curvature of that surface. Indeed, the curvature undercooling at the tip of the dendrite is $\Delta T_r = 2\Gamma/r$. Since $\Delta T_r = m \Delta C_r$, $\Delta C_r = 2\Gamma/(m r) = ct \cdot r^{-1}$. Curvature and local curvature differences must increase approximately proportionally with the inverse of the spacing λ. Thus. $r = ct \cdot \lambda$. It follows that $\Delta C_r = ct \cdot \lambda^{-1}$, and also $\Delta C_L = C_{r1} - C_{r2} = ct \cdot \lambda^{-1}$.

Substituting in Eq. (8.42) yields $d\lambda/dt = ct \cdot \lambda^{-2}$. Rearranging and integrating between an initial arm spacing, λ_o, and λ_f, and between zero and the final local solidification time (the difference between the times when the liquidus isotherm and the solidus isotherm pass the particular microvolume), t_f, gives:

$$\lambda_f^3 - \lambda_o^3 = \mu_o \cdot t_f$$

(8.43)

Assuming that $\lambda_o << \lambda_f = \lambda_2$, results in a final secondary arm spacing of:

$$\lambda_2 = \mu_o^{1/3} \cdot t_f^{1/3}$$

(8.44)

where t_f is the local solidification time.

Using the experimental data presented Figure 8.16 it can be calculated that, for Al-4.5% Cu alloys, the constant in the coarsening law has a value of 10^{-16} m/s^3.

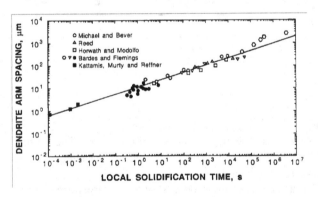

Figure 8.16. Relation between SDAS and solidification time for Al-4.5% Cu alloys (Flemings *et al.*, 1991). Copyright 1991 American Foundry Soc., used with permission.

The constant in Eq. (8.44) has been derived by a number of authors, mostly for the case of spherical particles (see for example the derivation in the inset). Some typical formulations and their basic assumptions are given in Table 8.3. Isothermal coarsening assumes that the only driving force is solute diffusion and that the fraction solid is constant. However, during solidification the temperature decreases, the fraction solid increases, and an additional driving force, thermal diffusion, must be considered. This is dynamic coarsening.

Table 8.3. Coarsening constants.

Model	Coarsening constant	Basic assumptions
Kattamis-Flemings, 1965	$\mu_o = \dfrac{20 D_L \Gamma \ln(C_E / C_o)}{m_L C_L (k-1)(C_E - C_o)}$	isothermal coarsening, of spheres; see inset for derivation
Ardell, 1972	$\mu_o = \dfrac{v_m^2 \gamma_{SL} C_o D_L}{RT}(1 - C_o)\dfrac{1-f}{f}$ C_o in fraction	dynamic coarsening of spheres, diffusion of solute depends on a characteristic distance defined by the main free path
Voorhees-Glicksman, 1984	$\mu_o = \dfrac{8}{9}\dfrac{v_m^2 \gamma_{SL} C_o D_L}{RT}\dfrac{\alpha^3}{1-f^{1/3}}$ α: fct. of f_S given in tabulated form	solution of dynamic multiparticle diffusion problem; random pattern of precipitates generated by Monte Carlo simulation
Mortensen, 1991	$\mu_o = \dfrac{27 D_L \Gamma}{2 m_L C_L (k-1) f_S^{2/3}\left(1 - f_S^{1/3}\right)}$	dynamic coarsening of spheres
	$\mu_o = \dfrac{27 D_L \Gamma}{4 m_L C_L (k-1) f_S \left(1 - f_S^{1/2}\right)}$	dynamic coarsening of cylinders

---

Derivation of coarsening constant - Kattamis/Flemings model.
The basic assumptions of the model are as follows: isotropic growth of two spherical dispersoids of constant total volume; constant concentration gradient; unidirectional diffusion; the radius of the larger dispersoid is much larger than that of the smaller dispersoid ($r_2 \gg r_1$).

Both spheres are at the same undercooling and separated by a distance λ. As discussed before, the difference in the interface concentration of the two spheres is:

$$\Delta C = C_L^{r_2} - C_L^{r_1} = \frac{\Gamma}{m}\left(\frac{1}{r_1} - \frac{1}{r_2}\right)$$

The flux of solute from r_2 to r_1 is: $J = -D\dfrac{\Delta C}{\lambda} = -\dfrac{D\Gamma}{\lambda m}\left(\dfrac{1}{r_1} - \dfrac{1}{r_2}\right)$

The flux of solvent from r_1 to r_2 is: $J = \left((1 - C_L^{r}) - (1 - C_S)\right)\dfrac{dr}{dt} = -C_L^{r}(1 - k)\dfrac{dr}{dt}$

Flux balance gives: $\dfrac{dr}{dt} = \dfrac{D\Gamma}{m C_L^{r}(1-k)\lambda}\left(\dfrac{1}{r_1} - \dfrac{1}{r_2}\right)$

It is further assumed that the interface concentration of the large particle is equal to the average liquid concentration due to segregation, $C_L^{r} \approx C_L$, and that the liquid concentration is a linear function of time, from an initial concentration C_o to the final eutectic concentration C_E:

$$C_L = C_o + (C_E - C_o)\frac{t}{t_f}$$

If it is also assumed that since $r_2 \gg r_1$, $r_2 = $ ct. and $r_1 = f(r)$, the last equation can be integrated between the initial time (0) when the small particle has a radius $r_1 = r_o$, and the time tf when the particle has vanished:

$$\int_{r_1=r_o}^{0} \frac{r_1 r_2}{r_2 - r_1} dr = \int_{0}^{t_f} \frac{D\Gamma}{mC_L(1-k)\lambda} dt$$

The final coarsening time is then: $t_f = \dfrac{mC_L(k-1)(C_E - C_o)}{D\Gamma \ln(C_E/C_o)} \lambda r_2^2 \left(\dfrac{r_o}{r_2} + \ln\left(1 - \dfrac{r_o}{r_2}\right) \right)$

It is not surprising that the coarsening time depends on the initial size of the particles. To obtain an order of magnitude of the coarsening constant it can be further assumed that $r_2 = 0.5\lambda$ and $r_o = 0.5\, r_2$. Then:

$$\lambda_f^3 = \frac{20 D_L \Gamma \ln(C_E/C_o)}{m_L C_L (k-1)(C_E - C_o)} t_f$$

The majority of models for the temporal evolution of λ_2 assume that the dendrite arms have a cylindrical shape (*e.g.* Kattamis *et al.* 1967). Calculations then predict the axial remelting of the thinner arm.

Other models consider the dendrite arms to be tear-shaped. If a tear-shaped arm is surrounded by two cylindrical arms (Chernov, 1956) material is transported from the base where the radius of curvature r_1 is small, to the tip, where the radius r_2 is large (Figure 8.17a). Eventually r_2 becomes zero and the arm detaches from the stem of the dendrite as demonstrated experimentally by Papapetrou (1935) (see Figure 8.18).

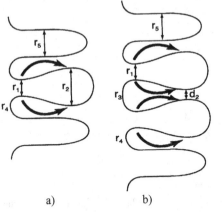

a)   b)

Figure 8.17. Material transport for (a) one tear-shaped arm surrounded by cylindrical arms, or (b) two tear-shaped arms surrounded by cylindrical arms (Mendoza *et al.* 2003).
With kind permission of Springer Science and Business Media.

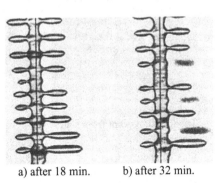

a) after 18 min.   b) after 32 min.

Figure 8.18. Separation of dendrite arms in NH4Cl (Papapetrou, 1935). After separation the detached arms move out of focus.

If case (b) in Figure 8.17 is considered (Young and Kirkwood, 1975), material transported away from the base accumulates at the tip and the distance d_2 decreases until coalescence occurs. Experimental evidence of this mechanism is shown in Figure 8.19.

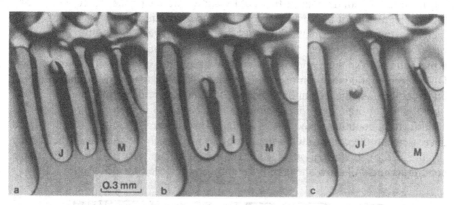

Figure 8.19. Coalescence of arms J and I in succinonitrile (Huang and Glicksman, 1981). Reprinted with permission from Elsevier.

Experimental data on secondary arm spacing have also been reported to fit an equation similar to Eq. (8.40) proposed for primary spacing. Analysis of 60 experimental data on two alloys by Bouchard and Kirkaldy (1997) produced the following relationship:

$$\lambda_2 = ct \cdot \left(\dot{T} \right)^{-0.34 \pm 0.02}$$

For alloys solidifying with equiaxed structure, it is not possible to define a primary arm spacing. To evaluate the length scale of the microstructure the average grain size (average diameter of grains on the metallographic sample), or volumetric grain density (number of grains per unit volume) are used. These numbers are primarily functions of the nucleation potential of the melt. Secondary arm spacing can also be used to evaluate the fineness of equiaxed structures. However, as noted above, it represents thermal conditions during solidification, not nucleation conditions. Thus, it is possible to have a coarse grained casting with fine secondary arm spacing.

8.5   The columnar-to-equiaxed transition

In many applications, either a columnar or an equiaxed structure is desired for the casting. If, because of lack of adequate process control, a sudden columnar-to-equiaxed transition (CET) occurs in these castings, they will be rejected having an unacceptable structure. Therefore, it is important to understand the conditions under which a CET can occur in a given casting.

An example of such a transition in an ingot is shown on Figure 8.20. This is not a schematic drawing but a modeled microstructure, which correctly describes the real microstructure. The details of the model will be discussed in Section 13.1.1. The results of the calculations are very realistic. Three different structural regions are shown on the middle ingot: a chill zone, made of small equiaxed grains resulting from rapid cooling against the mold wall; a columnar zone; an equiaxed zone toward the middle of the casting. The figure also indicates that as the undercooling increases the structure in the bulk of the ingot changes from fully columnar to mixed columnar /equiaxed. Thus, a CET occurs as the undercooling is increased. A typical structure showing the CET in an Al-5%Cu ingot is presented in Figure 8.21. One possible rationalization of the occurrence of CET is in terms of constitutional undercooling. At the beginning of solidification, the temperature gradient in the liquid is rather high, and constitutional undercooling is limited (Figure 8.22a). As solidification continues, the mold is heated. The temperature gradient in the liquid decreases and the constitutional undercooling may reach the middle of the casting (Figure 8.22b). If nucleation of equiaxed grains occurs they will have favorable conditions and will grow ahead of the columnar interface (see also Figure 7.13).

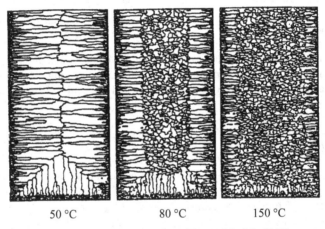

50 °C   80 °C   150 °C

Figure 8.20. Typical structural regions in castings (Zhu and Smith, 1992). Reprinted with permission from Elsevier.

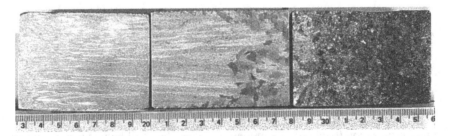

Figure 8.21. CET occurring in an Al-5 wt% Cu ingot solidified against a chill placed at its base (Guo and Stefanescu, 1992). Solidification is from right-to-left. The CET occurred when the temperature gradient in the melt ahead of the interface decreased in the range of 113 to 234K/m. In the equiaxed zone the number of grains was $5 \cdot 10^6 m^{-3}$. The average solidification velocity at the CET was measured to be $3.5 \cdot 10^{-4} m/s$. Copyright 1992 American Foundry Soc., used with permission.

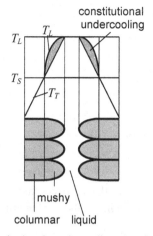 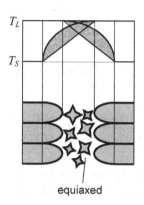

a) no constitutional undercooling in the middle of the casting - only columnar growth is possible

b) constitutional undercooling in the middle of the casting - if nucleation occurs, equiaxed growth is possible

Figure 8.22. Occurrence of CET because of increased constitutional undercooling resulting from lower temperature gradient.

While the role of constitutional undercooling in the CET is not under dispute, it does not seem to be the only mechanism responsible for it. Chalmers (1962) has shown that when the center of a casting is isolated with a cylinder, fewer and coarser grains grow than in the absence of isolation. This is in spite of the fact that the center is constitutionally undercooled in both cases. Chalmers then proposed the *"Big Bang Mechanism"* which postulates that equiaxed grains result from the nuclei formed during pouring by the initial chilling action of the mold. The grains are then carried into the bulk liquid. If they survive until the superheating is removed a CET occurs. However, assuming that only the big bang is responsible for CET cannot account for equiaxed zone formation in the absence of a chilled mold.

Jackson *et al.* (1966) noticed that increased convection during solidification of organic alloys produced a large number of nuclei in the liquid. They postulated that dendrite arms remelt because of recalescence, detach from the dendrite stem and then float into the center of the casting where they serve as nuclei for equiaxed grains. Another argument for the *dendrite detachment mechanism* is that the mechanical strength of the dendrite is negligible close to its melting point, and thus convection currents can simply break the dendrite (O'Hara and Tiller, 1967).

Hunt (1984) has proposed a 1D analytical model for the CET based on the following assumptions: equiaxed grains are formed by heterogeneous nucleation and do not move with the liquid, steady state is possible for a fully columnar, columnar + equiaxed, or fully equiaxed growth. It was further assumed that a fully equiaxed structure results when the fraction of equiaxed grains is higher than a critical fraction of solid, $f_S^e > f_S^{cr} = 0.49$, and that a fully columnar structure is produced when $f_S^c < 10^{-2} f_S^{cr} = 0.0049$. Using a model of hemispherical dendrite growth, and assuming small thermal undercooling the following criteria were derived:

- fully equiaxed growth occurs when:

$$G_T < 0.49 \left(\frac{N}{f_S^{cr}} \right)^{1/3} \left[1 - \frac{(\Delta T_N)^3}{(\Delta T_c)^3} \right] \Delta T_c \qquad (8.45a)$$

- fully columnar growth develops when:

$$G_T > 0.49 \left(\frac{100 \cdot N}{f_S^{cr}} \right)^{1/3} \left[1 - \frac{(\Delta T_N)^3}{(\Delta T_c)^3} \right] \Delta T_c \qquad (8.45b)$$

where N is the volumetric nuclei density, ΔT_N is the undercooling required for heterogeneous nucleation, and ΔT_c is the undercooling at the columnar front calculated as:

$$\Delta T_c = \left[-8\Gamma m_L (1-k) C_o V/D \right]^{1/2} \qquad (8.46)$$

The selection of the thresholds in the derivation of the above relationships is debatable. It seems difficult to accept that columnar grains can grow beyond the point when dendrite coherency is established, f_S^{coh}. Accordingly, a more reasonable upper limit is $f_S^{cr} = f_S^{coh}$. Typically $f_S^{coh} = 0.2$-0.4. Also, mixed equiaxed-columnar structures are seldom observed in castings. In most cases an abrupt CET is seen. Indeed, microgravity work performed by Dupouy *et al.* (1998) on Al-4% Cu alloys demonstrated that while a smooth (mixed structure) CET is obtained in microgravity (no thermo-solutal convection), an abrupt CET is seen on the same sample solidified under terrestrial conditions. Thus, it is reasonable to assume that the CET occurs simply when coherency is reached.

Another weak assumption is that of stationary equiaxed grains. Indeed, because of the thermo-solutal and shrinkage convection, the equiaxed grains will move with the liquid, unless coherency is reached. As discussed previously, one of the main reasons for the CET is the presence of thermosolutal convection.

Based on the preceding discussion a new model is proposed in the following paragraphs. Consider a volume element of length l, that extends from some arbitrary point in the columnar region to a region where the temperature is equal to the nucleation temperature, ΔT_N that is very close to the liquidus temperature (Figure 8.23). Consequently, the grains moving away out of the volume element in the bulk liquid in the x-direction will not survive, and no grains are advected from the bulk liquid into the element in the x-direction. It is further assumed that the net contribution of the flow in the y-direction to the number of grains is zero. Within the volume element the CET occurs if the equiaxed grains can reach f_S^{coh} before the columnar front traverses the element. Thus, condition for the CET is:

$$t_c \geq t_e^{coh} \qquad (8.47)$$

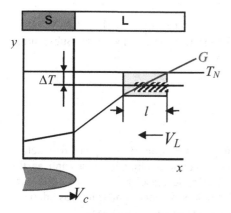

Figure 8.23. Volume element for calculation of CET.

where t_c is the time required for the columnar front to move across the volume element, and t_e^{coh} is the time required for the equiaxed grains to reach coherency. Then:

$$t_c = l/V_c = 2\Delta T/(GV_c) = 2\Delta T/(G\mu_c \Delta T^2)$$

where G and ΔT are the average thermal gradient and undercooling in the volume element, respectively, V_c is the growth velocity of the columnar dendrites, and μ_c is the growth coefficient of the columnar dendrites.

Similarly, since when assuming spherical grains, the fraction of solid in the volume element can be calculated as $f_S = (4/3)\pi \bar{r}^3 \bar{N}$, where $\bar{r}$ and $\bar{N}$ are the average grain radius and the average volumetric grain density, respectively:

$$t_e^{coh} = \frac{\bar{r}^{coh}}{V_e} = \left(\frac{3}{4\pi} \frac{f_S^{coh}}{\bar{N}} \right)^{1/3} \left(\mu_e \Delta T^2 \right)^{-1}$$

where V_e is the growth velocity of the equiaxed dendrites, and μ_e is the growth coefficient of the equiaxed dendrites.

Introducing the last two equations the CET criterion Eq. (8.47), the CET will occur when:

$$G_T \leq 3.22 \left(\frac{\bar{N}}{f_S^{coh}} \right)^{1/3} \frac{\mu_e}{\mu_c} \Delta T \tag{8.48}$$

In turn, the average volumetric grain density can be written as the difference between the active heterogeneous nuclei, N, and the grains entering the volume element because of fluid flow into the volume element:

$$\overline{N} = N + (V_L/V)N$$

where V_L is the flow velocity in the x-direction, and V is the solidification velocity. Note that if $V_L = V$, $\overline{N} = 0$, and equiaxed solidification is impossible. Substituting in Eq. (8.48) the final CET condition becomes:

$$G_T \le 3.22 \left[\frac{\overline{N}}{f_S^{coh}} \left(1 + \frac{V_L}{V} \right) \right]^{1/3} \frac{\mu_e}{\mu_c} \Delta T \qquad (8.49)$$

This equation suggests that the probability of formation of an equiaxed structure increases as the nucleation potential, and the undercooling increase, and as the coherency solid fraction and liquid convection decrease. Note that for the case when convection is ignored ($V_L = 0$), this equation is very similar to those derived by Hunt. However, it was obtained using less restrictive assumptions.

This model could be further developed to include grain transport into the volume element, by assuming that the outer limit of the element is at a temperature below the nucleation temperature, or if the number of grains to be advected in the volume element can be calculated.

8.6   Applications

Application 8.1

Compare the solutal and thermal undercooling for an Al-4.5% Cu alloy and for an Fe-0.09% C alloy.

Answer:
Since the expressions for undercooling include both velocity and tip radius it is not possible to calculate the undercooling without additional data. However, a comparison can be made by calculating the ratio $\Delta T/Vr$ for the two cases. From Eqs.(8.7)) and (8.6a), and using data in Appendix B:

$\Delta T_c/Vr = m(k-1)C_o/2kD = 1.03\cdot10^{10}$ K·s·m^{-2} for the Al-Cu alloy, and $= 8.9\cdot10^8$ K·s·m^{-2} for the Fe-C alloy

$\Delta T_T/Vr = \Delta H_f/2\alpha c = 4.94\cdot10^6$ K·s·m^{-2} for the Al-Cu alloy, and $= 2.53\cdot10^7$ K·s·m^{-2} for the Fe-C alloy

It is obvious that the thermal undercooling is very small as compared with the solutal undercooling for the Al-Cu alloy, but within an order of magnitude for the Fe-C alloy. Thus, the thermal undercooling cannot always be neglected.

Application 8.2

Compare the tip radius - growth velocity correlation for solutal dendrites and solutal-thermal dendrites for a Fe-0.09% C alloy using the Nastac-Stefanescu (NS) (1993) and the Trivedi-Kurz TK (1994) models.

Answer:
Combining Eqs. (8.23) and (8.24) and assuming steady-state, we obtain the *V-r* correlation
for the NS model as follows:

$$V r^2 = 8\pi^2 \Gamma \left(\frac{m(k-1)C_o}{k D_L} + \frac{\Delta H_f}{c \alpha_L} \right)^{-1}$$

(8.50)

For the TK model, assuming steady-state and low Péclet number (<<1) we can obtain a
similar equation from Eq. (8.19):

$$V r^2 = 4\pi^2 \Gamma \left(\frac{m(k-1)C_o}{D_L} + \frac{\Delta H_f}{2c_L \alpha_L} \right)^{-1}$$

(8.51)

For the solutal dendrite, that is a dendrite whose growth is controlled solely by the solutal field, only the first term in the parenthesis is used. Using data in Appendix B we obtain
the graph in Figure 8.24. It is seen that there is no difference between the solutal NS and TK
in the range of velocities used in the calculation. When the effects of thermal undercooling
are also used in the NS model, a slight decrease in tip radius is calculated.

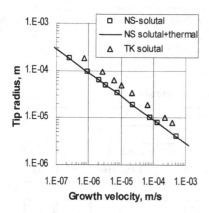

Figure 8.24. Calculated dendrite tip radii
for solutal and solutal-thermal dendrites
for a Fe-0.09% alloy.

Application 8.3

Compare the dendrite tip radius - growth velocity relationship calculated with the Trivedi-
Kurz (1994) and Nastac-Stefanescu (1993) models, with the experimental data on the Fe-
3.08% C-2.01% Si - 0.104% Mn - 0.016% S - 0.029% P alloys obtained by Tian and Stefanescu (1992).

Answer:
For the NS model the solutal part of Eq. (8.50) will be used. For the TK model the model for
columnar dendrite, Eq. (8.21), will be used. Again, assuming steady state ($C_L^* = C_o/k$) and
low Péclet number ($\xi_c = 1$) the TK columnar dendrite growth velocity becomes:

$$V = \left(\frac{4\pi^2 \Gamma}{r^2} + G_T \right) \frac{D_L}{m(k-1)C_o}$$

(8.52)

Since the alloy is a multicomponent alloy, the average composition C_o must be expressed as a carbon equivalent to reduce the multicomponent alloy to a binary one. The following relationship is used:

$$C_o = \%C + 0.31 \cdot \%Si + 0.33 \cdot \%P - 0.27 \cdot \%Mn + 0.4 \cdot \%S = 3.72$$

The predicted and experimental results are plotted in Figure 8.10. Note that while the NS model fails at growth velocities smaller than ~1μm/s, it describes growth reasonable well within the range of velocities typical for castings. It is also the simplest one to implement in a numerical code. The thermal gradient is important only in the cellular solidification range.

Application 8.4

Calculate the amount of eutectic that will solidify in the interdendritic regions of an Al-4% Cu alloy, for three different solidification velocities: $3 \cdot 10^{-5}$, $1 \cdot 10^{-7}$, $6 \cdot 10^{-8}$ m/s. Assume a constant temperature gradient of 2000 K/m.

Answer:
The required materials constant are obtained from Appendix B and listed in the Excel spreadsheet in column A. The maximum solubility of Cu in Al is 5.65%. When the composition reaches this value the rest of the liquid solidifies as eutectic.

The calculation of the solid composition is performed in columns D, E, and F based on the fraction solid listed in column C (Table 8.4). First the parameter a is calculated for the three velocities. Cells D3, E3, and F3 include the equation for a, i.e., Eq. (8.35). Columns D, E, and F starting with cells D6, E6, and F6 include the equation for C_S, i.e. Eq. (8.38). The calculation is run until $C_S = 5.65$.

Table 8.4. Organization of spreadsheet.

	A	B	C	D	E	F
1	Constants	Data				
2	D_L	2.8E-9	V	3.00E-05	1.00E-07	6.00E-08
3	G_L	2000	a	-1.15E-02	-3.46E+00	-5.76E+00
4	m	-3.6				
5	C_o	4.5	f_S	C_S	C_S	C_S
6	k	0.14	0	0.64	2.81	4.26
7	C_{max}	5.65	0.1	0.70	2.83	4.26
8			0.2	0.77	2.87	4.27
9			0.3	0.86	2.91	4.27
10			0.915	5.25	4.83	4.55
11			0.922	5.65	5.00	4.57
12			0.9405		5.65	4.66
13			0.9845			5.63

The calculation results are shown in Figure 8.25. The fraction of eutectic is calculated as $f_E = 1 - f_S$. It is seen that for $V = 3 \cdot 10^{-5}$, for which a is very small, the calculated fraction of eutectic is 0.078. As V increases to $1 \cdot 10^{-7}$, f_E decreases to 0.059, and to 0.016 for $V = 6 \cdot 10^{-8}$. It will become zero for equilibrium solidification when the velocity is so small that $a = (k - 1)/k$.

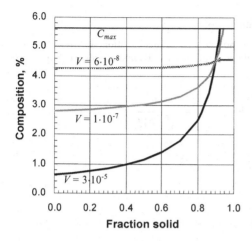

Figure 8.25. Calculation of amount of eutectic as a function of solidification velocity.

Application 8.5

Calculate the temperature and solid fraction evolution during the solidification of an equiaxed dendrite of an Fe-0.6%C alloy that has a volumetric grain density of 1 grain/mm^3. Assume that the alloy is cooled at constant heat extraction rate of $\dot{Q} = 3 \cdot 10^8$ J·m^{-3}·s^{-1}, and an initial temperature of 1520°C.

Answer:
The governing heat transport equation is: $\dot{Q} = \rho \Delta H_f \, df_S / dt - \rho c \, dT / dt$. Rearranging and discretizing for time-stepping:

$$T^{new} = T^{old} - \frac{\dot{Q}}{\rho c} \Delta t + \frac{\Delta H_f}{c} \Delta f_S^{new} \tag{a}$$

Note that this equation is independent of volume. Assuming a spherical equiaxed dendrite, $f_S = (4/3) \, \pi \, r_S^3 \, N$, and $df_S / dt = 4\pi \, r_S^2 \, N \, dr_S / dt = 4\pi \, r_S^2 \, N \, V_S$. The number of nuclei, N, is equal to one. The time-discretized equation is:

$$\Delta f_S^{new} = 4 \, \pi \left(r_S^{new} \right)^2 V_S^{new} \Delta t \tag{b}$$

Further, the grain size is:

$$r_S^{new} = r_S^{old} + V_S^{ewn} \Delta t \tag{c}$$

The solidification velocity is $V_S = \mu \cdot \Delta T^2$. The growth coefficient can be calculated with Eq. (8.22). Assuming a solutal dendrite, the discretized equation for the solidification velocity is:

$$V_S^{new} = \frac{D_L}{2\pi^2 \Gamma m(k-1)\langle C_L \rangle^{new}} \left(\Delta T^{new} \right)^2 \tag{d}$$

where $\langle C_L \rangle$ is the average liquid composition.

To calculate the average liquid composition needed in this equation, a diffusion model must be used. We will compare the Scheil and equilibrium diffusion models. The time discretized equations for the two models are:

Scheil: $C_L^{new} = C_o (1 - f_S^{old})^{k-1}$ Equilibrium: $C_L^{new} = \dfrac{C_o}{1-(1-k)f_S^{old}}$ \qquad (e)

The evolution of the fraction solid is: $f_S^{new} = f_S^{old} + \Delta f_S^{new}$ \qquad (f)

Finally, the undercooling is calculated as:

$$\Delta T = \Delta T_c + \Delta T_T = m\left(C_o - C_L\right) + T^* - T_{bulk} = T_f + mC_o - T_{bulk}$$

where T_{bulk} is the average temperature in the volume element (the macro-temperature). C_o is the average liquid composition, $\langle C_L \rangle$. In discretized form this is:

$$\Delta T^{new} = T_f + m\langle C_L \rangle^{new} - T^{old} \tag{g}$$

Table 8.5. Program implementation on the Excel spreadsheet.

time	$\langle C_L \rangle$	ΔT	V_S	r_S	Δf_S	f_S	T
Eq.	(e)	(g)	(d)	(c)	(b)	(f)	(a)
0				1.00E-07			1520
0.01	0.600	00	0.00E+00	1.00E-07	0.00E+00	0.00	1519.5
0.59	0.6	0.394	2.62E-05	3.62E-07	4.31E-19	4.31E-19	1489.1
1.47	0.602	44.7	3.35E-01	1.07E-01	4.82E-04	0.01	1444.8
4.63	0.891	19.4	4.29E-02	4.90E-01	1.29E-03	0.50	1447.2
9.17	1.746	17.6	1.79E-02	6.19E-01	8.62E-04	1.00	1380.5

Since above the liquidus temperature there is no undercooling, an IF statement must be included which allows this equation to become effective only at $T^{new} < T_L$. These equations are then implemented in the Excel spreadsheet, for example as shown in Table 8.5 fro the case of equilibrium. An initial radius at time zero is assumed. The initial temperature is 1520°C.

The calculated results are plotted in the Figure 8.26. It is seen that significant differences exist in both temperature and solid fraction evolution as a function of the chosen diffusion model. When the *Scheil* model is used, longer time is needed for completion of solidification, and a lower solidus temperature is reached (1187°C as compared to 1381°C for equilibrium). Dendritic solidification ends when the liquid composition becomes $C_L = C_E = 4.26$.

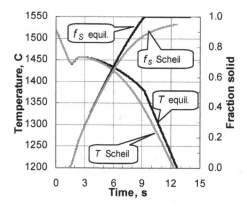

Figure 8.26. Evolution of temperature and solid fraction during the solidification of a condensed dendrite.

Application 8.6

Calculate the critical gradient for CET in the Al-5% Cu ingot presented in Figure 8.21 ($N = 5 \cdot 10^6$ m^{-3} and $V = 3.5 \cdot 10^{-4}$ m/s). Assume $f_S^{coh} = 0.3$. The other data required for calculation are given in Appendix B.

Answer:

Let us use the Hunt model first. From Eq. (8.46)) it is calculated that $\Delta T_c = 1.93$K. Then, assuming that $\Delta T_N \ll \Delta T_c$ (which is not necessarily true), and using Eq. (8.45a) it is calculated that the critical gradient for CET is $G_T = 205$K/m for the $f_S^{cr} = 0.49$ postulated but Hunt. This is within the range determined experimentally, which was 113 to 234 K/m.

Let us now use the model described by Eq. (8.49). The steady-state growth coefficient for columnar growth can be calculated from Eq. (8.26)) as $\mu_c D_L \left(\pi^2 \Gamma m (k-1) C_o \right)^{-1}$. We obtain $\mu_c = 7.64 \cdot 10^{-5}$. Then, from the same equation, assuming that all undercooling is constitutional, the undercooling is $\Delta T = \sqrt{V/\mu} = 2.14$ K. The growth coefficient for the equiaxed grains can be calculated with Eq. (8.22). When the thermal undercooling is ignored we obtain $\mu_e = 5.35 \cdot 10^{-5}$. Introducing these values in Eq. (8.49), it is obtained that $G_T = 123$ K/m, which is in range of the experimental data.

References

Akamatsu S., G. Faivre, and Th. Ihle, 1995, *Phys. Rev.* **E51**:4751–4773

Ananth R. and Gill W.N., 1988, *J. Crystal Growth* **91**:587

Ardell A.J., 1972, *Acta Metall.* **20**:61

Barbieri A. and Langer J.S., 1989, *Physical Review A* **10**:5314-5325

Bensimon D., Pelce P., and Shraiman B. I., 1987, *J. Phys. A* **48**:2081

Biloni H. and Boettinger W.J., 1996, in: *Physical Metallurgy*; fourth edition, R.W. Cahn and P. Haasen eds., Elsevier Science BV p.670

Boettinger W. J., Coriell S.R., and Trivedi R., 1988, in: *Rapid Solidification Processing: Principles and Technologies*, R. Mehrabian and P.A. Parrish eds., Claitor's Publishing, Baton Rouge, LA p.13

Bouchard D. and Kirkaldy J.S., 1997, *Metall. and Mater. Trans.* **28B**:651

Bowers T.F., Brody H.D. and. Flemings M.C, 1966, *Trans. AIME* **236**:624

Burden N. H. and Hunt J.D., 1974, *J. Cryst. Growth* **22**

Chalmers B., 1962, *J. Aust. Inst. Met.* **8**:225

Chernov A.A., 1956, *Kristallographya* **65**:583

Cohen M. and Flemings M.C., 1985, in: *Rapidly Solidified Crystalline Alloys*, S.K. Das, B.H. Kear and C.M. Adam eds., TMS, Warrendale, PA p.3

Dupouy M.D., Camel D., Botalla F., Abadie J., and Favier J.J., 1998, *Microgravity sci. technol.* **XI**(1):2

Elliott R., 1983, *Eutectic Solidification Processing*, Butterworths, London

Flemings M.C., Kattamis T.Z., and Bardes B.P., 1991, *AFS Trans.* **99**:501

Glicksman M.E., Koss M.B., Bushnell L.T., Lacombe J.C. and Winsa E.A., 1995, *ISIJ International* **35**:604

Guo X. and Stefanescu D.M., 1992, *AFS Trans.* **100**:273

Huang S.C. and Glicksman M.E., 1981, *Acta Metall.* **29**:701

Hunt J. D., 1979, in: *Solidification and Casting of Metals*, The Metals Society, London p.3

Hunt J. D., 1984, *Mat. Sci. and Eng.* **65**:75

Hunt J.D. and Jackson K.A, 1966, *Trans. Met. Soc. AIME* **236**:843

Ivantsov G. P., 1947, *Doklady Akademii Nauk SSSR* **58**:695

Jackson K.A., Hunt J.D., Uhlmann D., and Seward T.P., 1965, *Trans. AIME* **236**:149

Kattamis T.Z. and Flemings M.C., 1965, *Trans. Met. Soc. AIME* **233**:992

Kattamis T.Z., Coughlin J.C., and Flemings M.C., 1965, *Trans. TMS-AIME* **239**:1504

Kessler D.A and Levine H., 1986, *Phys. Rev. Lett.* **57**:3069

Koseki T. and Flemings M.C., 1995, *ISIJ International* **35**:611

Kurz W., Giovanola B. and Trivedi R., 1986, *Acta Metall.* **34**:823

Kurz W. and Fisher D.J., 1989, *Fundamentals of Solidification*, 3rd ed., Trans Tech Publications, Switzerland

Langer J. S. and Müller-Krumbhaar H., 1978, *Acta Metall.* **26**:1681

Laxmanan V., 1985, *Acta Metall.* **33**:1023

Lipton J., Glicksman M.E. and Kurz W., 1984, *Mat. Sci. Eng.* **65**:57

Lu S.Z. and Liu S., 2007, *Metall. and Mater. Trans.* **38A**:1378-1387

Lux B., Minkoff I., Mollard F., and Thury E., 1975, in: *The Metallurgy of Cast Iron*, B. Lux, I. Minkoff and F. Mollard eds., Georgi Publ. Co., St Saphorin, Switzerland p.497

Mendoza R., Alkemper J., and Voorhees P.W., 2003, *Metall. and Mater. Trans.* **34A**:481

Miyata Y., 1995, *ISIJ International* **35**:600

Morris L.R. and Winegard W.C., 1969, *J. Crystal Growth* **6**:61

Mortensen A., 1991, *Metall. Trans.* **22A**:569

Nash G.E. and Glicksman M.E., 1974, *Acta Metall.* **22**:1283

Nastac L. and Stefanescu D.M., 1993, *Metall. Trans.* **24A**:2107

Nastac L., Chou J.S., and Pang Y., 1999, in: *Symp. on Liquid Metal Processing and Casting*, Santa Fe, New Mexico

O'Hara S. and Tiller W.A., 1967, *Trans. AIME* **239**:497

Papapetrou A., 1935, *Z. Kristall.* **92**:89

Trivedi R. and Kurz W., 1988, in: *Metals Handbook Ninth Edition vol. 15 Casting*, D.M. Stefanescu ed., ASM International, Metals Park, Ohio p.114

Trivedi R. and Kurz W., 1994, *International Materials Reviews* **39**,2:49

Voorhees P.W. and Glicksman E.M., 1984, *Metall. Trans.* **15A**:1081

Warren J.A. and W.J. Boettinger, 1995, *Acta metall. mater.* **43**:689

Young K.P. and Kirkwood D.H., 1975, *Metall. Trans.* **6A**:197

Zhu P. and Smith R.W., 1992, *Acta metall. mater.* **40**:683 and 3369

9

EUTECTIC SOLIDIFICATION

Some of the most important casting alloys in terms of tonnage as well as applications, such as cast iron and aluminum-silicon alloys, are essentially two-phase alloys. The eutectic has a fixed composition in terms of species A and B, and is in fact a two-phase solid ($\alpha + \beta$). Solidification of a liquid of eutectic composition proceeds by transformation of the liquid into a two-phase solid.

9.1 Classification of Eutectics

Many eutectic classifications have been proposed, based on different criteria. A first classification of eutectics is based on their growth mechanism:

- *cooperative growth*: the two phases of the eutectic grow together as a diffusion couple;
- *divorced growth*: the two phases of the eutectic grow separately; there is no direct exchange of solute between the two solid phases and no trijunction.

Another classification was proposed (Hunt and Jackson, 1966) based on the interface kinetics of the component phases of the eutectic. As discussed earlier, phases having low entropies of fusion solidify with a non-faceted interface, while phases having high entropy of fusion solidify with faceted interface. Thus, the following classification was proposed:

- non-faceted / non-faceted eutectics (nf /nf);
- non-faceted / faceted (nf /f);
- faceted / faceted (f /f).

The first two categories are common and have commercial applications. The f /f eutectics are less studied. Croker *et al.* (1973) suggested that in addition to the entropy of fusion, the volume fraction of the two phases plays a significant role in the resulting microstructure. Depending on the ratio between the fractions of the two phases of the eutectic, f_α and f_β, and on the morphology of the liquid - solid, several types of cooperative eutectics may form (Figure 9.1). The nondimensional entropy of fusion, $\Delta S_f/R$, where R is the gas constant, is used to distinguish between faceted and non- faceted morphologies. The classification in Figure 9.1 is a simplification of the rather extensive one proposed by Croker *et al.* (1973), which includes a large number of irregular (anomalous) structures.

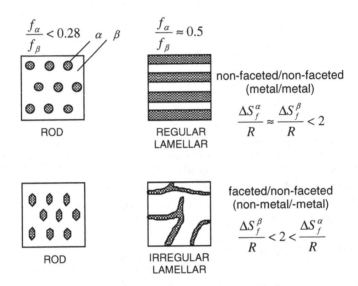

Figure 9.1. Types of cooperative eutectics (Kurz and Fisher, 1989).

Alloys such as Pb-Sn and Al-Al2Cu (Figure 9.2a), where there are approximately equal volume fractions of non-faceted phases solidify as regular, lamellar eutectics. If one of the phases is non-faceted the morphology becomes irregular, because the faceted phase grows preferentially in a direction determined by specific atomic planes. A typical example is the Mg-Mg$_2$Sn eutectic shown in Figure 9.2b.

When the volume fraction of one phase is significantly lower than that of the other (typically lower than 0.25), a fibrous structure will result (example the Ni-NbC eutectic shown in Figure 9.2c). This is because of the tendency of the system to minimize its interfacial energy by selecting the morphology that is associated with the smallest interfacial area. Fibers have smaller interfacial area than lamellae. However, when the minor phase is faceted, a lamellar structure may form even at a very low volume fraction, because specific planes may have the lowest interfacial energy. The minor phase will then grow such as to expose these planes even when lamellae rather than fibers are formed. The two commercially most significant eutectics, Al-Si and Fe-graphite (Gr) fall into this category. Note that in the Fe-Gr eutectic $f_{Gr} = 0.07$. The Fe-*Gr* eutectic can be either cooperative, irregular, as is the case for lamellar graphite cast iron, or divorced, as for spheroidal graphite cast iron (Figure 9.2d). In this last case, at the beginning of solidification the two phases, graphite and austenite dendrites, grow independently from the liquid without establishing a diffusion couple.

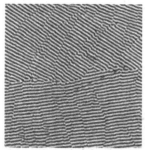

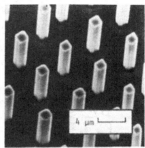

a) regular nf /nf eutectic (Al-Al$_2$Cu) (Magnin and Kurz, 1988a)

b) irregular f /nf eutectic (Mg-Mg$_2$Sn). The dark phase is the faceted Mg$_2$Sn (Magnin and Kurz, 1988a)

c) rod f /nf eutectic (Ni-NbC) (Mc Lean, 1983)

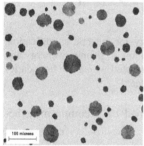

Figure 9.2. Eutectic microstructures.

d) divorced eutectic (Fe-spheroidal graphite)

9.2   Cooperative Eutectics

During the solidification of single-phase alloys, the solute is rejected from the growing tip of the dendrite and from the dendrite sides, and accumulates in the interdendritic regions. During the solidification of two-phase eutectic alloys, two solutes are rejected. Solute A accumulates in front of the β phase, while solute B accumulates in front of the α phase. The solute only needs to diffuse along the S/L interface from one phase to the other. Accordingly, the diffusion boundary layer for eutectics should be much smaller than for single-phase alloys (see Application 9.1). Each of these solutes is then incorporated in the growing solid solutions. Sideways diffusion is thus the reason for eutectic solidification. Thus, the morphology of the eutectic is made of alternative lamellae of phases α and β (Figure 9.3a). The smaller is the lamellar spacing, the smaller the solute buildup. In other words, solute diffusion tends to decrease the lamellar spacing.

Since excess free energy is associated with grain boundaries, a lower equilibrium temperature will exist for the tri-junction $\alpha/\beta/L$. At the three-phase junction $\alpha/\beta/L$, the surface energies must be balanced to insure mechanical equilibrium. This imposes fixed contact angles, and in turn induces a curvature of the S/L interface of each lamella (Figure 9.3b). Because the contact angles are material constants, this curvature is smaller when the lamellar spacing is higher. Accordingly, curvature acts to increase the lamellar spacing.

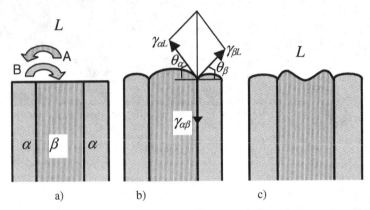

Figure 9.3. Schematic representation of interface morphology for eutectic solidification, explaining the effects of boundary energy and solute accumulation: a) boundary energy and solute accumulation at interface ignored; b) excess energy associated with grain boundaries considered; c) boundary energy and solute accumulation considered.

The scale of the eutectic structure is therefore determined by a compromise between two opposing factors:

    solute diffusion which tends to decrease the spacing
    surface energy (interface curvature) which tends to increase the spacing.

The dominant variables of eutectic solidification are undercooling, ΔT, growth velocity, V, and interlamellar spacing, λ. Their quantitative evaluation is performed starting again with the equation of the interface undercooling:

$$\Delta T = \Delta T_k + \Delta T_r + \Delta T_c + \Delta T_T$$

where the kinetic and the thermal undercooling are neglected. Since $\Delta T_r = 2\Gamma/r$ and, $r = ct\cdot\lambda$, as discussed for secondary DAS, we have $\Delta T_r = \mu_r \lambda^{-1}$, where μ_r is a material constant depending on curvature at the tip of the lamella. The constitutional undercooling is a function of composition, $\Delta T_c = m\Delta C_c$, as derived from Figure 9.4 The solidification velocity will increase with diffusivity and difference in composition, but decrease with higher lamellar spacing: $V = ct \cdot D\Delta C_c/\lambda$. Thus, $\Delta C_c = ct \cdot \lambda V$ and the constitutional undercooling becomes $\Delta T_c = \mu_c \lambda V$, where μ_c is a material constant depending on composition. Substituting ΔT_r and ΔT_c in the equation for the total undercooling results in:

$$\Delta T = \mu_r/\lambda + \mu_c \lambda V \qquad (9.1)$$

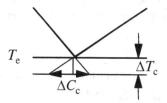

T_e

ΔT_c

ΔC_c

Figure 9.4. Correlation between under-cooling and composition.

To obtain the specific values of μ_r and μ_c a more detailed analysis must be conducted.

It must be further noted that the solute concentration is higher at the middle than at the edge of a lamella. Consequently, the liquidus temperature will be lower in the middle, and a lamella may have a negative curvature in the middle, as shown in Figure 9.3c for phase β. To demonstrate this, let us consider the interface composition (Figure 9.5a) and the undercooling (Figure 9.5b) ahead of the eutectic. The interface undercooling, T^*, is the same for both lamellae. Thus, the interface is maintained isothermal at T^* by the lamellae adjusting their radii of curvature. If ΔT_r becomes negative, as is the case for the β lamella, curvature and the radius of curvature becomes negative (Figure 9.5c). This means that the β lamella will have three curvatures rather than one.

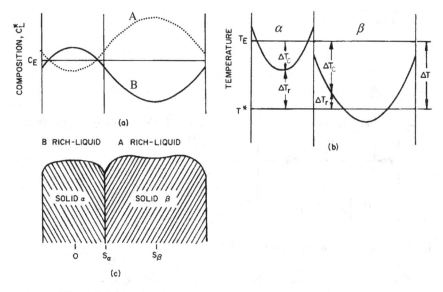

Figure 9.5. Morphology of eutectic front (Hunt and Jackson, 1966).

9.2.1   Models for regular eutectic growth

A formal theory of directionally solidified eutectics was proposed as early as 1946 by Zener and then modified in 1958 by Tiller. It is based on the free energy change for eutectic solidification of unit volume of liquid written as $\Delta G_v^\lambda = \Delta G_v^\infty + 2\gamma_{\alpha\beta}/\lambda$, where the free energy change for ∞ spacing is $\Delta G_v^\infty = \Delta H_f \, \Delta T_c/T_e$, ΔH_f is the latent heat, T_e is the equilibrium temperature, λ is the lamellar spacing, $2/\lambda$ is the α/β interface per unit volume, $\gamma_{\alpha\beta}$ is the α/β interface energy, and the significance of other quantities are given in Figure 9.4. The minimum spacing λ_{min} is obtained for $\Delta G_v^\lambda \to 0$, as $\lambda_{min} = 2\gamma_{\alpha\beta}/(\Delta H_f \, \Delta T_c)$.

The growth velocity can be calculated as $V = ct. D_L \, \Delta C_c/\lambda$, where D_L is the liquid diffusivity. Since from Figure 9.4 it is clear that $\Delta C_c = ct. \, \Delta T_c$, a relationship between V, λ and ΔT_c is obtained. Assuming now that solidification occurs at the maximum rate or minimum undercooling (extremum criterion), correlations between the various quantities are obtained:

$$\lambda_{min}^2 \, V_{max} = ct. \quad \text{and} \quad V_{max} = ct. \Delta T_c^2 \tag{9.2}$$

A more complete mathematical analysis was done by Jackson and Hunt (1966). For steady state growth, with the coordinate system moving with velocity V in the x direction (Figure 9.6), the diffusion equation to be solved is:

$$\nabla^2 C + \frac{V}{D}\frac{\partial C}{\partial x} = 0$$

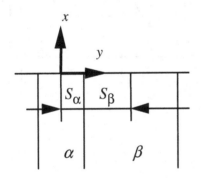

Figure 9.6. Coordinate system for the Jackson/Hunt model.

The following boundary conditions apply:

B.C.1	at $x \to \infty$	$C = C_o = C_E + \Delta C_\infty$
B.C.2	at $y = 0$ and at $y = S_\alpha + S_\beta$	$\partial C/\partial y = 0$

where ΔC_∞ is the difference between the initial liquid composition and the eutectic composition, C_E. The general solution of this diffusion equation is (see derivation in inset):

$$C(x, y) = B_o \exp\left(-\frac{V}{D}x\right) + \sum_{n=1}^{\infty} B_n \cos\frac{n\pi y}{L}\exp\left\{\left[-\frac{V}{2D} - \left[\left(\frac{V}{2D}\right)^2 + \left(\frac{n\pi}{L}\right)^2\right]^{1/2}\right]x\right\} \tag{9.3}$$

where $L = S_\alpha + S_\beta$.

Derivation of the Jackson-Hunt eutectic model. The governing equation is:

$$\frac{\partial^2 u}{\partial x^2} + \frac{\partial^2 u}{\partial y^2} + \frac{V}{D}\frac{\partial u}{\partial x} = 0$$

Using the method of separation of variables, $u(x,y) = P(x) Q(y)$, which, after substitution in the governing equation and appropriate manipulations results in:

$$\frac{1}{P}\frac{\partial^2 P}{\partial x^2} + \frac{1}{Q}\frac{\partial^2 Q}{\partial y^2} + \frac{V}{D}\frac{1}{P}\frac{\partial P}{\partial x} = 0 \quad \text{or} \quad \frac{1}{P}\frac{\partial^2 P}{\partial x^2} + \frac{V}{D}\frac{1}{P}\frac{\partial P}{\partial x} = -\frac{1}{Q}\frac{\partial^2 Q}{\partial y^2} = \lambda$$

where λ is the separation variable.

The first equation, $\dfrac{\partial^2 P}{\partial x^2} + \dfrac{V}{D}\dfrac{\partial P}{\partial x} - P\lambda = 0$, has the general solution $P = C_1 \exp(-C_2 x)$, where C_2 must be the root of the equation $C_2^2 + (V/D)C_2 - \lambda = 0$, i.e., $C_2 = -(V/2D) - \left[(V/2D)^2 + \lambda\right]^{1/2}$. Note that we have kept only the root with negative sign to avoid having a constant zero at $\lambda = 0$. Thus, the solution of the P function is:

$$P = C_1 \exp\left\{\left[-\frac{V}{2D} - \left[\left(\frac{V}{2D}\right)^2 + \lambda\right]^{1/2}\right]x\right\}$$

The second equation, $\partial^2 Q/\partial y^2 + \lambda Q = 0$, has the solutions $Q = C_3 \cos(C_4\, y)$ and $Q = C_5 \cos(C_6\, y)$, which gives:

$$Q = C_3 \cos(\lambda\, y) + C_5 \sin(\lambda\, y) \quad \text{and} \quad dQ/dy = -C_3\sqrt{\lambda}\sin(\sqrt{\lambda}y) + C_5\sqrt{\lambda}\cos(\sqrt{\lambda}y)$$

The applicable boundary conditions are:

$$Q(0,t) = dQ/dy = 0 \quad \text{and} \quad Q(L,t) = dQ/dy = 0 \quad \text{where } L = S_\alpha + S_\beta$$

Applying these conditions in the expression of Q we have $C_5 = 0$ and $(dQ/dy)_L = -C_3\sqrt{\lambda}\sin(\sqrt{\lambda}\,L) = 0$

For $C_3 \neq 0$ we obtain $\sin(\sqrt{\lambda}\,L) = \sin(n\,\pi)$ which gives the eigenvalue $\lambda = (n\pi/L)^2$ and the eigenfunction $Q = C_n \cos(n\pi y/\lambda)$. Thus, a general solution of the governing equation is:

$$u(x, y) = P(x)Q(y) = C_1 \exp\left\{\left[-\frac{V}{2D} - \left[\left(\frac{V}{2D}\right)^2 + \lambda\right]^{1/2}\right]x\right\}\cos\frac{n\pi x}{L}$$

and the complete solution is:

$$
u(x, y) = \sum_{n=0}^{\infty} B_n \cos\frac{n\pi y}{L} \exp\left\{\left[-\frac{V}{2D} - \left[\left(\frac{V}{2D}\right)^2 + \left(\frac{n\pi}{L}\right)^2\right]^{1/2}\right]x\right\}
$$

$$
= B_o \exp\left(-\frac{V}{D}x\right) + \sum_{n=1}^{\infty} B_n \cos\frac{n\pi y}{L} \exp\left\{\left[-\frac{V}{2D} - \left[\left(\frac{V}{2D}\right)^2 + \left(\frac{n\pi}{L}\right)^2\right]^{1/2}\right]x\right\}
$$

The exponential term contains two quantities: $V/2D$ which is the inverse of the primary phase boundary layer, and $n\pi/L$ which is the inverse of the eutectic boundary layer For the case of eutectic solidification $L = S_\alpha + S_\beta = \lambda/2$. Thus, the eutectic boundary layer is $\lambda/2\pi$, and the primary phase boundary layer is $2D/V$. Now $V/2D \ll 2\pi/\lambda$ because $\delta = D/V$ is large at slow growth rates, while λ is small (see Application 9.1). Thus, $V/2D$ will be neglected.

The solution given by Eq. (9.3) satisfies BC2, but not BC1. Since the solution also has to satisfy BC1, it becomes:

$$
C_L = C_E + \Delta C_\infty + B_o \exp\left(-\frac{V}{D}x\right) + \sum_{n=1}^{\infty} B_n \cos\frac{2n\pi y}{\lambda} \exp\left(-\frac{2n\pi x}{\lambda}\right) \tag{9.4}
$$

To evaluate the Fourier coefficients in Eq. (9.4), conservation of mass at the interface is applied:

$$
\left(\frac{\partial C}{\partial x}\right)_{x=0} = -\frac{V}{D}(C_E - C_{\alpha M}) \quad \text{for} \quad 0 \le y \le S_\alpha
$$

$$
\left(\frac{\partial C}{\partial x}\right)_{x=0} = -\frac{V}{D}(C_E - C_{\beta M}) \quad \text{for} \quad S_\alpha \le y \le S_\alpha + S_\beta
$$

The Fourier coefficients are:

$$
B_o = \overline{C}_0^\alpha f_\alpha - \overline{C}_0^\beta f_\beta \quad \text{and} \quad B_n = \frac{\lambda}{(n\pi)^2}\frac{V}{D}\left(\overline{C}_o^\alpha + \overline{C}_o^\beta\right)\sin\left(n\pi f_\alpha\right)
$$

where $f_\alpha = S_\alpha/(S_\alpha + S_\beta)$, $f_\beta = S_\beta/(S_\alpha + S_\beta)$, and $\overline{C}_o^\alpha$ and $\overline{C}_o^\beta$ are defined in Figure 9.7.

If the liquid is of exact eutectic composition Eq. (9.4) simplifies to:

$$
C_L = C_E + \sum_{n=1}^{\infty} B_n \cos\frac{2n\pi y}{\lambda} \exp\left(-\frac{2n\pi x}{\lambda}\right) \tag{9.5}
$$

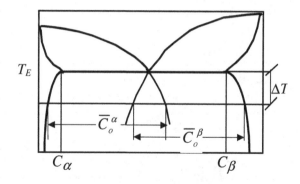

Figure 9.7. Definition of $\overline{C}_o^{\alpha}$ and $\overline{C}_o^{\beta}$.

Eq. (9.4) is valid when it is assumed that the density difference between the two eutectic phases is negligible. A more complete analysis that included the density difference was done by Magnin and Trivedi (1991).

At a distance ahead of the interface $x = \lambda/2$, exp($-n\pi$) is: exp($-\pi$)=0.04, exp(-2π)=0.001, and exp(-3π)=0.00008. Thus, from Eq. (9.5), $C_L \approx C_E$. This means that composition deviation of liquid from eutectic is damped out fast ahead of the interface. The eutectic diffusion layer is thus $\lambda/2$. Calculations show that maximum composition deviation from eutectic is small, ~0.1%.

To find the particular values of the constants in Eq. (9.1) it is necessary to formulate both ΔT_c and ΔT_r. In turn, ΔT_c can be obtained from Eq. (9.5) by specifying V and λ, since $\Delta T_c = m(C_L - C_E) = m \cdot f(\lambda, V)$.

To obtain ΔT_r it is necessary to use the contact angles at the tri-junction (Figure 9.3). From the definition of curvature in Cartesian coordinates, it can be shown that $\Delta T_r = \Gamma K = 2\sin\theta_i / f_i \lambda$, where the subscript i stands for the phase α or β. By manipulating these last two equations, and Eq. (9.5), and substituting in Eq. (9.1), the values of the constants are found to be:

$$\mu_c = \frac{m_\alpha m_\beta}{m_\alpha + m_\beta} \frac{C_\beta - C_\alpha}{D} F(f) \quad \text{and} \quad \mu_r = \frac{2m_\alpha m_\beta \delta}{m_\alpha + m_\beta} \left(\frac{\Gamma_\alpha \sin\theta_\alpha}{f_\alpha m_\alpha} + \frac{\Gamma_\beta \sin\theta_\beta}{f_\phi m_\beta} \right)$$

where C_α and C_β are the compositions of the α and β phases, respectively, m_α and m_β are the slopes of the liquidus lines of the α and β phases, respectively, $F(f) = 0.335\left(f_\alpha f_\beta\right)^{0.65}$, f_α and f_β are the volume fraction of solids of the eutectic phases, respectively, Γ_α and Γ_β are the Gibbs-Thomson coefficients of the two phases, θ_α and θ_β are the contact angles, and δ is a parameter equal to 1 for lamellar eutectics. This equation is valid for lamellar eutectics. For α-fibrous eutectics the equation is:

$$F(f_\alpha) \approx 4.908 \cdot 10^{-3} + 0.3122 f_\alpha + 0.6918 f_\alpha^2 - 2.604 f_\alpha^3 + 3.238 f_\alpha^4 - 1.619 f_\alpha^5$$

A comparison between calculated and experimental results is shown in Figure 9.8.

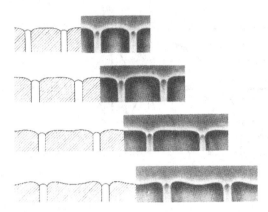

Figure 9.8. Comparison of calculated and observed shapes for a transparent organic eutectic (Jackson and Hunt, 1966).

Eq. (9.1) is represented graphically in Figure 9.9. It is seen that the curve representing the total undercooling, ΔT, is the result of the sum of the solutal and curvature undercoolings. The total undercooling is high for small and large values of the lamellar spacing, and reaches a minimum at a certain value of the spacing, which we will call extremum, λ_{ex}. For $\lambda < \lambda_{ex}$, capillarity effects are controlling the process ($\Delta T_c < \Delta T_r$). For $\lambda > \lambda_{ex}$ diffusion is the controlling mechanism ($\Delta T_c > \Delta T_r$). It is also seen that no unique solution is available. Indeed, pairs of λ and ΔT will satisfy the equation. Nevertheless, experiments demonstrate that for a given system solidifying at constant velocity, for regular eutectics, the lamellar spacing varies within a very narrow range. In the case of irregular eutectics, the range of variation of the spacing is considerably larger. It is generally accepted that the lamellar spacing will vary between λ_{ex} and the branching spacing, λ_{br}.

For the case of regular eutectics, the most common approach is to use the extremum criterion, proposed by Zener. It states that the eutectic will grow at the minimum undercooling possible. This criterion has been found to yield correct results for regular eutectics such as Pb-Sn.

When the extremum criterion is applied, the first derivative of Eq. (9.1) with respect to λ is equated to zero. This gives:

$$\lambda_{ex}^2 V = \mu_r / \mu_c \quad \text{or} \quad \lambda_{ex} = \mu_\lambda V^{-1/2} \tag{9.6}$$

with $\mu_\lambda = \sqrt{\mu_r / \mu_c}$. This equation indicates that, as the growth velocity increases the lamellar spacing decreases. Since growth velocity increases with cooling rate, it is clear that cooling rate can be used to control the lamellar spacing in castings poured from eutectic alloys.

It can also be shown by substituting Eq.(9.6) in Eq. (9.1) that:

$$\Delta T_{min} V^{1/2} = 2 (\mu_c \mu_r)^{1/2} \quad \text{or} \quad V = \mu_V \Delta T_{min}^2 \tag{9.7}$$

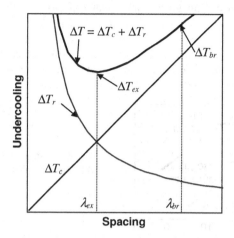

Figure 9.9. Extremum and branching lamellar spacing in eutectics.

In these equations μ_λ and μ_V are material constants. Some typical values for these constants are given in Appendix B, Tables B1-B4.

However, as demonstrated by Seetharaman and Trivedi (1988), even for regular eutectics, for a given velocity the spacing is not fixed, but rather occupies a range whose lower limit is indeed λ_{ex}, but whose upper limit is about 20% higher than the extremum value. In other words, an average spacing exists, that is higher than λ_{ex}.

A more recent theory by Catalina, Sen and Stefanescu (2003) that will be discussed in the next section explains this departure from λ_{ex} and calculates λ_{ex} without the use of the arbitrary extremum criterion.

For irregular eutectics such as Al-Si, Fe-C, Fe-Fe$_3$C, the experiments revealed that the average spacing is much higher than predictions made by the JH model (Toloui and Hellawell 1976, Elliot and Glenister 1980, Jones and Kurz 1981, Hogan and Song 1987, Magnin *et al.* 1991). Typical values are shown in Figure 9.10.

9.2.2 Models for irregular eutectic growth

The irregular faceted /non-faceted structure occurs because the specific surface energy between phases is very anisotropic. Certain lowest-energy crystallographic orientations develop between phases to minimize the interfacial energy. The α phase in Figure 9.11 grows faster than the β phase, and it can overgrow it.

The lamellar spacing of irregular eutectics does not obey Eqs. (9.6) or (9.7). In other words, the extremum criterion does not explain the mechanism of occurrence of lamellar spacing, or rather is not sufficient to explain it. While local spacing corresponding to the extremum condition can be found, the mean spacing is considerably larger. This results from the fact that in these eutectics the crystallographic growth directions of the two phases do not coincide. Consequently, the smallest spacing will be the one dictated by the extremum criterion, λ_{ex}, but a larger spacing will also exist, dictated by a branching condition, λ_{br} (Kurz and Fisher,

1989). The two phases can grow convergent (one toward the other) until λ_{ex} is reached, then they will grow divergently until λ_{br} is reached, as shown in Figure 9.9 and Figure 9.11. Because of this growth, irregular eutectics solidify with a non-isothermal interface. The lamellar graphite (L_{Gr}) - austenite eutectic formed in gray iron, or the silicon plates - aluminum eutectic are typical examples.

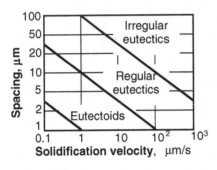

Figure 9.10. λ - V correlation for various diffusion couples.

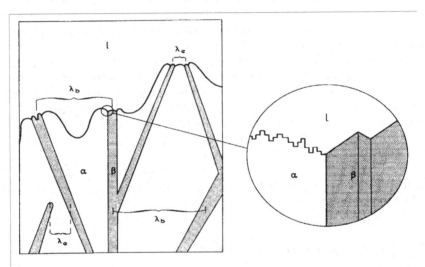

Figure 9.11. Growth of irregular eutectics (Kurz and Fisher, 1998). Note: non-isothermal S/L interface; the diffuse α interface can grow easily, while the faceted β phase grows more difficult. λ_b and and λ_e are the branching and the extremum spacing, respectively.

The irregular nature of the lamellar graphite- austenite eutectic was not confronted until 1974, when Sato and Sayama introduced the concept of partially cooperative growth of the nonfaceted and faceted phases. Although the calculated spacings were larger than the JH predictions, they still failed to approach the even larger experimental value.

Fisher and Kurz (1980) and then Magnin and Kurz (1987) proposed their irregular faceted/nonfaceted eutectic model. They used the following main assumptions: i) non-isothermal interface; ii) the γ phase that has a diffuse interface grows faster than the graphite phase that is faceted; iii) branching occurs when a depression forms on the faceted phase. To impose a non-isothermal coupling condition over the interface, they ascribed a cubic function. They demonstrated that the smallest spacing of the lamellar eutectic is dictated by the extremum condition, but that a larger spacing will also exist, λ_{br}, dictated by a branching condition. λ_{br} can be calculated as the product between a function of the physical constants of the faceted phase and a material constant. This constant must be postulated to match the theoretical predictions with the experimental measurements, which limits the generality of the model. The model predicts that the average spacing in eutectics with non-isothermal SL interface obeys relationships similar to those developed by JH for the isothermal interface:

$$\langle\lambda\rangle^2 V = \phi^2 \mu_r / \mu_c \quad \text{or} \quad \langle\lambda\rangle\langle\Delta T\rangle = (\phi^2 + 1)\mu_r \quad \text{or} \quad V^{-1}\langle\Delta T\rangle^2 = (\phi + 1/\phi)^2 \mu_r \mu_c \quad (9.8)$$

where the average spacing, $\langle\lambda\rangle$, and the operating point factor, ϕ, are defined as follows:

$$\langle\lambda\rangle = (\lambda_{ex} + \lambda_{br})/2 = \varphi\lambda_{ex} \quad \text{and} \quad \varphi = (1 + \lambda_{br}/\lambda_{ex})/2$$

Note that since $\lambda_{br} > \lambda_{ex}$, $\varphi > 1$, and $\langle\lambda\rangle > \lambda_{ex}$. This is shown in Figure 9.10 where the λ-V line for irregular eutectics is positioned above the one for regular eutectics. For eutectoids, diffusion occurs only through solid phases and it is slower. Thus, the diffusion distance will be lowered by decreasing the spacing.

Eq. (9.8) describes the growth of both regular ($\varphi = 1$) and irregular ($\varphi > 1$) eutectics. Assuming that β is the faceted phase, for irregular eutectics:

$$\varphi = 0.5 + \left[F'(f_\beta)\left(1 + \frac{f_\beta m_\beta}{f_\varepsilon m_\alpha}\frac{\Gamma_\alpha \sin\theta_\alpha}{\Gamma_\beta \sin\theta_\beta}\right)\right]^{-1/2} \quad (9.9)$$

where $F'(f_\beta) \approx 0.03917 + 0.6047 f_\beta - 1.413 f_\beta^2 + 2.171 f_\beta^3 - 1.236 f_\beta^4$.

9.2.3 The unified eutectic growth model

Catalina et al. (2003) proposed a modified Jackson-Hunt model for eutectic growth (the CSS model) that is applicable to both regular and irregular eutectics. The model allows for local variation of composition and curvature, which results in different local undercoolings in front of the two eutectic phases (Figure 9.12). After calculations that account for the density difference between the liquid and the two solid phases (as suggested by Magnin and Trivedi, 1991) and following the same procedure as JH, two expressions for the undercooling of each eutectic phases can be derived:

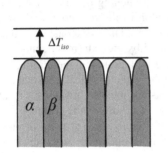
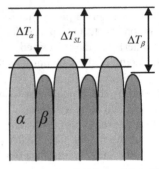

Figure 9.12. S/L interface assumptions: left - JH model; right - CSS model. ΔT_{SL} is the average interface undercooling.

$$\Delta T_\alpha = \mu_c^\alpha \lambda V + \mu_r^\alpha / \lambda \quad \text{and} \quad \Delta T_\beta = \mu_c^\beta \lambda V + \mu_r^\beta / \lambda \tag{9.10}$$

These equations are similar with the JH equation for isothermal S/L interface. In addition, the average undercooling of the S/L interface is:

$$\Delta T_{SL} = f_\alpha \Delta T_\alpha + f_\beta \Delta T_\beta \tag{9.11}$$

The undercoolings described by Eqs. (9.10) exhibit a minimum for certain spacings λ_α and λ_β, respectively:

$$\lambda^i \sqrt{V} = \sqrt{\mu_r^i / \mu_c^i} \tag{9.12}$$

where, for lamellar eutectics, $\mu_c^i = -m_i \left(1 + \xi / f_i \Omega\right)\left(C_E \Omega^2 / D_L\right) \cdot P$, $\mu_r^i = 2\Gamma_i \sin \theta_i / f_i$, $i = \alpha$ or β, $\xi = 1$ if $i = \alpha$ and $\xi = -1$ if $i = \beta$, $\Omega = \left(\rho_\alpha / \rho_L\right)\left(1 - k_\alpha\right) - \left(\rho_\beta / \rho_L\right)\left(1 - k_\beta\right)$, and $P = \sum_{n=1}^\infty \sin^2 \left(n\pi f_\alpha\right) / \left(n\pi\right)^3$.

If an isothermal interface exists, that is if $\Delta T_\alpha = \Delta T_\beta = \Delta T_{iso}$, a third characteristic spacing, λ_{iso}, can be obtained from Eq. (9.10) without invoking the extremum criterion:

$$\lambda_{iso} \sqrt{V} = \sqrt{-\left(\mu_r^\alpha - \mu_r^\beta\right) / \left(\mu_c^\alpha - \mu_c^\beta\right)} \tag{9.13}$$

Furthermore, using Eq. (9.11) a spacing λ_{SL} for which the average undercooling reaches a minimum can be derived:

$$\lambda_{SL} \sqrt{V} = \sqrt{-\left(f_\alpha \mu_c^\alpha + f_\beta \mu_c^\beta\right) / \left(f_\alpha \mu_r^\alpha + f_\beta \mu_r^\beta\right)} \tag{9.14}$$

This analysis reveals the existence of four different spacings that characterize eutectic growth. They all obey relationships similar to those obtained by JH for isothermal interface. However, this analysis shows that for a given growth velocity there is a unique spacing, $\lambda_{iso} = \lambda_{ex}$, at which the SL interface is isothermal without invoking the extremum criterion. However, isothermal growth is only possible in eutectic systems for which Eq. (9.12) has rational solutions (the quantity under the square root on the RHS is positive). Calculations demonstrate that such a solution exits for Al-Si but not for the Fe-Gr system. Indeed, for the Al-Si system all the curves, including ΔT_{iso}, intersect at a point characterized by a spacing λ_{iso} (Figure 9.13). This means that growth with isothermal interface is possible for the Al-Si system. For the Fe-Gr system such an intersection does not occur.

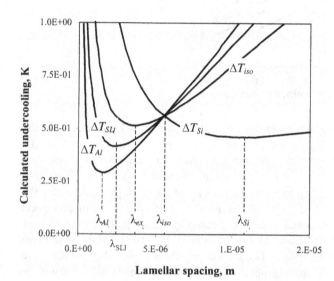

Figure 9.13. Predicted undercooling *vs* lamellar spacing for the Al-Si system (V = 10 μm/s) (Catalina *et al.*, 2003). With kind permission of Springer Science and Business Media.

The question is now which of the four possible spacings will be selected by the system? The answer to this question can be provided by comparison with experiments.

The various theoretical spacings in the Al-Si system are compared to experimental results in Figure 9.14. The extremum spacing, λ_{ex}, predicted by JH model modified by Magnin and Trivedi (1991) is also presented. It is apparent from this figure that the experimental measurements are in close agreement only with the theoretical spacing λ_{iso} and λ_β predicted by the CSS model. That is, the minimum measured spacing corresponds to the theoretical λ_{iso}, while the average experimental spacing corresponds to λ_β.

The ratio $\lambda_\beta / \lambda_{ex}$, takes a value of 3.147 which is very close to the value of 3.2 attributed to the operating factor, φ, of the Al-Si eutectic. A mathematical expression for φ can be easily derived, but we must point out that the true existence of φ depends on the existence of λ_{ex}. According to the CSS approach, it appears that λ_{ex} does not have a real physical correspondent, at least for the Al-Si system.

Consequently, the application of an operating factor, φ, as a spacing selection mechanism for irregular eutectics is questionable.

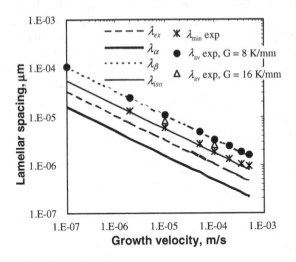

Figure 9.14. Various theoretical (Catalina *et al.*, 2001) and measured (Magnin *et al.*, 1991) lamellar spacing vs. growth velocity for the AlSi system.

As the Fe- Gr eutectic system cannot grow with an isothermal interface only the spacing λ_γ (austenite), λ_{Gr} (graphite) and λ_{SL} remain to be compared to the experimental spacings. From the comparison presented in Figure 9.15 it is apparent that the theoretical spacing λ_{SL} is in good agreement with the minimum experimental spacing (λ_{min}) up to a growth velocity of about 10 µm/s. Also, for the same velocity range (*i.e.*, $V < 10$ µm/s) the theoretical spacing λ_{Gr} is in very good agreement with the experimental values of the average spacing. This suggests that the spacing selection mechanism in this system is governed by the minimum undercooling of the graphite phase. For growth velocities higher than 10 µm/s the experiments show that the spacing deviates from the relationship $\lambda^2 V = const.$ as austenite dendrites begin to appear in the microstructure and the graphite lamellae

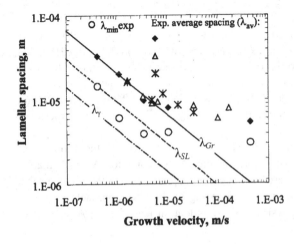

Figure 9.15. Comparison of minimum and average experimental spacings (Jones and Kurz 1981, Magnin and Kurz 1988) of Fe-Gr eutectic with theoretical predictions of the CSS model.

exhibit a relatively random orientation.

For a complete discussion, as well as for a discussion on the Fe-Fe₃C eutectic the reader is referred to the original paper (Catalina *et al.*, 2001).

The JH model as well as the CSS analysis discussed so far, do not account for the influence of the temperature gradient, G_T, because the problem was treated independently of the thermal field. Fisher and Kurz (1980) have included the effect of G_T in their non-isothermal coupling condition. Their results for Al-Si, obtained through a numerical method, led to the following relationships:

$$\lambda^2 V = 1.24 \cdot 10^{-5} G_T \quad \text{and} \quad \Delta T / \sqrt{V} = 156 \cdot G_T^{-0.41} \tag{9.15}$$

These and other proposed equations were found to overestimate the effect of the gradient (Elliot and Glenister, 1980). Attempting to correct this problem, CSS proposed a different type of relationship, which allows restrictions to be imposed on the effect of the gradient, as follows:

$$\lambda_{av} \sqrt{V} = \left[4.3185 \cdot \exp\left(-G_T / G_T^*\right) + 1.74805\right] \cdot 10^{-8} \quad (m^3 / s)^{1/2} \tag{9.16}$$

where G_T^* is a neutral temperature gradient at which the selection of the average spacing is dictated by the minimum undercooling of the faceted (β) phase. For the Al-Si system $G_T^* = 8.405$ K/m.

9.3   Divorced eutectics

In the case of divorced eutectics, the two phases of the eutectic grow independently from one another. This may happen when in a eutectic forming at the end of solidification of a two-phase system the fraction of remaining liquid is so small that the width is comparable to the eutectic spacing. In such a situation, the second phase may form as single particle or layer between the dendrites.

Another case is that of the spheroidal graphite - austenite eutectic (ductile iron). Here, eutectic austenite dendrites, independent graphite spheroids (in the initial stage) and austenite-graphite aggregates are formed during eutectic solidification. While the austenite grows in contact with the liquid, the graphite grows in contact with the liquid only at the beginning of solidification. Then, the graphite is enveloped in an austenite shell and grows through solid diffusion of carbon through this shell (Figure 9.16).

The first analytical model to describe growth of the eutectic in spheroidal graphite (SG) iron was proposed in 1956 by Birchenall and Mead. Later, Wetterfall *et al.* (1972) made calculations of the diffusion-controlled steady-state growth of graphite through the austenite shell (Figure 9.17) based on Zener's growth equation for an isolated spherical particle in a matrix of low supersaturation. The growth velocity of the γ shell was derived to be (see derivation in inset):

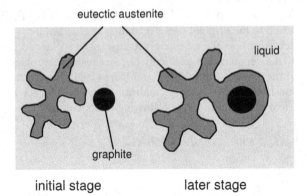

Figure 9.16. Growth of the divorced spheroidal graphite - austenite eutectic.

$$\frac{dr_\gamma}{dt} = D_C^\gamma \frac{r_G}{(r_G - r_\gamma)r_\gamma} \frac{C^{G/\gamma} - C^{L/\gamma}}{C^{\gamma/L} - C^{L/\gamma}}$$ (9.17)

where D_C^γ is carbon diffusivity in austenite. This equation can be further simplified if it is assumed that, as demonstrated experimentally, the ratio between the radius of the austenite shell and that of the graphite spheroid remain constant: $r_\gamma = 2.4 \, r_{Gr}$. The simplified growth rate is (Svensson and Wessen, 1998):

$$\frac{dr_{Gr}}{dt} = 2.87 \cdot 10^{-11} \frac{\Delta T}{r_{Gr}}$$ (9.18)

Derivation of growth law for solidification of SG iron. The basic assumptions are:

- growth controlled by carbon diffusion through the austenite shell
- steady-state diffusion of austenite through the austenite shell.

Solutal balance at the S/L interface ($r = r_\gamma$ in Figure 9.17) gives:

$$\frac{d}{dt}\left[\frac{4}{3}\pi r_\gamma^3 \rho_\gamma \left(C^{\gamma/L} - C^{L/\gamma}\right)\right] = 4\pi r_\gamma^2 \rho_\gamma D_C^\gamma \left(\frac{\partial C}{\partial r}\right)_{r=r_\gamma} \quad \text{or} \quad V = \frac{dr_\gamma}{dt} = \frac{D_C^\gamma}{C^{\gamma/L} - C^{L/\gamma}}\left(\frac{\partial C}{\partial r}\right)_{r=r_\gamma}$$

To calculate the carbon flux at the S/L interface, the steady-state diffusion equation assuming constant diffusivity, no advection and no source term is used:

$$D\nabla^2 C = 0 \quad \text{or, in spherical coordinates:} \quad D_L \frac{\partial}{\partial r}\left(r^2 \frac{\partial C}{\partial r}\right) = 0$$

The solution of this equation can be obtained by successive integration, as follows:

$\partial C/\partial r = C_1/r^2$ and $C = -C_1/r + C_2$ with the boundary conditions

BC1:   at $r = r_G$ $C = C^{G/\gamma}$ and   BC2:   at $r = r_\gamma$ $C = C^{L/\gamma}$

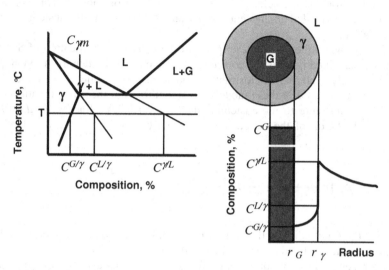

Figure 9.17. Schematic diagram showing the concentration profile of carbon throughout the liquid and the austenite shell based on the binary phase diagram.

After evaluating the constants, the flux at the interface can be obtained from the above equation as:

$$\left(\frac{\partial C}{\partial r}\right)_{r=r_\gamma} = \frac{r_G\left(C^{L/\gamma} - C^{G/\gamma}\right)}{r_\gamma\left(r_\gamma - r_G\right)}$$

Substituting in the velocity equation we obtain the growth velocity of the austenite shell:

$$V_\gamma = \frac{dr_\gamma}{dt} = D_C^\gamma \frac{r_G}{\left(r_G - r_\gamma\right)r_\gamma} \frac{C^{G/\gamma} - C^{L/\gamma}}{C^{\gamma/L} - C^{L/\gamma}}$$

The radius of the austenite shell is calculated from the velocity equation. Once r_γ is known, r_G is calculated from solute balance (Chang *et al.*, 1992):

$$\rho_G \frac{4}{3}\pi\left[\left(r_G^{i+1}\right)^3 - \left(r_G^i\right)^3\right]C_G = \rho_\gamma \frac{4}{3}\pi\left[\left(r_\gamma^{i+1}\right)^3 - \left(r_\gamma^i\right)^3\right]C^{L/\gamma} \tag{9.19}$$

The radius of the austenite shell is calculated from the velocity equation. Once r_γ is known, r_G is calculated from solute balance (Chang *et al.*, 1992):

$$\rho_G \frac{4}{3}\pi\left[\left(r_G^{i+1}\right)^3 - \left(r_G^i\right)^3\right]c_G = \rho_\gamma \frac{4}{3}\pi\left[\left(r_\gamma^{i+1}\right)^3 - \left(r_\gamma^i\right)^3\right]c^{L/\gamma} \tag{9.20}$$

This model has survived the test of time and is used today in most computational models for microstructure evolution in one form or another. This approach has also been extended to the eutectoid transformation for the austenite-ferrite transformation.

Although the mathematical description of the eutectic solidification of SG iron previously described ignores the contribution of the austenite dendrites, it can predict surprisingly well the microstructural outcome. It is believed that the mass balance equations used in these models are smoothing out the error, by attributing the dendritic austenite to the austenitic shell. Nevertheless, Lacaze *et al.* (1991) included calculation of the off-eutectic austenite by writing the mass balance as follows:

$$\rho_G \int_0^{r_G} r^2\,dr + \rho_\gamma \int_{r_G}^{r_\gamma} r^2\,dr + \left[\rho_L g_L + \rho_\gamma(1 - g_L)\right]\int_{r_\gamma}^{r_0} r^2\,dr = \rho_L \frac{r_0^3}{3}$$

Here, r_o is the final radius of the volume element. In previous models, only the first two terms on the left hand side were included.

More recently, Fredriksson *et al.* (2005) argued on the basis of observations on samples quenched during solidification that spheroidal graphite may be in contact with γ without being surrounded by a shell. They contended that when the nodules are small the interface kinetics produces a high growth rate. An austenite shell will not develop in this case because of lack of sufficient driving force for the plastic deformation of austenite. After the graphite reaches a certain size, the interface kinetics of graphite growth slows down, an austenite shell is formed, and γ/Gr equilibrium can be assumed. At this time, there are no mathematical models that describe this growth sequence.

9.4   Interface stability of eutectics

Contrary to what the equilibrium diagrams suggest, eutectic-like structures (composites) may be obtained even with off-eutectic compositions. Let us assume that an alloy of composition $C_o < C_E$ is grown in such a way that plane front is maintained. The distribution of solute in the liquid ahead of the interface is given by Eq. (4.5):

$$C_L = C_o + (C_L - C_o)\exp\left(-\frac{V}{D_L}x\right) = C_o\left[1 + \frac{1-k}{k}\exp\left(-\frac{V}{D_L}x\right)\right]$$

At steady state $C_L = C_o/k = C_E$. Thus, the above equation becomes:

$$C_e = C_o + (C_e - C_o)\exp(-(V/D_L)x)$$

Substituting in the Jackson-Hunt equation for eutectic solidification, Eq. (9.5):

$$C_E = C_o + (C_E - C_o)\exp\left(-\frac{V\,x}{D_L}\right) + \sum_{n=1}^{\infty} B_n \cos\frac{2n\pi\,y}{\lambda}\exp\left(-\frac{2n\pi\,x}{\lambda}\right) \qquad (9.21)$$

From this equation, it is seen that there are two characteristic distances in the x-direction:

- $\cong \lambda/2$: transverse transport of solute occurs within this layer
- $\cong D_L/V$: as for single-phase crystals; within this layer $C_L > C_o$

For usual solidification velocities, $\lambda/2 \ll D_L/V$. This means that perturbations in the y-direction dampens out much quicker than those in the x-direction. When $\lambda/2 \ll D_L/V$ the third RHT vanishes, and Eq.(9.21) reduces to the equation for growth of single phase alloys.

To find the influence of processing conditions on the composition of the result-ing solid we will use mass balance at the eutectic front. It gives $(\partial C_L/\partial x)_{x=0} = -(C_E - \overline{C}_S)(V/D)$, where $\overline{C}_S$ is the average two-phase composition. Thus, since $\partial C_L/dx = G_L/m$:

$$\overline{C}_S = C_E + (D/V)(G/m) \qquad (9.22)$$

For very small G/V ratios the composition of the solid is $\overline{C}_S = C_E$. For high G/V ratios $\overline{C}_S = C_o < C_E$

Let us try to evaluate the morphology of this solid. In the absence of convection the criterion for interface stability is:

$$G_T \geq \Delta T_o V/D = -m(C_E - C_o)V/D \qquad (9.23)$$

This is the same as Eq. (9.22) for high G/V when $\overline{C}_S = C_o$ If the stability crite-rion is not satisfied (small G/V), a dendritic structure will result (Figure 9.18). Eutectic will still form at the end of solidification. If the interface is stable (high G/V), a composite structure will form, having a eutectic-like morphology.

From the eutectic phase diagram, it is apparent that a eutectic structure can be obtained only when the composition is exactly eutectic. Nevertheless, experimental observations, as well as the preceding theoretical analysis show that, depending on the growth conditions, eutectic microstructures can be obtained at off-eutectic compositions. This is possible because the eutectic grows faster than the dendrites, since diffusion-coupled growth is much faster than isolated dendritic growth. Ac-cordingly, even in off-eutectic compositions, the eutectic may outgrow the individ-ual dendrites, resulting in a purely eutectic microstructure. On the other hand, at high growth velocities, dendrites can be found in alloys of eutectic compositions.

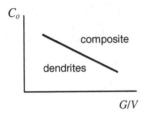

Figure 9.18. Dendritic /eutectic transition.

From the preceding analysis it is clear that, depending on the processing conditions of a two-component alloy, stable or unstable interfaces may be obtained. A third alloying element, which is partitioned between both solid phases will lead to two-phase instability. The influence of the G/V ratio and of composition on the solidification microstructure of off-eutectic alloys can be summarized as follows:

- high G/V, no third element -- composite (eutectic-like structure)
- low G/V, no third element -- dendrites + planar (lamellar) eutectic (Figure 9.19a); this is because one phase becomes heavily constitutionally undercooled
- low G/V, some third element -- two phase eutectic cells (colony) (Figure 9.19b), or dendrites + eutectic cells; this is because both phases are constitutionally undercooled
- lower G/V, some third element -- dendrites + eutectic grains; a long range diffusion boundary layer is established ahead of the S/L interface

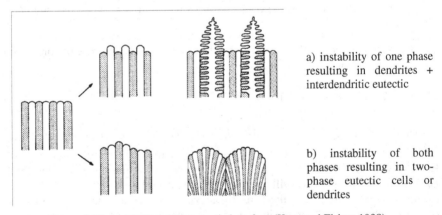

a) instability of one phase resulting in dendrites + interdendritic eutectic

b) instability of both phases resulting in two-phase eutectic cells or dendrites

Figure 9.19. Instability of the eutectic interface (Kurz and Fisher, 1998).

An analysis of the possible solidification microstructure of a binary alloy can be made based on the growth velocities of the competing phases. As shown in Figure 9.20a, the phase that will be present in the final microstructure at a given undercooling is that one which has the highest solidification velocity. Thus, this is a kinetic effect.

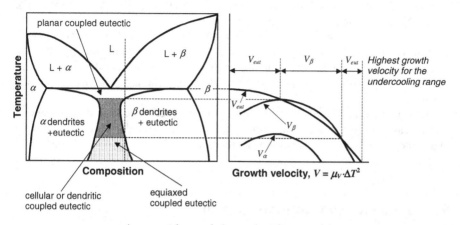

a) symmetric coupled zone (regular eutectic)

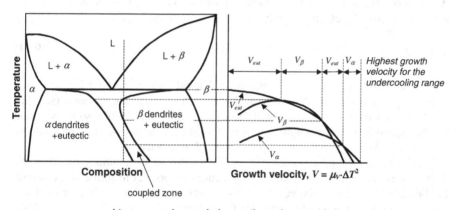

b) asymmetric coupled zone (irregular eutectics)

Figure 9.20. Coupled eutectic zones.

For regular eutectics, where the two primary dendritic phases have similar undercooling, the coupled zone is symmetric. Consider the solidification of the slightly hypereutectic alloy in Figure 9.20. At small undercooling, the eutectic has the highest growth velocity and a planar, coupled eutectic solidifies. At higher undercooling, the β phase will have higher growth velocity, and a structure made of eutectic and dendrites will result. At even higher undercooling, the eutectic velocity will become again the highest. However, because of the undercooling, a planar structure is not possible, and equiaxed coupled growth will result.

If one of the eutectic phases is faceted, the growth of this phase and consequently that of the eutectic is slowed down. Dendrites of the other phase may grow faster at a given undercooling than the eutectic, even for eutectic composition. Consequently, purely eutectic microstructures can be obtained only at hypereutectic compositions. This is exemplified in Figure 9.20b, for the case of faceted β phase. An asymmetric coupled zone results.

The coupled zone is thus the solidification velocity dependent composition region in which the eutectic grows more rapidly, or at a lower undercooling, than the α or β dendrites. Note that the widening of the coupled zone near the eutectic temperature is observed only in directional solidification, where the thermal gradient is positive.

The practical significance of the concept of coupled zone is that the composition of cast iron or of aluminum-silicon alloys, both irregular eutectics, must be hypereutectic at high solidification velocities, if it is desired to avoid primary dendrites.

9.5   Equiaxed eutectic grain growth

During the solidification of eutectic commercial casting alloys such cast iron and Al-Si alloys, the most common appearance of the eutectic phase is that of eutectic grains. The growth velocity of eutectic grains can be calculated as:

$$V = \mu_V \Delta T_{bulk}^2 \tag{9.24}$$

where the growth coefficient deviates from that calculated from the Jackson-Hunt theory. Typically, values that are calculated from experiments are used (see Appendix B). Example of calculation of the cooling curve and faction solid evolution for the equiaxed eutectic solidification of eutectic and hypoeutectic cast irons are presented in Application 9.2 and Application 9.3, respectively.

The morphology of eutectics can be altered by small additions of elements. This process is called *modification*. Its practical purpose is to refine and compact the brittle component of the eutectic phase, when such a phase exists, or to eliminate the occurrence of a brittle eutectic. Typical examples are modification of Al-Si alloys with strontium to refine the Si phase, or the modification of lamellar graphite cast iron with Mg to spheroidize the graphite.

Modification differs from *grain refinement* in that grain refinement affects only the nucleation of the primary dendritic phase, while modification controls the undercooling of the eutectic phase. As discussed, growth of the eutectic is controlled by its undercooling. Thus, it is possible to combine different modification treatments with the same grain refinement to obtain entirely different eutectic phases, and thus entirely different casting structures. The most common example of this is cast iron. In cast iron, the graphite eutectic morphology can be controlled by a modification treatment to produce either the flake form found in gray iron, or the spheroidal form found in ductile iron. The liquid treatment of lamellar graphite iron with ferro-silicon, which is called inoculation, is in fact a modification process since it results in the elimination of the formation of the brittle iron-carbide eutectic. Modification of the eutectic Al-Si alloys with Na or Sr to refine the silicon plates is another example.

As some of the most important casting alloys, cast iron and Al-Si alloys solidify with equiaxed eutectic grains, they will be discussed in more details in the following sections.

9.6   Solidification of Cast Iron

Primitive people worked with meteoric iron long before learning to extract iron from iron ore. The Sumerian word AN.BAR, the oldest word designating iron, is made up of the pictogram 'sky' and 'fire'. Similar terminology is found in Egypt ('metal from heaven') and with the Hittites ('black iron from sky'). In most ancient cultures, the metallurgist was believed to have a direct link to the divine, if not of divine origin himself. However, the beginning of this metallurgy on an industrial scale was not possible until the secret of smelting magnetite or hematite was discovered, followed by the art of hardening the metal through quenching (about 1200-1000 B.C.) in the mountains of Armenia (Eliade, 1978). The earliest dated iron casting is a lion produced in China in 502 B.C. Introduction of cast iron in Europe did not occur until about 1200-1450 A.D. Remarkable European cast iron artifacts include the sewer pipes in Versailles (1681) and the iron bridge near Coalbrookdale in England (1779).

Before the invention of microscope in 1860, only two types of iron were known, based on the appearance of their fracture: white and gray. In 1896, the first paper on cast iron to be published in the newly created Journal of the American Foundrymen's Association (1896) stated that "The physical properties of cast iron are shrinkage, strength, deflection, set, chill, grain and hardness. Tensile test should not be used for cast iron, but should be confined to steel and other ductile materials. Compression test should be made, but is generally neglected, from the common erroneous impression that the resistance of a small cube or cylinder, which is enormous, is always in excess of loads which can be applied". It took another 50 years for ductile iron to be discovered (1938-1940 independently by Adey, Millis and Morrogh). The major discoveries of cast iron ended in the '70s with the recognition of compacted graphite iron as a grade in its own merit.

Today, cast iron remains the most important casting material accounting for about 75% of the total world casting tonnage. The main reasons for cast iron longevity are the wide range of mechanical and physical properties associated with its competitive price.

When describing microstructure formation in cast iron during the liquid-solid transformation, we must explain two distinct stages: (I) solidification of the austenite dendrites and graphite crystallization from the liquid (before the beginning of eutectic solidification), and (II) solidification of the stable and metastable eutectics, including the various shapes of the carbon-rich phase (graphite or carbide).

9.6.1   Nucleation and growth of austenite dendrites

Liquid quenching experiments were conducted to evaluate the grain density of primary austenite grains for an iron having 2.98% C and 1.65% Si (Tian and Stefanescu, 1993). The experimental results are shown in Figure 9.21. No significant grain coalescence was observed. This is not surprising since the fraction of austenite is small and the grains do not come in contact. At a quenching temperature of 1180°C the correlation between the area grain density (N_γ in mm^{-2}) and cooling rate (instantaneous nucleation) was found to be:

$$N_\gamma = 48.12 + 5.33\,\dot{T} + 0.087\,\dot{T}^2 \qquad (9.25)$$

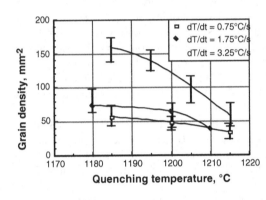

Figure 9.21. Austenite grain density as a function of quenching temperature and cooling rate (Tian and Stefanescu, 1993).

The number of austenite grains in lamellar graphite (LG) iron can also be expressed as a function of undercooling (continuous nucleation) as:

$$N_\gamma = \mu_\gamma \Delta T^n \qquad (9.26)$$

Tian and Stefanescu (1993) proposed $\mu_\gamma = 2.45$ and $n = 0.93$ for N_γ in mm^{-2}, based on the experimental work presented previously. Fras et al. (1993) suggested $\mu_\gamma = 500$ and $n = 2$ for N_γ in mm^{-3}.

To evaluate the austenite dendrites tip radius and spacing Tian and Stefanescu (1992) have conducted DS experiments combined with a liquid metal decanting technique. The method allowed separation of the solidification front from the liquid by means of gravitational and additional centrifugal forces. The typical morphology of dendrites in a Fe - 3.08 %C - 2.01 %Si alloy is shown in the SEM micrographs in Figure 9.22. It is apparent that the tip has the form of a paraboloid, which is consistent with observations on other systems. The experimental results plotted on Figure 8.10, as well as metallographic observations, indicated that at a growth velocity of 0.65μm/s a cellular-to-dendritic transition occurred.

Mampey (2001) compared the amount of austenite calculated with the lever rule from the Fe-C equilibrium diagram with that measured on 100mm diameter bars cast in green sand. As shown in Figure 9.23, for white irons the calculated equilibrium value is very close to the experimental one. This means that ledeburitic solidification is not influenced by the austenite dendrites. There is no measurable growth of the eutectic austenite on the austenite dendrites.

For gray irons, much more dendrites are present than predicted by the lever rule. There seem to be no difference between inoculated and uninoculated irons. Thus, this difference cannot be attributed to the difference in undercooling.

The difference in behavior between the white and gray iron can be explained through the difference in their coupled zone. A shown by Jones and Kurz (1980), the white eutectic remains in the coupled zone regarding of the undercooling, while for the gray eutectic the coupled zone is asymmetric, and below an undercooling of 7.3 °C austenite and eutectic will grow together. Thus more austenite than predicted by equilibrium will form.

a) array of dendrites   b) paraboloidal shape dendrite tip   c) cell

Figure 9.22. Interface morphology at decanted L/S interface in a Fe-3.08 %C-2.01 %Si alloy ($G = 50$ K/cm) (Tian and Stefanescu, 1992). With kind permission of Springer Science and Business Media.

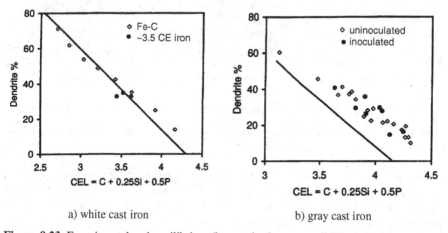

a) white cast iron   b) gray cast iron

Figure 9.23. Experimental and equilibrium (lever rule shown as solid line) dendrite volume in iron-base alloys (Mampey, 2001). CEL is the carbon equivalent liquidus. Copyright 2001 American Foundry Soc., used with permission.

9.6.2   Crystallization of graphite from the liquid

Cast iron is one of the most complex, if not the most complex alloy used in industry, mostly because it can solidify with formation of either a stable (austenite-graphite) or a metastable (austenite-Fe_3C) eutectic. Furthermore, depending on composition and cooling rate several graphite shapes can be obtained at the end of solidification as exemplified in Figure 9.24.

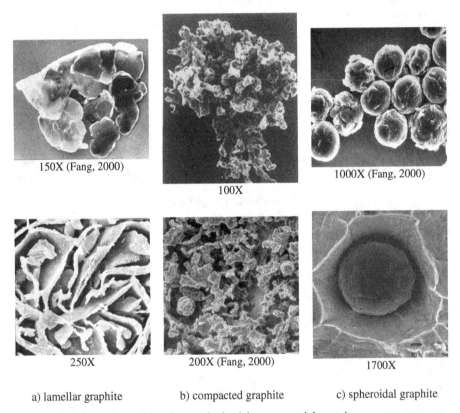

150X (Fang, 2000)

100X

1000X (Fang, 2000)

250X

200X (Fang, 2000)

1700X

a) lamellar graphite   b) compacted graphite   c) spheroidal graphite

Figure 9.24. Typical graphite shapes obtained in commercial cast iron: upper row - extracted graphite; lower row - deep etched microstructure.

Nucleation of SG iron has been described through empirical relationships by a good number of investigators (see Section 7.1.1). Recently, an attempt at better quantification of the nucleation law (Skaland *et al.*, 1993) resulted in an equation that includes the effect of fading time:

$$N = c \cdot \ln \frac{1.33 + 0.64 t_E}{1.33 + 0.64 t_S} \quad \text{in mm}^{-3} \tag{9.27}$$

where c is a kinetic constant to be evaluated experimentally, and t_S and t_E are the time intervals between inoculation and start and end of solidification, respectively.

The nature of nuclei was also addressed. It was concluded that SG nuclei are sulfide inclusions (MgS, CaS) that are covered by an outer shell of complex *Mg* silicates (*e.g.*, $MgO \cdot SiO_2$, $2MgO \cdot SiO_2$), or oxide inclusions ($MgO \cdot SiO_2$). These inclusions do not have a high potency because of a large disregistry between graphite and the substrate. After inoculation with (Me, Al)-containing ferro-silicon (Me = Ca, Sr, or Ba), hexagonal silicate phases of the $MeO \cdot SiO_2$ or the $MeO \cdot Al_2O_3 \cdot 2SiO_2$ type can form at the surface of the oxide inclusions. They allow for the formation of coherent/semicoherent low energy interfaces between the substrate and the graphite.

Many theories have been proposed over the years to describe the mechanisms of formation of various graphite shapes, and reviews of these theories have been periodically written (*e.g.*, Minkoff 1983, Elliott 1988, Stefanescu 1992). A new review is not the aim of this section. It will rather be attempted to evaluate the credibility of the most significant proposed models. The criterion used in this analysis is that if a given set of facts about a phenomenon that we try to describe can be explained by more than one theory, the most probable theory is that one that has the least number of independent assumptions. As an example, two such theories will be discussed.

The first one, proposed by Herfurth (1965), postulates that the change from lamellar to spheroidal graphite occurs because of the change in the ratio between growth on the $[10\bar{1}0]$ face (a direction) and growth on the [0001] face of graphite (c direction). For equilibrium conditions, the Gibbs-Curie-Wulf law states that the crystalline phase with the higher interface energy grows more rapidly in the normal direction. For graphite, $\gamma_{[10\bar{1}0]} = 7.7 \cdot \gamma_{[0001]}$. Thus, $V_{[10\bar{1}0]} > V_{[0001]}$. Bravais' rule stipulates that the growth rate in the direction normal to a plane is inversely proportional to the density of atoms located on the plane. Thus, again, the preferred growth direction should be the a direction. However, according to Herfurth, under the nonequilibrium conditions that prevail during the solidification of cast iron, kinetic considerations become important. Assuming growth by two-dimensional nucleation, the highest rate of growth will be experienced by the face with the higher density of atoms, where the probability for nucleation is higher. Therefore, in a pure environment (pure Fe-C-Si alloy), the highest growth rate will be in the [0001] direction of the graphite crystal, that is growth would be preferred in the c direction. This will result in the formation of unbranched single crystals (coral graphite), as shown in Figure 9.25a.

In a contaminated environment, surface-active elements such as sulfur or oxygen are adsorbed on the high-energy plane $[10\bar{1}0]$, which has fewer satisfied bonds. Subsequently, the $[10\bar{1}0]$ plane face achieves a lower surface energy and higher atomic density than the [0001] face, and growth becomes predominant in the a direction. Lamellar (plate-like) graphite results (Figure 9.25b). Finally, the reactive impurities (such as magnesium, cerium, and lanthanum) in the environment scavenge the melt of surface-active elements (sulfur, diatomic oxygen, lead, antimony, titanium, etc.), after which they also block growth on the $[10\bar{1}0]$ prism face. The preferred growth direction becomes the *c* direction, and *a* polycrystalline spheroidal graphite results (Figure 9.25c).

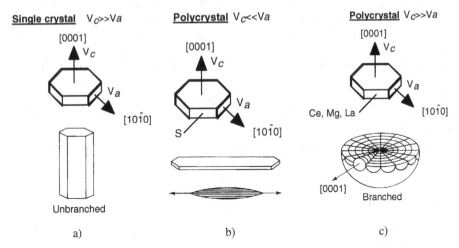

Figure 9.25. Change in the growth velocity of graphite due to adsorption of foreign atoms according to Herfurth (1965): (a) Pure environment. (b) Environment contaminated with surface-active elements such as S or O_2 adsorbed on the prism faces. (c) Spheroidizer (Mg, Ce, La) added as a reactive impurity.

Graphite crystals have a hexagonal habit, with ABAB stacking of semi-infinite hexagonal layers. These layers are called "graphene" sheets (Dresselhaus *et al.*, 1988). They are effectively two-dimensional polymeric sheets, to which atoms can be easily attached in the *a* direction, but more difficult in the *c* direction. Thus, it seems reasonable to assume that the predominant growth direction of graphite in liquid iron is in the *a* direction, with limited growth in the *c* direction, as stipulated by both Gibbs-Curie-Wulf theory and Bravais rule. Indeed, more recent models than the one previously discussed, suggest that the "natural" graphite morphology in pure Fe-C-Si alloys is spheroidal.

According to Sadocha and Gruzleski (1975), this shape results from the circumferential growth model of graphite (Figure 9.26), by movement of steps around the surface of the spheroid. These steps grow in the *a* direction by curved crystal growth, with the low energy basal plane of graphite exposed to the liquid. The growing steps run into one another forming boundaries on the surface. From these boundaries, new steps can develop and grow over the surface, producing a cabbage leaf effect.

In the presence of impurities such, as S or O_2, which decrease the surface tension, the spherical shape is deteriorated into a lamellar one. The major role of spheroidizing elements, such as Mg or Ce, is to act as scavengers, and to produce a high interface surface tension, which helps the curved growth of graphite.

According to Double and Hellawell (1995), bending and wrapping of the graphene layers is facilitated by the mono-layer structure of the graphite sheets. Such growth will result in significant mismatch between the graphite layers that must be accommodated by radial faults. These types of faults are characteristic of all microscopic spheroids, mineralogical or polymeric (Jaszaczak, 1994). This manner of growth requires the minimum activation energy.

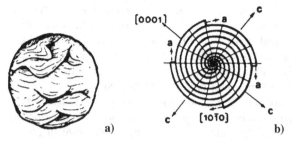

Figure 9.26. Circumferential growth of graphite spheroids: (a) Surface showing cabbage leaf effect. (b) Diametric section showing growth in *a* direction (Sadocha and Gruzleski, 1975).

The introduction of divalent oxygen or sulfur into a graphite ring produces a saturated site that cannot continue to grow by carbon addition to form a new hexagonal graphite ring (Figure 9.27). This will retard, without stopping, growth in the *a* direction. Growth in the *c* direction will become more probable with formation of new graphene sheets, and thus of graphite flakes. Such flakes have limited flexibility and can change growth direction mostly by tilting and twining.

This is why in gray cast iron, that is relatively rich in oxygen and sulfur, the graphite shape is flake. Magnesium, cerium, or lanthanide additions will scavenge the oxygen (MgO) and sulfur (MgS), thus promoting growth and wrapping of graphene sheets in the *a* direction, resulting in spheroidal growth.

It can be generally concluded that the spheroidal shape is the natural growth habit of graphite in liquid iron. Flake graphite is a modified shape, the modifiers

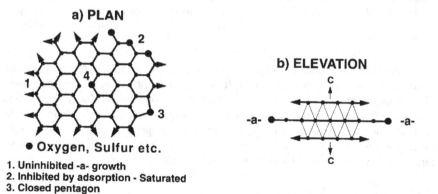

Figure 9.27. Plan of graphene sheet, doped by oxygen or sulfur to form heterocyclic rings (a). Growth in the *c* direction becomes more probable if that in the *a* direction is retarded by the adsorption into saturated heterocyclic rings (b) (Double and Hellawell, 1995). Reprinted with permission from Elsevier.

being sulfur and oxygen. They affect graphite growth through some surface adsorption mechanism.

Some facts that support this model are as follows. Auger analysis show concentrations of oxygen and sulfur in iron, but not in graphite, adjacent to the metal-flake graphite interfaces, of about 20at% O_2 and 5at% S, in some two or three atomic layers (Johnson and Smartt, 1979). Simple calculations (Double and Hellawell, 1995) show that there is more than sufficient oxygen and/or sulfur in a typical gray iron to cover all of the flake graphite surfaces. Since only limited amounts of these elements can be incorporated in the hexagonal rings, the excess solute will be rejected in the liquid, and accumulate at the graphite flake - iron interface.

When comparing the two models discussed so far, it becomes apparent that the model that assumes the natural growth habit of graphite in pure melts to be the spheroidal one (through circumferential growth) is more straight forward. It requires fewer assumptions. According to the selection principle introduced at the beginning of this section, we will lend it more credibility.

It was also proposed that graphite spheroids develop first by circumferential growth in the a direction, followed by pyramidal growth in the c direction (Figure 9.28) (Liu and Liu, 1993). The duplex structure of SG was used as the basic argument. However, the duplex structure may simply be the result of growth through the austenite shell, or during solid-state transformation.

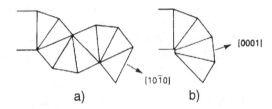

a)   b)

Figure 9.28. Model for duplex growth of SG (*Liu and Liu,* 1993). Copyright 1993 American Foundry Soc., used with permission.

Figure 9.29. Change of direction of the tilt of the boundary due to the amount of reactive impurity: a) insufficient spheroidization; b) sufficient spheroidization (*Zhu et al.,* 1985).

Other intermediate graphite shapes, such as compacted graphite, are probably the result of a mixed growth mechanism. It is proposed, for example, that the growth of compacted graphite occurs by the twin/tilt of boundaries (Zhu *et al.*, 1985). When insufficient reactive element is present, the tilt orientation of twin boundaries can alternate, resulting in typical compacted graphite (Figure 9.29a). On the other hand, when there is enough impurity in the melt, the tilt orientation is singular, and spheroidal graphite will occur (Figure 9.29b).

9.6.3   Eutectic Solidification

The basic parameters affecting the morphology of the eutectic are the G/V ratio and composition. The G/V ratio can be best controlled during directional solidification, when steady state conditions are achieved. It is possible to obtain a variety of graphite and matrix structures in cast iron when varying G/V and/or the level of impurities, such as Mg or Ce. Argo and Gruzleski (1986) have achieved transition from a SG through compacted graphite (CG) to LG structure by directional solidification of Mg containing SG iron. The cooling rates used in their work varied from 0.0015 to 0.013°C/s, which is relatively low. Consequently, complete structural transition, to include the stable/metastable transition, was not obtained.

The complete structural transition from metastable to stable, and for different graphite morphologies, has been documented for cast irons of hypoeutectic composition, as a function of growth velocity, temperature gradients at the solid/liquid interface, and cerium concentration (Figure 9.30). It was found that while the metastable-to-stable (white-to-gray) transition depends mostly on the G/V ratio, the transition between different graphite shapes (lamellar to compacted to spheroidal) depends mostly on the cerium concentration.

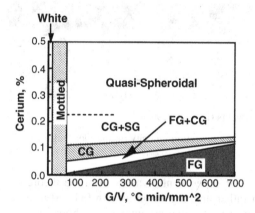

Figure 9.30. Influence of G/V ratios and percent cerium on structural transitions in cast iron (Bandyopadhyay *et al.*, 1989). FG stands for flake graphite, *i.e.*, lamellar graphite.

Based on this and other experimental work (*e.g.* Rickert and Engler, 1985) a sequence of changes in the eutectic morphology of directionally solidified cast iron was proposed, as shown in Figure 9.31. As the ratio between the temperature gradient at the S/L interface and the growth velocity, G/V, decreases, or the composition C_o (*e.g.*, Mg or Ce) increases, the S/L interface changes from planar, to cellular, and then to equiaxed, while graphite remains basically flake (lamellar). Cooperative growth of austenite and graphite occurs. Further change of G/V or of C_o brings about formation of an irregular interface, with austenite dendrites protruding in the liquid. Graphite becomes compacted and then spheroidal. Eutectic growth is divorced. From these structures, those that have practical importance are the first three that is the irregular SG, the irregular CG, and the eutectic colonies of lamellar graphite.

As shown in Figure 9.32, the solidification mechanisms of lamellar and spheroidal graphite cast iron during continuous cooling are quite different. LG iron solidifies with skin formation while SG iron is characterized by mushy solidification.

For solidification of regular eutectics, a number of relationships have been established between process and material parameters based on the extremum criterion (see section 9.2.1 for analytical models). Eutectic gray iron can solidify with a planar interface. However, the relationship $\lambda^2 \cdot V = ct.$ is not obeyed in the growth of the γ-LG eutectic. Some experimental results are compared with theoretical calcu-

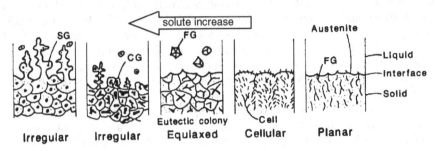

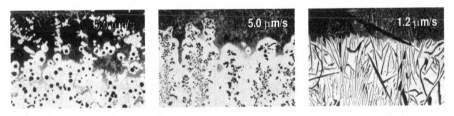

a) schematic representation (Stefanescu *et al.* 1986, Banerjee and Stefanescu 1991)

b) DS experiments (Li *et al.*, 1990)

Figure 9.31. Influence of composition and solidification velocity on the morphology of the S/L interface. Copyright 1991 and 1990 American Foundry Soc., used with permission.

lations in Figure 9.33. The departure of experimental values from the theoretical line is a clear indication of irregular rather than regular growth in LG cast iron. As discussed earlier, lamellar graphite iron is an irregular eutectic and both an extremum and branching spacing can be defined. The average spacing should be higher than the extremum one predicted form Jackson-Hunt theory. Other data for the growth coefficient μ_V of directionally solidified gray and white iron are given in Appendix B, Table B3. For continuously cooled irons μ_V is not a constant. Some typical values for gray and white iron are given in Appendix B, Table B4.

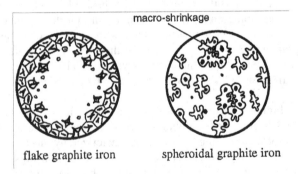

Figure 9.32. Schematic illustration of solidification mechanisms of continuously cooled lamellar and spheroidal graphite cast iron.

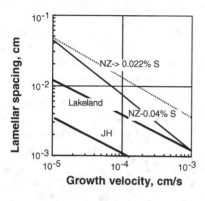

Figure 9.33. λ-V relationships in LG cast iron. Lakeland (1964) $\lambda = 3.8 \cdot 10^{-5} \, V^{-0.5}$ cm; NZ – Nieswaag and Zuithoff (1975) $\lambda = 0.56 \cdot 10^{-5} \, V^{-0.78}$ cm (0.004%S); NZ – Nieswaag and Zuithoff (1975) $\lambda = 7.1 \cdot 10^{-5} \, V^{-0.57}$ cm (>0.02%S); JH – Jackson-Hunt $\lambda = 1.15 \times 10^{-5} \, V^{-0.5}$ cm (theoretical).

The γ-SG eutectic is a divorced eutectic (see section 9.3 for analytical model). However, a number of reports (Rickert and Engler 1985, Banerjee and Stefanescu 1991) have demonstrated that primary austenite dendrites play a significant role in the eutectic solidification of SG. This contention is supported by microstructures found in the microshrinkage of thin SG iron plates presented in Figure 9.34. Microshrinkage can be rationalized as interrupted solidification. Two types of dendrites exhibiting non-similar morphologies can be identified: *primary austenite dendrites* and *eutectic SG grains*. The morphology of the primary austenite dendrites is typical for dendrites in metallic alloys. They exhibit clear primary and secondary arms (Figure 9.34a). The eutectic SG grains (Figure 9.34b) are thick and rounded, suggesting a cauliflower shape. While displaying branching, there is no clear distinction between primary and secondary arms. It appears that the SG grains are made of several graphite nodules surrounded by quasi-spherical austenite envelopes. This multi-nodular morphology of the SG eutectic grain may be visualized with other techniques, such as color etching metallography, first introduced by Rivera *et al.* (1997) (Figure 9.35).

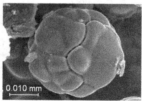
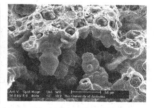

a) primary austenite dendrite   b) eutectic austenite dendrite   c) overall view of mi-
and SG aggregate   croshrinkage

Figure 9.34. Microstructures of SG iron found in the same microshrinkage cavity from a SG iron plate (Ruxanda *et al.*, 2001). Copyright 2001 American Foundry Soc., used with permission.

Based on previous work and on the new experimental evidence presented here, the solidification sequence of eutectic SG iron may be summarized as shown in Figure 9.36. At the beginning of eutectic solidification, graphite nodules and austenite crystals are nucleating independently in the liquid. Natural convection will move both phases and no significant interaction between graphite and austenite dendrites is expected, as shown in Figure 9.36a. Once nucleated, the austenite dendrites will grow freely in the liquid. This is possible even though the composition is eutectic, as cast irons solidify following an asymmetric phase diagram. While the austenite dendrites grow, carbon is continuously rejected ahead of the interface, the viscosity of the S/L mixture increases, and therefore the convection velocity decreases.

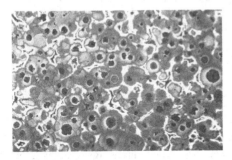
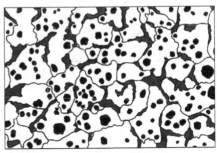

a) microstructure of a 7 mm thick plate after color etching

b) grain and segregation delineation

Figure 9.35. Microstructure of SG iron plates after color etching (Ruxanda *et al.*, 2001).
Copyright 2001 American Foundry Soc., used with permission.

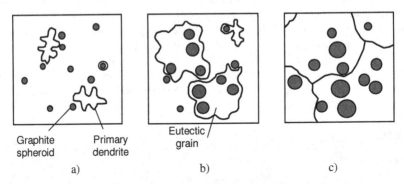

Graphite   Primary
spheroid   dendrite

Eutectic
grain

a)   b)   c)

Figure 9.36. Schematic illustration of the sequence of formation of eutectic grains in SG iron for the case of continuous cooling solidification.

Based on limited measurements on directionally solidified SG iron samples, it is considered that the graphite-austenite interaction occurs when the fraction of solid reaches ~0.3 (dendrite coherency), because natural convection is suppressed (Figure 9.36b). Because of graphite particle-austenite dendrite interactions, several nodules may be incorporated in one austenite grain.

The graphite particles grow isotropically by solid-state diffusion of carbon from the liquid through the austenite shell, to the graphite particle. The first graphite particles to be incorporated will have the longest time to grow and will become the largest nodules. Smaller graphite spheroids will be found in the austenite closer to the interface.

To understand the interaction between austenite dendrites and graphite nodules in the early stages of solidification, the concepts developed for particle engulfment and pushing (PEP), described in Section 13.1, may be used. As demonstrated for transparent organic materials, a particle may be engulfed by the growing dendrite into the side of the dendrite arm, or through the splitting of the dendrite tip. Alternatively, the particle may be pushed in the interdendritic region and entrapped in the last regions to solidify, process that will result in particle alignment. A similar behavior may be inferred for the austenite-SG eutectic system.

The austenite growing into the liquid will tend to grow anisotropically in its preferred crystallographic orientation. However, restrictions imposed by isotropic diffusion growth will impose an increased isotropy on the system. Consequently, the dendritic shape of the austenite will be altered and the γ-L interface will exhibit only small protuberances instead of clear secondary arms. This process is dominant in the last regions to solidify.

This interpretation is consistent with recent phase-field modeling of dendritic growth. Karma and Rappel (1998), have compared 3D growth of dendrites with cubic symmetry for the extreme cases of high anisotropy, Figure 9.37a, and no anisotropy, Figure 9.37b. It is seen that a metal with a cubic lattice will grow in a "cauliflower" shape in the absence of anisotropy. The SEM investigation showed remarkable similarities between the simulated and real austenite morphology for the case of isotropic growth (compare Figure 9.37b with Figure 9.37c).

a) simulated growth with high anisotropy. Reprinted with permission from Phys. Rev. Copyright 1998 by the American Physical Soc. www.aps.org

b) simulated growth without anisotropy. Reprinted with permission from Phys. Rev. Copyright 1998 by the American Physical Soc. www.aps.org

c) SG eutectic dendrites where the loss of anisotropy lead to an effective modification of the dendrite morphology

Figure 9.37. Simulated growth of the dendrite morphology (Karma and Rappel, 1998) and real dendrite morphology (Ruxanda *et al.*, 2001).

9.6.4   The gray-to-white structural transition

Of particular interest to the cast iron manufacturer is the stable-to-metastable microstructure transition (also know as the gray-to-white transition, GWT), as it signals the occurrence of unwanted iron carbides in the gray iron. In a binary Fe-C alloy, the difference between the stable (T_{st}) and metastable (T_{met}) eutectic temperatures is only 3K. Thus, during cooling of the casting the temperature of the melt may become smaller than T_{met} before any stable structure nucleates and grows. This may happen even at very low cooling rates. In the Fe-C-Si system the T_{st} -T_{met} interval is much larger, and stable solidification may occur before the temperature reaches T_{met}.

The first rationalization of the GWT was based on the influence of cooling rate on the stable (T_{st}) and metastable (T_{met}) eutectic temperatures. As the cooling rate increases, both temperatures decrease (Figure 9.38). However, since the slope of T_{st} is steeper than that of T_{met}, the two intersect at a cooling rate which is the critical cooling rate, $(dT/dt)_{cr}$, for the GWT. At cooling rates smaller than $(dT/dt)_{cr}$ the iron solidifies gray, while at higher cooling rates it solidifies white.

Magnin and Kurz (1985) further developed this concept by using solidification velocity rather than cooling rate as a variable. As shown on Figure 9.39, the curves describing the growth velocities of the two eutectics intersect at V_{cr}. Since above this velocity the metastable growth velocity is higher than the stable one at any undercooling ($V_{met} > V_{st}$), it can be assumed that above this velocity (or under this temperature) only white iron will solidify, while under V_{cr} the structure is completely gray. Thus, if only growth considerations are invoked, a clear GWT exists.

These arguments have ignored nucleation. It is well accepted that nucleation of white iron is more difficult than that of gray iron. Consequently, additional undercooling, in excess of that predicted from growth velocity considerations, is required for a complete GWT. This is shown in the figure as ΔT_n^{met}, that is the nucleation temperature of the metastable cementite. Thus, a complete white iron is obtained only at growth velocities larger than V_{g-w}. A similar argument holds for the white-to-gray transition, when a complete transition cannot occur unless the undercooling is smaller than ΔT_n^{st}, which is the nucleation temperature of the stable gray eutectic. Thus, a region of mixed structure, gray and white, will exist at growth velocities between V_{w-g} and V_{g-w}. This is the mottled region.

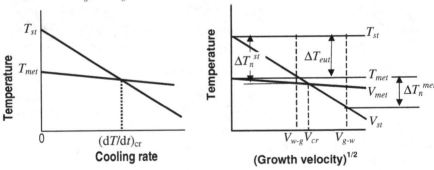

Figure 9.38. Critical cooling rate for the GTW transition.

Figure 9.39. Structural transitions in cast iron as function of undercooling and growth velocity (Magnin and Kurz, 1985).

Based on this discussion it is clear that a model that predicts the GWT must describe nucleation and growth of both the gray and white eutectic. An analytical model was proposed by Fras and Lopez (1993). For eutectic iron the *chilling equivalent* is:

$$E = \frac{T_{st}^{1.08}}{1.9\Delta T_{pour}^{0.5}} \left(\frac{1}{\mu_1 \mu_2^{\ 3} c^5 \left(\Delta T_{eut} + \Delta T_n^{\ met} \right)^{10}} \right)^{1/6}$$

where ΔT_{pour} is the superheating above the eutectic temperature, μ_1 and μ_2 are nucleation and growth coefficients, respectively, and c is the specific heat. It is seen that the chilling tendency (chilling equivalent) decreases as the number of eutectic grains and their growth rate increase (μ_1 and μ_2 increase), as the eutectic interval, ΔT_{eut}, and the pouring temperature increase.

9.7   Solidification of aluminum-silicon alloys

Out of the about 238 compositions for casting aluminum alloys registered with the Aluminum Association of America, 46% consists of aluminum-silicon alloys. However, 90% of all aluminum alloy shaped castings are made of Al-Si alloys (Granger and Elliott, 1992). Most of these alloys are hypoeutectic. The length scale of the primary and eutectic phases is controlled through grain refinement and modification.

9.7.1   Nucleation and growth of primary aluminum dendrites

Most properties of practical interest depend on the length scale of the primary aluminum dendrites (grain size) of the aluminum alloy. Grain refinement is obtained by the heterogeneous nucleation of the α-aluminum phase. While it is possible to achieve this goal by dynamic nucleation, in metal casting practice this is mostly done by addition of chemicals. An efficient grain refiner must provide stable nuclei at the liquidus temperature of the alloy. The most common grain refiner is a master alloy containing titanium and boron. When adding more than 0.15% titanium to the aluminum alloy, $TiAl_3$ forms, as can be seen on the phase diagram in Figure 9.40. The $TiAl_3$ is solid at a temperature higher than the melting point of the Al-Si alloy, and serves as heterogeneous nuclei. If boron is also added complex borides, $(Al,Ti)B_2$, serve as nuclei.

The evaluation of grain refinement efficiency can be performed through cooling curve analysis. As shown in Figure 9.41 grain refinement decreases the amount of undercooling observed in the initial stages of the solidification. However, the undercooling alone is not enough to evaluate the degree of grain refinement. Indeed, from Figure 9.42 it is seen that both very fine and very coarse grain structures may have minimal undercooling. Thus, other features of the cooling curve must be used, such as the time difference between the maximum undercooling and the recalescence or elements of the first derivative curve.

9.7.2   Eutectic solidification

Typical microstructures resulting during the solidification of *Al-Si* eutectic alloys are shown in Figure 9.43. A rod-angular morphology of the silicon phase is typical for the unmodified eutectic (type D). This type of structure is very brittle. Modification is used to change the shape of the silicon phase to fibrous.

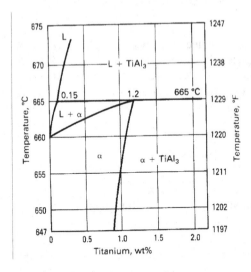

Figure 9.40. The aluminum-titanium phase diagram.

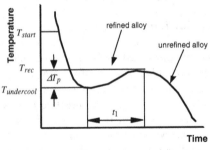

Figure 9.41. Schematic cooling curve showing primary solidification of an aluminum alloy and role of grain refinement.

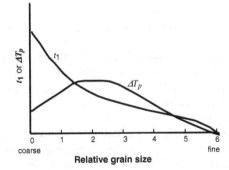

Figure 9.42. Correlation between grain size and time and temperature of undercooling in hypoeutectic aluminum alloys (Charbonier, 1984).

Two types of modifications are possible:

- through rapid solidification (quench modification)
- impurity modification (chemical modification): elements in groups I and IIa of the periodic table and lanthanides can be used, but the most efficient are Na and Sr.

The effect of growth velocity and composition on microstructure is also shown in Figure 9.43. Typical morphologies occurring in the coupled zone as a function of growth velocity are shown in Figure 9.44. Impurity modification produces similar structures.

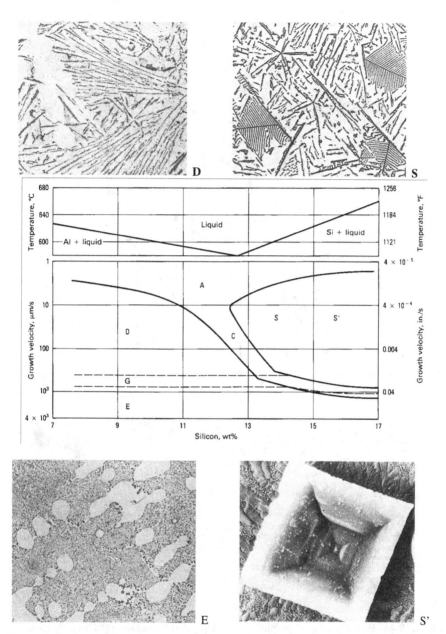

Figure 9.43. Coupled zone diagram for Al-Si alloys obtained through directional solidification under a thermal gradient of 125°C/cm; D: unmodified structure of flake eutectic Si and Al dendrites, 100X; S: complex regular and star-like primary Si and flake eutectic Si, 100X; E: chill modified structure showing fibrous eutectic Si and Al dendrites, 100X; S': SEM showing a (100) section through an octahedral primary Si particle revealing four {111} planes, 1500X (Granger and Elliott, 1988). Reprinted with permission of ASM International. All rights reserved. www.asminternational.org

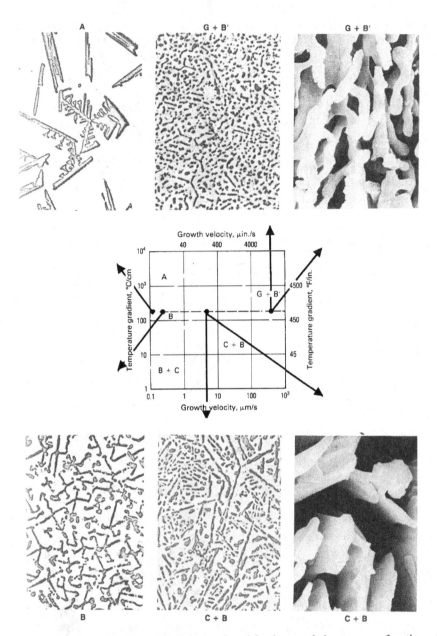

Figure 9.44. Eutectic silicon morphologies found in the coupled zone as a function of growth velocity and temperature gradient in the liquid at the solid/liquid interface; A: massive faceted silicon and Al dendrites (100X); B: rod eutectic silicon (100X); B+C: angular and some flake eutectic silicon; C+B: flake and some angular eutectic silicon (100X and 1500X; G+B': quench modified fibrous silicon (100X and 1500X). (Hellawell 1973)

Modification was originally attributed to the repeated nucleation of the eutectic silicon phase at a reduced temperature. According to Crossley and Mondolfo (1966), the mechanism of Na modification is as follows. In unmodified melts, Si nucleates on AlP precipitates. Excessive P results in large eutectic Si particles. Na restricts both nucleation and growth of the Si phase because when Na is added, NaP rather than AlP is formed. Thus, nucleation of the Si phase is retarded, and the eutectic temperature is depressed. If too much Na is added AlNaSi which nucleates Si is formed (Sigworth, 1983). This is over modification.

However, because of the similarities between structures obtained through quench and impurity modification, modification is now considered the result of a change in the growth mechanism of the silicon phase, from faceted growth to a more isotropic growth (Figure 9.45). Although the eutectic temperature is depressed, silicon grows continuously, without repeated nucleation (Elliott, 1983). The aluminum phase is not affected. There is experimental evidence that both sodium and strontium are concentrated in the silicon phase.

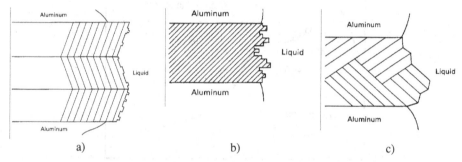

a)   b)   c)

Figure 9.45. Interface morphology of the eutectic liquid/solid interface (Granger and Elliott, 1988): a) unmodified flake silicon, growth through propagation of steps on widely spaced twins across the interface; b) quenched-modified fibrous silicon, continuous growth on the atomically rough interface; c) impurity modified fibrous silicon, growth through twin-plane reentrant edge of finely spaced twins.

Cooling curve analysis can also be used for evaluation of modification efficiency. Modification results in higher undercooling (Figure 9.46). This is similar to the results of Mg-treatment of cast iron, which is modification, but opposed to the FeSi treatment of lamellar graphite iron, which decreases undercooling because it increases the number of eutectic grains. However, just as in the case of grain refinement, evaluating only the undercooling is not enough to decide whether the required degree of undercooling has been achieved. Indeed, as shown in Figure 9.47, the highest undercooling is obtained for the under-modified eutectic having an acicular lamellar structure.

The Jackson-Hunt parabolic law of irregular eutectic growth, as modified by Fisher and Kurz (1980), has been found to be suitable to model the growth of both modified and unmodified equiaxed eutectic grains of irregular eutectics like Al-Si (Degand et al., 1996). The modifiers restrain the growth of Si in favorable directions, increasing the twins density and thus, the Si kinetic undercooling. Including

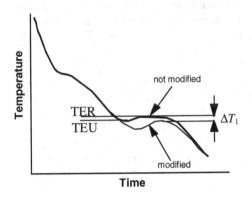

Figure 9.46. Schematic diagram of cooling curves for non modified and modified hypoeutectic Al-Si alloys.

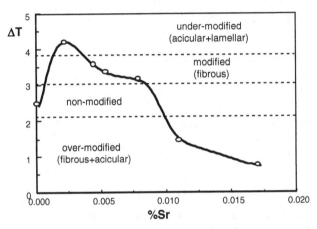

Figure 9.47. Influence of Sr content on ΔT = TER - TEU for an Al-7% Si alloy (based on data from Argyropoulos *et al.*, 1983).

the kinetic undercooling of the Si phase, and using weighted averages for the two phases leads to:

$$\overline{\Delta T} = \frac{m_{Si}\overline{\Delta T^{Al}} + |m_{Al}|\overline{\Delta T^{Si}}}{|m_{Al}| + m_{Si}} = \Delta T_c + \Delta T_r + \Delta T_k \qquad (9.28)$$

with $\Delta T_k = |m_{Al}|\overline{\Delta T_k^{Si}} / (|m_{Al}| + m_{Si})$, where m_{Al} and m_{Si} are the slopes of the liquidus in the Al-Si phase diagram. To obtain an equation uniquely relating ΔT to the growth velocity V, one needs to know ΔT_k^{Si} as a function of V. Magnin and Trivedi (1991) have derived some values of ΔT_k^{Si} from experimental data. When ΔT_k^{Si} was plotted versus V, it was noticed that a parabolic law fits better the experimental values than the exponential law suggested by Tiller (1969). This can be written as:

$$V = \mu_{Si}\left(\Delta T_k^{Si}\right)^2 \tag{9.29}$$

This means that a parabolic dependency of V on ΔT is still valid in the case of modified irregular eutectics, with a different growth parameter:

$$\mu = \left[\sqrt{K_1 K_2}\left(\phi + \frac{1}{\phi}\right) + \frac{|m_{Al}|}{\left(|m_{Al}| + m_{Si}\right)}\mu_{Si}^{-1/2}\right]^{-2} = \left[\mu_{ex}^{-1/2} + A\mu_{Si}^{-1/2}\right]^{-2} \tag{9.30}$$

where ϕ is a coefficient relating the lamellar spacing of regular and irregular eutectics: $\lambda_{ir} = \phi\lambda_{ex}$. For Al-Si alloys ϕ has values between 2.2 for rod morphology and 3.2 for lamellar morphology. This law has been verified experimentally by several authors (*e.g.*, Magnin and Trivedi, 1991) by using directional solidification experiments.

The growth parameter was evaluated experimentally (Degand *et al.*, 1996). It was found that it decreases as the degree of modification increases (Figure 9.48). During solidification of sodium-modified Al-Si alloys, the grain morphology changes from the center to the outside. As shown in Figure 9.49, a coarse microstructure surrounds the nucleus while a finer fibrous structure is observed at the periphery of the grain. This suggests that a variable growth parameter should be used to simulate the eutectic grain growth. μ is therefore supposed to be initially large and then decreases because of sodium rejection by the solidification interface. Such a growth parameter will impose less undercooling and recalescence on the cooling curve.

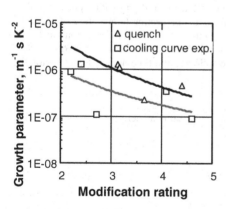

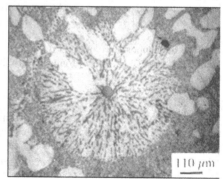

Figure 9.48. Influence of modification on the average value of the growth parameter (modification rating: 2 – low; 5 – high) (Degand *et al.*, 1996).

Figure 9.49. Microstructure of an Al-11%Si eutectic grain core in a sodium modified casting (Degand *et al.*, 1996).

This micro-level analysis can be included in a macro-model by using an Avrami-type equation for the growth parameter:

$$\mu = \mu_{rod} + \left(\mu_{lam} - \mu_{rod}\right)\exp\left(-\frac{Vt^{1/2}}{D^{1/2}}\right)$$

(9.31)

where μ_{lam} is the initial growth parameter (lamellar eutectic $\approx 5{\cdot}10^{-5}$ m s^{-1} K^{-2}) and μ_{rod} the growth parameter of the fibrous eutectic ($\approx 5.10^{-7}$m s^{-1} K^{-2}). The factor $V/D^{1/2}$ has been used to account for the opposite effects of interface velocity and diffusion in the liquid on the sodium segregation ahead of the liquid-solid interface.

9.8   Applications

Application 9.1

Demonstrate that the diffusion characteristic length is much smaller for the eutectic than for the primary phase. Perform the calculation for a Sn-15% Pb alloy solidifying at $2{\cdot}10^{-6}$ m/s.

Answer:
For the eutectic, the diffusion characteristic length is $\delta = \lambda/2$. In turn, $\lambda = \mu_\lambda{\cdot}V^{0.5} = (\mu_r/\mu_c)^{0.5}{\cdot}V^{0.5}$. Using the data in Appendix B Table B1, $\delta = 5.9{\cdot}10^{-9}/2/\sqrt{2{\cdot}10^{-6}}$ $= 2{\cdot}10^{-6}$. For the primary phase, $\delta = D_L/(2V) = 1.1{\cdot}10^{-9}/2/(2{\cdot}10^{-6}) = 2.75{\cdot}10^{-4}$. Thus, the diffusion characteristic length of the eutectic is two orders of magnitude smaller than that of the primary phase.

Application 9.2

Calculate the final solidification time, the cooling curve and the time evolution of the solid fraction of a eutectic cast iron cube of volume 0.001 m^3 solidifying with equiaxed grains. The casting is poured into a silica sand mold, from a superheating temperature of 1350 °C. Use a time-stepping analysis assuming resistance in the mold. Use transformation kinetics to calculate the temporal evolution of the eutectic solid fraction. Assume instantaneous nuclea-tion with $N_E = 3.32{\cdot}10^7$ (m^{-3}). Compare with the solidification time obtained when using the Chvorinov's equation.

Answer:
The governing equation in time-stepping format is the same as derived in Application 5.2, that is:

$$T^n = T^{n-1} - \frac{A}{\nu\rho c}\sqrt{\frac{k_m \rho_m c_m}{\pi t^i}}\left(T^{n-1} - T_o\right)\Delta t + \frac{\Delta H_f}{c}\Delta f_S^n$$

(a)

where T^n and T^{n+1} are the temperatures at time n and $n + 1$ respectively, and T_o is the ambi-ent temperature. The solid fraction at time n is:

$$f_S^n = f_S^{n-1} + \Delta f_S^n$$

(b)

Assuming instantaneous nucleation, the fraction solid at the end of solidification is given by $f_S = (4/3)\pi N r_E^3$, where r_E is the final radius of the eutectic grain and N is their number.

The evolution in time of the fraction solid is then $df_S/dt = 4\pi N r_E^2 dr_E/dt$, or in discretized format, since $dr_E/dt = V_E$:

$$\Delta f_S^n = 4\pi \left(r_E^{n-1}\right)^2 NV_E^n \left(1 - f_S^{n-1}\right)\Delta t \tag{c}$$

Then:

$$r_E^n = r_E^{n-1} + V_E^n \Delta t \left(1 - f_S^{n-1}\right) \tag{d}$$

Note that both the fraction solid and the radius have been decreased by the factor $(1 - f_S)$ to account for grain impingement. The solidification velocity is calculated with Eq. (9.24).

$$V_E^n = \mu_E^n \left(\Delta T_E^n\right)^2 \tag{e}$$

For eutectic the growth coefficient, μ_E, can be assumed constant (see Appendix B). The eutectic undercooling is:

$$\Delta T_E^n = T_E - T^{n-1} \tag{f}$$

Some comments are necessary regarding the use of nucleation laws. According to the present calculations, the cooling rate before the beginning of solidification is 0.4 °C/s. Using this value and the numbers in Appendix B, it is calculated that $N_E = 3.32 \cdot 10^7$ m^{-3}. Note that the number of gains per unit area was transformed in number of grains per unit volume with the relationship $N_V = 0.87 (N_A)^{1.5}$. Since eutectic solidification occurs in a very narrow range, it is reasonable to use an instantaneous nucleation law. This algorithm is implemented in an Excel spreadsheet that is organized as shown in Table 9.1. In the second row of the table, the equations used in the various cells are shown. In addition, some calculation results at different times are also included. It is seen that the solidification time, corresponding to fraction solid 1, is 1953 s. Chvorinov's equation gave 1817 s.

Table 9.1. Organization of Excel spreadsheet.

t	ΔT_E	V_E	r_E	Δf_S	f_S	T
Eq.	(f)	(e)	(d)	(c)	(b)	(a)
1						1350
2	0	0	0	0	0	1345
10	0	0	0	0	0	1323
286	0.04	4.0E-11	4.0E-11	2.7E-23	2.7E-23	1154
1953	19.6	1.1E-05	1.9E-03	0	1	1134

As the fraction solid is calculated only during solidification, two IF statements must be introduced as follows:

for Δf_S: $IF(f_S^{n-1} > 0.99, 0, Eq.(c))$ $\Delta f_S = 0$ when $f_S^{n-1} > 0.99$

for f_S: $IF(f_S^{n-1} > 0.99, 1, f_S^{n-1} + \Delta f_S^n)$ $f_S^{n-1} = 1$ when $f_S^{n-1} > 0.99$

The calculated cooling curve and fraction of solid evolution are given in Figure 9.50. A slight recalescence is seen at the beginning of eutectic solidification.

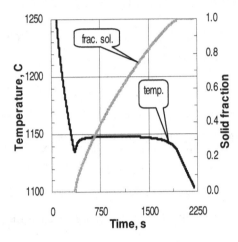

Figure 9.50. Temperature and fraction of solid evolution for cast iron solidifying in a sand mold.

Application 9.3

Calculate the final solidification time, the cooling curve and the time evolution of the solid fraction of a 3.9% C cast iron cube of volume 0.001m³, poured into a silica sand mold, from a superheating temperature of 1350°C. Use a time-stepping analysis assuming resistance in the mold. Use transformation kinetics to calculate the temporal evolution of both the primary and eutectic solid fraction. Assume instantaneous nucleation with $N_\gamma = 1 \cdot 10^7$ and $N_E = 3.32 \cdot 10^7$ (m⁻³). Compare with the solidification time obtained when using the Chvorinov's equation.

Answer:
The difference compared with Application 9.2 is that this is a hypoeutectic iron. Consequently the solidification microstructure will include both a primary phase (austenite) and eutectic. The governing equation in time-stepping format is the same as Eq. (a) in Application 9.2, while the solid fraction at time *n* is given by equation (b). However, bceause the composition is hypoeutectic, the solid fraction is the sum of the fraction of austenite dendrites and of the fraction of eutectic:

$$\Delta f_S^n = \Delta f_{S\gamma}^n + \Delta f_{SE}^n \tag{c}$$

The fraction of austenite or eutectic is calculated with an equation similar to Eq. (c) in Application 9.2:

$$\Delta f_{S\phi}^n = 4\pi \left(r_\phi^{n-1}\right)^2 N_\phi V_\phi^n \left(1 - f_S^{n-1}\right) \Delta t \tag{d}$$

where the subscriptϕ is either the austenite, γ, or the eutectic, *E*. N_ϕ is the number of grains per unit volume. Then:

$$r_\phi^n = r_\phi^{n-1} + V_\phi^n \Delta t \left(1 - f_S^{n-1}\right) \tag{e}$$

Note that both the fraction solid and the radius have been decreased by the factor $(1 - f_S)$ to account for grain impingement. The growth velocity of the phases is:

$$V_\phi^n = \mu_\phi^n \left(\Delta T_\phi^n\right)^2 \tag{f}$$

For austenite, as shown in Application 8.5:

$$\mu_\gamma^n = \frac{D_L}{2\pi^2 \Gamma m(k-1)C_L^n} \tag{g}$$

$$\Delta T_\gamma^n = T_f + mC_L^n - T^{n-1} \tag{h}$$

$$C_L^n = C_o\left(1 - f_S^{n-1}\right)^{k-1} \qquad \text{(the Scheil model is assumed for diffusion)} \tag{i}$$

For eutectic the growth coefficient, μ_E, can be assumed constant (see Appendix B). The eutectic undercooling is:

$$\Delta T_E^n = T_E - T^{n-1} \tag{j}$$

This algorithm is implemented in an Excel spreadsheet that is organized as shown in Table 9.2. In the second row of the table, the equations used in the various cells are shown. In addition, some calculation results at different times are also included. It is seen that the solidification time, corresponding to fraction solid 1, is 1954 s. Chvorinov's equation gave 1817 s.

The calculated cooling curve and fraction of solid evolution are given in Figure 9.50. A slight recalescence is seen at the beginning of both primary and eutectic solidification. The fraction solid curve shows two regions: the first one when only primary grains are formed and a second one where mostly eutectic solidification occurs.

Table 9.2. Organization of Excel spreadsheet.

t	C_L	μ_γ	ΔT_γ	V_γ	r_γ	Δf_γ	ΔT_E	V_E	r_E	Δf_E	Δf_S	f_S	T
	(i)	(g)	(h)	(f)	(e)	(d)	(h)	(f)	(e)	(d)	(c)	(b)	(a)
1	3.9												1350
2	3.9	1.97 E-6	0	0	0	0	0	0	0	0	0	0	1345
10	3.9	1.97 E-6	0	0	0	0	0	0	0	0	0	0	1323.0
172	3.9	1.97 E-6	7.84	1.21 E-4	1.21 E-4	1.79 E-4	0	0	0	0	1.79 E-04	1.79 E-04	1198.3
533	4.51	1.71 E-6	0.97	1.59 E-6	1.94 E-3	4.51 E-4	0.03	2.72 E-11	2.05 E-11	3.60 E-24	4.51 E-04	2.47 E-01	1153.9
1954	40.9	1.88 E-7	0	0	2.08 E-3	0	21.67	1.41 E-05	1.70 E-03	1.69 E-04	0	1	1132.2

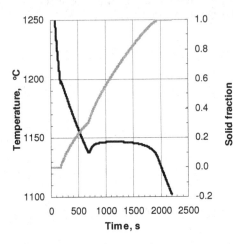

Figure 9.51. Temperature and fraction of solid evolution for cast iron solidifying in a sand mold.

Some comments are necessary regarding the use of nucleation laws. The number of eutecitc nuclei was calculated as in Application 9.2. Using the same procedure, it is calculated that the number of grains for primary solidification is $N_\gamma = 5.03 \cdot 10^8$ m^{-3}. If this number in conjunction with instantaneous nucleation is used the model becomes unstable (the temperature oscillates back and forth around the liquidus temperature at the beginning of solidification). This is because in reality, nucleation occurs over the whole interval of primary solidification, and only a limited number of nuclei are available for growth at the beginning of solidification. Since implementation of a continuous nucleation law is more difficult on the Excel spreadsheet, an instantaneous nucleation law, with a smaller number of nuclei was proposed ($N_\gamma = 10^7$ m^{-3}).

References

Argyropoulos S., B.Closset, J.E.Gruzleski and H.Oger, 1983, *Trans. AFS* **91**:351

Argo D. and J.E. Gruzleski, 1986, *Mat. Sci. Tech.* **10**(2):1019

Bandyopadhyay D.K., D.M. Stefanescu, I. Minkoff and S.K. Biswal, 1989, in: *Physical Metallurgy of Cast Iron IV*, G. Ohira, T. Kusakawa and E. Niyama eds., Tokyo, Mat. Res. Soc. Proc., Pittsburgh, Pa. p.27

Banerjee D. and D.M. Stefanescu, 1991, *AFS Trans.* **99**:747

Birchenall C.E.and Mead H.W, 1956, *J. of Metals* 1104

Catalina A.V., Sen S., and Curreri P.A., 2001, in: *The Science of Casting and Solidification*, D. M. Stefanescu, R. Ruxanda, M. Tierean, and C. Serban eds., Editura Lux Libris, Brasov p.44

Catalina A.V., Sen S. and Stefanescu D.M., 2003, *Metall. Mat. Trans.* **34A**:383-394

Chang S., Shangguan D. and Stefanescu D.M., 1992, *Metall. Trans.* **23A**:1333

Charbonier J., 1984, *Trans. AFS* **92**:907

Croker M.N., Fidler R.S. and Smith R.W., 1973, *Proc. Roy. Soc. London,* **A335**:15

Crossley P.B. and L.F. Mondolfo, 1966, *Modern Casting* **49**:53

Degand C., D.M. Stefanescu and G. Laslaz, 1996, in: *Solidification Science and Processing*, I. Ohnaka and D.M. Stefanescu eds., TMS, Warrendale Pa. p:55

Double D.D. and A. Hellawell, 1995, *Acta metall. mater.* **43**(6):2435

Dresselhaus M. S., G. Dresselhaus, K. Surihara, I. L. Spain and H. A. Goldberg, 1988, in: *Graphite Fibers and Filaments*, Springer, Berlin
Eliade M., 1978, *The Forge and the Crucible*, The University of Chicago Press
Elliott R., 1983, *Eutectic Solidification Processing*, Butterworth, London
Elliott R., 1988, *Cast Iron Technology*, Butterworth, London
Elliot R. and Glenister S.M.D., 1980, *Acta Metallurgica* **28**:1489-94
Fang K., Wang G.C, Wang X.J., Huang L. and Deng G.F., 2006, *Proc. Eighth Intern. Symp. on Sci. and Processing of Cast Iron*, Li Y.X, Shen H.F, Xu Q.Y and Han Z.Q. eds., Tsinghua Univ. Press, Beijing, China, p.181
Fisher D.J. and Kurz W., 1980, *Acta Metallurgica* **28**:777-94
Fras E. and H. F. Lopez, 1993, *Trans. AFS* **101**:355
Fredriksson H., Stjerndahl J., and Tinoco J., 2005, *Mat. Sci. Eng A* **413-414**:363-372
Granger D. and R. Elliott, 1988, in: *Metals Handbook Ninth Edition, Vol.15, Casting*, D.M. Stefanescu ed., ASM International, Metals Park, Ohio p.159
Hellawell A., 1973, *Progr. Mater. Sci.* **15**:1
Herfurth K., 1965, *Freiberg Forschungs* **105**:267
Hogan L.M. and Song H., 1987, *Metall. Trans. A* **18A**:707-13
Hunt, J.D. and Jackson K.A., 1966, *Trans. AIME* **236**:843
Jackson K.A. and Hunt J.D., 1966, *Trans. Met. Soc. AIME* **236**:1129
Jaszaczak, J.A., 1994, in: *Mesomolecules: Molecules to Materials*, D. Mendenhall, J. Liebman and A. Greenberg eds., Chapman and Hall p.161
Johnson W. C. and H. B. Smartt, 1979, in: *Solidification and Casting of Metals*, The Metal Society, Book No. 192 p.129
Jones H. and Kurz W., 1980, *Metall. Trans.* **11A**:1265
Jones H. and Kurz W., 1981, *Z. Metallkde.***72**:792-97
Journal of the American Foundrymen's Assoc., 1896, **1**
Karma, A. and W-J. Rappel, 1998, *Physical Revue E* **57**(4):4323
Kurz W. and Fisher D.J., 1998, *Fundamentals of Solidification*, 4th ed., Trans Tech Publications, Switzerland
Lacaze J., M. Castro, C. Selig and G. Lesoult, 1991, in: *Modeling of Casting, Welding and Advanced Solidification Processes - V*, M. Rappaz et al. eds., TMS, Warrendale Pa. p.473
Lakeland K.D., 1964, BCIRA J. **12**:634
Li Y.X., B.C. Liu and C.R. Loper Jr., 1990, *AFS Trans.* **98**:483-488
Liu Q. and Q. Liu, 1993, *AFS Transactions* **101**:101
Magnin P. and Kurz W., 1985, in: *The Physical Metallurgy of Cast Iron*, H. Fredriksson and M. Hillert, eds., North Holland, New York p.263
Magnin P. and Kurz W., 1987, *Acta Metall.* **35**:1119
Magnin P. and Kurz W., 1988a, *Metall. Trans. A* **19A**:1955-63
Magnin P. and Kurz W., 1988, in: *Metal Handbook vol. 15 Casting*, D.M. Stefanescu ed., ASM International, Metals Park, OH p.119
Magnin P., Mason J.T., and Trivedi R., 1991, *Acta metall. mater.* **39**(4):469-80
Magnin P. and Trivedi R., 1991, *Acta metal.mater.* **39**:453
Magnin, P., Mason, J.T., Trivedi, R., 1991, *Acta metal.mater.* **39**:469
Mampey F., 2001, in: *Proceedings of Cast Iron Division*, AFS 105th Casting Congress, American Foundry Soc., Des Plaines IL p.51
McLean, M., 1983, *Directionally Solidified Materials for High Temperature Service*, The Metals Soc., London
Minkoff I., 1983, *The Physical Metallurgy of Cast Iron*, John Wiley & Sons
Mullins, W.W., Sekerka, R.F., 1964, *J. Appl. Phys.* **35**:444
McLean, M., 1983, *Directionally Solidified Materials for High Temperature Service*, The Metals Soc., London

Nieswaag H. and A.J. Zuithoff, 1975, in: *The Metallurgy of Cast Iron*, B. Lux, I. Minkoff and F. Mollard, Eds., Georgi Publishing, Switzerland p.327

Rickert A. and S. Engler; 1985, in: *The Physical Metallurgy of Cast Iron*, H. Fredriksson and M. Hillert, Ed., Proceedings of the Materials Research Society, North Holland **34**:165

Rivera, G. Boeri, and R. Sikora, J., 1997, *Advanced Materials Research* **4-5**:169

Ruxanda R., L. Beltran-Sanchez, J. Massone, and D.M. Stefanescu, 2001, in: *Proceedings of Cast Iron Division*, AFS 105th Casting Congress, American Foundry Soc., Des Plaines IL p.37

Sadocha J.P. and J.E. Gruzleski, 1975, in *The Metallurgy of Cast Iron*, B. Lux, I. Minkoff and F. Mollard, Eds., Georgi Publishing, Switzerland p.443

Sato T. and Sayama Y., 1974, *J. Crystal Growth*, **22**:259-71

Seetharaman V. and Trivedi R., 1988, *Metall. Trans.* **19A**:2955

Sigworth G.K., 1983, *Trans. AFS* **91**:7

Skaland T., F. Grong and T. Grong, 1993, *Metall. Trans.* **24A**:2321 and 2347

Stefanescu D.M, 1992, in: *ASM Handbook Vol. 15 Casting*, D.M. Stefanescu ed., ASM International, Metals Park, Ohio p.168

Stefanescu D. M, P.A. Curreri and M.R. Fiske, 1986, *Metall. Trans.* **17A**:1121

Svensson I.L. and Wessen M., 1998, in *Modeling of Casting, Welding and Advanced Solidification Processes-VII*, B.G. Thomas and C. Beckermann eds., TMS, Warrendale Pa. p.443

Tian H. and D.M. Stefanescu, 1992, *Metall. Trans.* **23A**:681

Tian H. and D.M. Stefanescu, 1993, in: *Modeling of Casting, Welding and Advanced Solidification Processes-VI*, T. S. Piwonka, V. Voller and L. Katgerman editors, TMS, Warrendale Pa. p.639

Tiller W.A., 1958, in *Liquid Metals and Solidification*, ASM, Metals Park, OH p276

Tiller W.A., 1969, in: *Solidification*, American Society for Metals p.84

Toloui B. and Hellawell A., 1976, *Acta Metallurgica* **24**:565-73

Wetterfall S.E., Fredriksson H. and Hillert M., 1972, *J. Iron and Steel Inst.* p.323

Zener C., 1946, *Trans. AIME* **167**:550

Zhu P., R. Sha, and Y. Li, 1985, in: *The Physical Metallurgy of Cast Iron*, H. Fredriksson and M. Hillert eds., Proceedings of the Materials Research Society, North Holland, **34** p.3

10

PERITECTIC SOLIDIFICATION

Peritectic solidification is very common in the solidification of metallic, ceramic and organic materials. Typical examples of systems with peritectic solidification include Fe-C (steel), Fe-Ni, Cu-Sn (bronze), Cu-Zn (brass), Al-Ti, lanthanide magnets (Nd—Fe-B) (Umeda et al., 1996), and ceramic superconductors (Y-Ba-Cu-O) (Izumi and Shiohara, 2005). Some of these materials have exotic applications. For example, the naphthalene-capric acid system is a potential latent heat storage material (Jin and Xiao, 2004).

10.1 Classification of peritectics

Three different types of peritectic systems can be considered (Figure 10.1). In the first type (left), the β-solidus and the β-solvus lines have slopes of the same sign. At the peritectic invariant α +L → α + β. In the second type (middle), the β-solidus and the β-solvus have opposite sign slopes. At the peritectic invariant α +L → β solid solution. The third type (right), has a very narrow or no solubility region. At the peritectic invariant α +L → β compound. The first two diagrams are typical for metallic alloys such as steel, while the third one is typical for rare earth permanent magnet alloys (Nd-Fe-B) and ceramic superconductors.

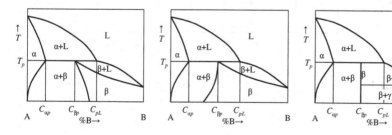

Figure 10.1. Types of peritectics.

Depending on the G/V ratio, a planar or dendritic interface can be obtained. If the G/V ratio is sufficiently high, an $\alpha + \beta$ composite of uniform composition will solidify with planar interface even for off-peritectic compositions.

The volume fraction of each phase will be given by the lever rule if the alloy solidifies under equilibrium conditions. A simple Scheil-model has also been proposed (Flemings, 1974). An example of such a calculation is given in Application 10.1. Nevertheless, since kinetics and the diffusion rate in the solid phases are determining the time for reaching equilibrium, in most cases, neither the lever rule nor the Scheil model will give the correct volume fraction of the different phases.

Control of the α to β transition is essential for the quality of the product in steel as well as in many other alloys. The phase diagram of the region of interest for steel is shown in Figure 10.2. At temperatures less than 1498°C the solidification microstructure of low-carbon steel is single phase austenite. Yet, depending on the carbon content the solidification path can be quite different. At carbon contents less than 0.16%, at the end of peritectic solidification both δ ferrite and γ phase coexist, while over 0.16% C liquid and γ coexist. At carbon contents higher than 0.53% only austenite solidifies from the liquid. Also shown on the figure are the metastable extensions of the γ phase that could form at any composition directly from the liquid if nucleation of the δ phase is suppressed. Such suppression of the pro-peritectic phase has been demonstrated in a number of peritectic binary melts, including Fe–Mo, Co–Si and Al–Co alloys (Löser *et al.* 2004). An extensive review of peritectic solidification and its mechanism has been done by Stefanescu (2006).

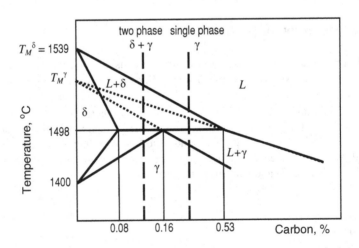

Figure 10.2. Schematic phase diagram of the peritectic region of carbon steel.

10.2 Peritectic microstructures and phase selection

Direct evidence of the mechanisms of peritectic solidification was provided recently through *in situ* dynamic observation of the progress of peritectic reactions and transformations of Fe-C alloys made with a combination of a confocal scanning laser microscope and an infrared image furnace (Shibata *et al.*, 2000). Selected micrographs are presented in Figure 10.3. It is observed that as the thermal gradient decreases from 22 K/mm (left column) to 4.3 K/mm (middle column) the δ solid/liquid interface becomes unstable and changes from planar to cellular. Further increase in solidification velocity from 2.5 to 19.3 μm/s showed that γ phase starts growing at the boundaries of the δ cells. Upon further increase of velocity to 38.7 μm/s, island like δ crystals appeared (Figure 10.3, right column, a) which then underwent peritectic reaction and transformation (Figure 10.3, right column, b). The wrinkles observed on the γ crystals that transformed form the δ crystals are thought to be due to the volume contraction of the transformation.

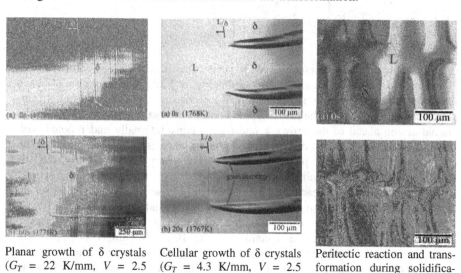

Planar growth of δ crystals (G_T = 22 K/mm, V = 2.5 μm/s)

Cellular growth of δ crystals (G_T = 4.3 K/mm, V = 2.5 μm/s)

Peritectic reaction and transformation during solidification at 1768 K (G_T = 4.3 K/mm, V = 38.7 μm/s)

Figure 10.3. Typical microstructures of peritectics in Fe-0.14%C alloys (Shibata *et al.*, 2000). With kind permission of Springer Science and Business Media.

Similar observations were also made for the Fe-Ni system (McDonald and Sridhar, 2003). The two stages of the peritectic transition involving the reaction (austenite growing along the liquid-ferrite interface) and the transformation (direct solidification of austenite from the liquid) were observed.

In general, a variety of microstructures can result from peritectic solidification, mostly depending on the temperature gradient/solidification velocity (G_T/V) ratio and nucleation conditions. The possible structures include cellular, plane-front, bands, eutectic-like structures.

Simultaneous growth of two phases as oriented fibers and lamellae has been observed in some peritectic alloys when the composition was on the tie-line of the two solid phases and the G_T/V ratio was close to the limit of constitutional undercooling for the stable phase having the smaller distribution coefficient (Vandyoussefi et al., 2000). An example of such a structure for a Fe-Ni alloy is presented in Figure 10.4.

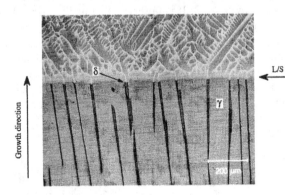

Figure 10.4. Quenched S/L interface of simultaneous two-phase growth in peritectic Fe–Ni alloy (Vandyoussefi et al., 2000). Reprinted with permission from Elsevier.

Banded structures have been observed in peritectic alloys at low growth rates (Tokieda et al. 1999). An example is provided in Figure 10.5a. The formation of bands is explained by nucleation and growth of the second phase during the initial transient of planar growth of the primary phase and vice versa. This occurs because the liquid at and ahead of the growing interface is constitutionally undercooled with respect to the other phase. As the second phase nucleates and grows ahead of the primary phase, the former phase cannot reach the steady state. Similarly, the primary phase nucleates again during the transient growth regime of the second phase, preventing it for reaching the steady state. Consequently, a cycle is set up leading to the layered microstructure (Trivedi 1995, Yasuda et al. 2003).

Several other structures can be obtained depending on the relative importance of nucleation diffusion and convection (Figure 10.5) (Boettinger et al., 2000). Theoretical models and experimental studies in thin samples suggest that the structures (a)-(e) can form under diffusive regime, while microstructure (f) requires the presence of oscillatory convection in the melt.

In an attempt to rationalize this plethora of peritectic microstructures, prediction of phase and microstructure selection was attempted by generating microstructure selection diagrams for peritectics. The main variables controlling microstructure evolution include interface velocity (V), thermal gradient (G_T), alloy composition (C_o) and nucleation potential.

Assuming that the leading phase that is the phase that growth at the highest interface temperature is the kinetically most stable one, Umeda et al. (1996) developed an equation that describes the transition velocity from δ dendrites to γ dendrites in directionally solidified alloys:

$$V_{tr}^{\delta-\gamma} = \frac{D_L}{4\pi^2}\left(\frac{\Delta T_m^{\delta-\gamma} - C_o \Delta m_L^{\delta-\gamma}}{\left(C_o m_L^{\delta}(k_{\delta}-1)\Gamma_{\delta}\right)^{1/2} - \left(C_o m_L^{\gamma}(k_{\gamma}-1)\Gamma_{\gamma}\right)^{1/2}}\right) \tag{10.1}$$

where D_L is the liquid diffusivity, $\Delta T_m^{\delta-\gamma}$ and $m_L^{\delta-\gamma}$ are the melting point difference and the liquidus slope difference between δ and γ, respectively, and k_i and Γ_i are the partition coefficient and the Gibbs-Thomson coefficient of the γ or δ phase. The numerator in the parenthesis represents the difference in liquidus temperature between the two phases (effect of phase equilibria on V_{tr}), and the denominator represents the difference in growth kinetics between the dendrites of the two phases. This equation was plotted in Figure 10.6b as a function of composition for Fe-Ni alloys whose phase diagram is presented in Figure 10.6a. It is noticed that for a given composition of 4.2at% Ni, as the interface velocity increases, a transition from δ to γ dendrites occurs at about $8 \cdot 10^{-2}$m/s.

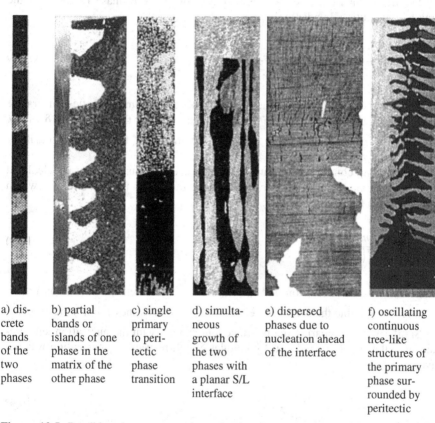

a) discrete bands of the two phases

b) partial bands or islands of one phase in the matrix of the other phase

c) single primary to peritectic phase transition

d) simultaneous growth of the two phases with a planar S/L interface

e) dispersed phases due to nucleation ahead of the interface

f) oscillating continuous tree-like structures of the primary phase surrounded by peritectic

Figure 10.5. Possible microstructures in peritectic alloys with compositions in the two-phase region. Solidification direction –upward (Boettinger *et al.*, 2000). Reprinted with permission from Elsevier.

This model cannot explain band formation, which is apparently the result of nucleation and growth of the second phase during the initial transient of planar solidification of the primary phase and vice-versa. According to Trivedi (1995), the liquid ahead of the growing interface is constitutionally undercooled with respect to the other phase. As the second phase nucleates and grows ahead of the primary phase, the former phase cannot reach steady state. Then, the primary phase nucleates ahead of the growing second phase preventing it from reaching steady state. Thus, a cycle leading to banded microstructure is set up.

By combining the maximum growth temperature criterion with nucleation considerations, Hunziker *et al.* (1998) developed a microstructure selection diagram for peritectic alloys close to the limit of constitutional undercooling. The diagram, presented in Figure 10.6c for Fe-Ni alloys, assumes negligible nucleation undercooling for both δ and γ phases and allows prediction of planar front, cellular, dendritic, and band solidification. The transition lines on the G/V –%Ni graph are calculated with the equations presented in the following text and plotted on Figure 10.6c.

The transition from planar-to-cellular growth of the δ phase is given by the limit of constitutional undercooling:

$$\frac{G}{V} \geq \frac{-m_L^\delta (C_L - C_o) - m_S^\delta (C_o - C_\delta)}{D_L} \tag{10.2}$$

where C_L is the composition of the liquid at the equilibrium peritectic temperature and m_S^δ is the solidus slope of the δ phase. By substituting the superscript δ with γ, the equation can be adapted to describe the transition from planar-to-cellular growth of the γ phase.

The stability condition for planar δ with respect to γ nucleation was calculated by comparing the interface temperature with the nucleation temperature of γ, which resulted in the following equation:

$$C_o < C_\delta - \frac{m_L^\delta \Delta T_N^\gamma}{m_S^\delta (m_L^\delta - m_L^\gamma)} \tag{10.3}$$

where ΔT_N^γ is the nucleation undercooling. Note that when the nucleation undercooling is negligible the condition reduces to $C_o < C_\delta$, as shown in Figure 10.6c.

Similarly, by comparing the interface temperature with the nucleation temperature of γ, the stability condition for cellular δ with respect to γ nucleation was derived as:

$$\frac{G}{V} < \frac{m_L^\delta}{D_L}(C_o - C_L) + \frac{\Delta T_N^\gamma}{m_L^\delta - m_L^\gamma} \tag{10.4}$$

The stability condition for planar γ with respect to δ is given by an equation similar to (10.3):

$$C_o > C_\gamma - \frac{m_L^\gamma \Delta T_N^\delta}{m_S^\gamma (m_L^\gamma - m_L^\delta)} \tag{10.5}$$

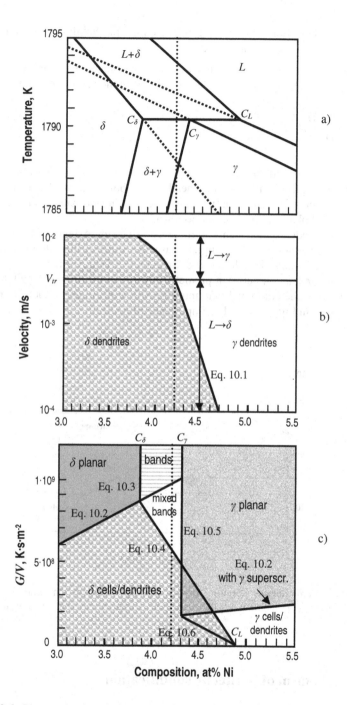

Figure 10.6. Phase selection during directional solidification of peritectic Fe-Ni alloys (Umeda *et al.* 1996, Hunziker *et al.* 1998). Reprinted with permission from Elsevier.

When the nucleation undercooling is negligible this stability condition reduces to $C_o > C_\gamma$, as shown in Figure 10.6c.

Additional stability conditions that will not be presented here are invoked to derive the stability condition between the two phases in the cellular regime:

$$\frac{G}{V} > \frac{\left(m_L^\delta - m_L^\gamma\right)\left(C_o - C_L\right) - \Delta T_N^\delta}{D_L \ln\left(m_L^\delta / m_L^\gamma\right)} \tag{10.6}$$

In the region marked "bands" on Figure 10.6, neither δ nor γ are stable at steady state, and either phase can nucleate ahead of the other's plane front. This is the condition for bands formation. In the region marked "mixed bands" alternate layers of cellular δ and planar γ are expected to form.

Nucleation undercooling can significantly affect the extent of the bands. Indeed, as the nucleation undercooling increases the stability lines defined by (10.3) and (10.4 move to the right, while that defined by (10.5) moves to the left.

Finally, a comparison between predictions with the maximum growth temperature criterion and the combined maximum growth temperature – nucleation model are summarized in Figure 10.7. The importance of nucleation is quite clear.

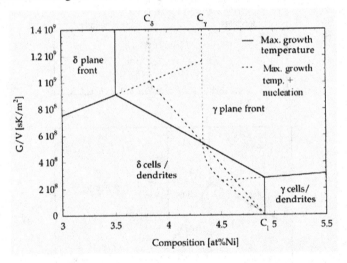

Figure 10.7. Microstructure selection maps using the maximum temperature prediction criterion alone or in combination with nucleation considerations (Hunziker *et al.*, 1998). Reprinted with permission from Elsevier.

10.3 Mechanism of peritectic solidification

Two different mechanisms are involved in peritectic solidification, namely *peritectic reaction* and *peritectic transformation* (Kerr *et al.*, 1974). These mechanisms are presented schematically in Figure 10.8. The peritectic solidification starts wit a *peritectic reaction* in which all three phases, α, β, and liquid are in contact with each other. The peritectic β phase will grow along the solid/liquid (S/L) interface α/L, driven by liquid super-saturation. Solute rejected by the β phase will diffuse through the liquid to the α phase contributing to its dissolution. The β phase will also thicken in the direction perpendicular to its growth, by direct growth in the liquid and at the expense of the α phase by solid state diffusion. Once the reaction is completed and all the α/L interface is covered by β, the *peritectic transformation* starts. The liquid and the primary α phase are isolated by the β phase. The transformation α ▯ β takes place by long-range solid-state diffusion through the peritectic β phase. The β phase grows by direct solidification in the liquid.

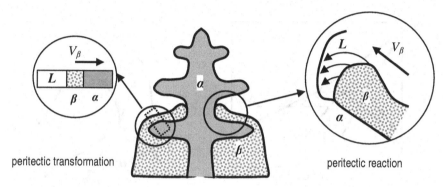

peritectic transformation   peritectic reaction

Figure 10.8. Mechanisms of peritectic solidification (Stefanescu, 2006).

10.3.1   The rate of the peritectic reaction

Depending on surface energy conditions, two different types of the peritectic reactions can occur (Fredriksson, 1988):

- Nucleation and growth of the *β* crystals in the liquid without contact with the *α* crystals. Following nucleation, the secondary phase grows freely in the liquid, while the primary phase will dissolve.
- Nucleation and growth of the *β* crystals in contact with the primary α phase. This is the most common. Following nucleation, lateral growth of the *β* phase around the *α* phase takes place (Figure 10.8).

The second type of peritectic reaction, which is the propagation of the triple point L/β/α along the L/α boundary of planar α crystals, consists of the dissolution of the α phase and growth of the β phase. It is controlled by the growth of β since dissolution is the fastest process. Bosze and Trivedi (1974) simplified an earlier equation developed by Trivedi to describe the relative contributions of diffusion, surface energy and interface kinetics during the growth of parabolic shape precipitates. Using their model it appears that the peritectic reaction is controlled by undercooling and liquid diffusivity according to the equations:

$$V_\gamma = \frac{9}{8\pi} \frac{D_L}{r} \frac{\Omega^2}{\left(1 - 2\Omega/\pi - \Omega^2/2\pi\right)^2} \quad \text{with} \quad \Omega = \frac{C_{L\beta} - C_{L\alpha}}{C_{L\beta} - C_{\beta L}} \qquad (10.7)$$

where r is the radius of the leading edge (the plate will have a thickness of $2r$), and C_{ij} are interface concentrations (see Figure 10.9 for definitions).

Using the maximum growth rate theory (Fredriksson and Nylen 1982, Hunziker et al. 1998) it can be shown that the thickness of β increases with lower solidification velocity and with larger surface energy difference $\Delta\gamma = \gamma^{L\beta} + \gamma^{\alpha\beta} - \gamma^{L\alpha}$ (see Figure 10.10).

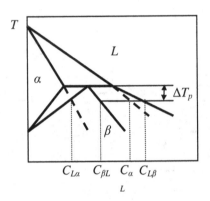

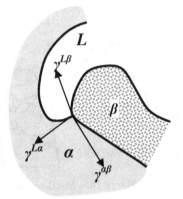

Figure 10.9. Definition of concentration terms in Eq. (10.7).

Figure 10.10. Surface energy of phases involved in the peritectic reaction.

However, *in situ* observation in Fe-C systems (Shibata *et al.*, 2000) showed that the experimental velocities were much higher than the ones predicted with this model. This suggests that the peritectic reaction is not controlled by carbon diffusivity in the liquid, but perhaps by either massive transformation of α into β, or direct solidification of β from the liquid. Support of theses hypotheses was brought recently by the experimental work of Dhindaw *et al.* (2004) who studied the peritectic reaction in medium–alloy steel (0.22mass%C, 1.3mass%Cr, 2.6mass%Ni). Microsegregation measurements on directionally and isothermally solidified samples showed that when the segregation ratio for Ni is higher than that fro Cr a peritectic reaction has occurred. However, when the segregation ratio for Cr was higher than for Ni, the liquid was transformed directly into β without undergoing a peritectic reaction. Based on the evaluation of the energy of transformation through differential thermal analysis the authors concluded that the transformation is a difusionless transformation α ⮀ β.

10.3.2 The rate of the peritectic transformation

The thickness of the β layer will normally increase during subsequent cooling through diffusion through the β layer, precipitation of β directly from the liquid, and precipitation of β directly from the α phase.

Both the precipitation of β directly from the liquid and the solid, and the diffusion process through the β layer, depend on the shape of the phase diagram and the cooling rate. In addition, the diffusion process through the β layer depends on the diffusion rate.

Assuming isothermal transformation, the growth of the β layer is controlled by the diffusion rate through the layer at a temperature just below the peritectic temperature. The notations on the phase diagram and the concentration profile are shown in Figure 10.11. Under these assumptions, the mass balance at the interface can be written as (Hillert, 1979):

$$D\left(\frac{dC}{dx}\right)_{x=0} = -V \cdot \Delta C \tag{10.8}$$

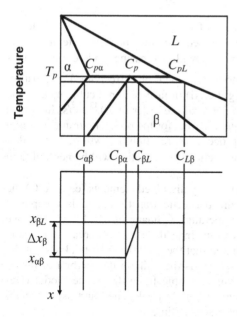

Figure 10.11. Definition of quantities in Eq. (10.9).

Then, for the $\alpha\beta$ interface and for the βL interface we have, respectively:

$$D_\beta \frac{\partial C_\beta}{\partial x}\bigg|_{x=x_{\alpha\beta}} = -\frac{dx_{\alpha\beta}}{dt}\left[C_{\alpha\beta} - C_{\beta\alpha}\right] \quad \text{and} \quad D_\beta \frac{\partial C_\beta}{\partial x}\bigg|_{x=x_{\beta L}} = -\frac{dx_{\beta L}}{dt}\left[C_{\beta L} - C_{L\beta}\right] \tag{10.9}$$

where D_β is the diffusion coefficient in the β phase. All other terms are concentrations that are defined in Figure 10.11. Assuming that the concentration gradient through the β phase is constant, it can be expressed as:

$$\frac{\partial C_\beta}{\partial x} = \frac{C_{\beta L} - C_{\beta \alpha}}{x_{\beta L} - x_{\beta \alpha}} = \frac{C_{\beta L} - C_{\beta \alpha}}{\Delta x}$$

Substituting in the above two equations and adding the equations one obtains:

$$\frac{d(\Delta x)}{dt} = \frac{D_\beta}{C_{\beta \alpha} - C_{\alpha \beta}} \frac{C_{\beta L} - C_{\beta \alpha}}{\Delta x} + \frac{D_\beta}{C_{\beta L} - C_{L\beta}} \frac{C_{\beta L} - C_{\beta \alpha}}{\Delta x} = \frac{D_\beta}{\Delta x} \left[\Omega_{\alpha \beta} + \Omega_{\beta L} \right]$$

where $\quad \Omega_{\alpha \beta} = \dfrac{C_{\beta L} - C_{\beta \alpha}}{C_{\beta \alpha} - C_{\alpha \beta}} \quad$ and $\quad \Omega_{\beta L} = \dfrac{C_{\beta L} - C_{\beta \alpha}}{C_{L\beta} - C_{\beta L}}$. Integrating:

$$\Delta x_\beta = \left[2 D_\beta \left(\Omega_{\alpha \beta} + \Omega_{\beta L} \right) t \right]^{1/2} \tag{10.10}$$

This equation shows that the thickness of the β layer and the growth rate increase with increasing undercooling. Indeed, at the peritectic temperature, $C_{\beta L} - C_{\beta \alpha} = 0$, but the difference increases with increasing undercooling. This equation also shows that the growth rate is dependent on diffusivity. For substitutionally dissolved alloying elements (e.g. Fe-Ni) in face-centered cubic metals, the diffusion coefficient near the melting point is of the order of 10^{-13} m²/s. In such a case, the growth rate will be very low and the time for the peritectic transformation will be very large. In a normal casting process, the reaction rate will be so low that the amount of β phase formed by the peritectic transformation will be negligible in comparison with the precipitation of β from the liquid.

For body-centered cubic metals interstitially dissolved elements (e.g., Fe-C) the diffusion rates are much higher and the peritectic transformation is completed within 6 to 10 K of the equilibrium temperature (Chuang et al., 1995). The rate controlling phenomenon is carbon diffusion. Indeed, in situ observation in Fe-C systems (Shibata et al., 2000) demonstrated that the growth of the thickness of the γ phase follows a parabolic law which supports the opinion that carbon diffusion determines growth rate. Calculations with a simple finite difference model (Ueshima et al. 1986) showed good agreement between calculated and experimental migration distances of the γ/δ and L/γ interfaces in time.

Fredriksson and Stjerndahl (1982) expanded Hillert's model to continuous cooling by assuming that the boundary conditions change during cooling and that the cooling rate is constant. From the isothermal equation they calculated that the thickness of the γ layer is given by:

$$\Delta x = ct.\cdot\left(\frac{dT}{dt}\right)^{-1}\cdot\Delta T_p \qquad\qquad (10.11)$$

This equation was used to plot the peritectic temperature range as a function of the carbon content (see Figure 10.12). It is seen that the reaction is relatively fast and is finished at maximum 6 or 10K below the peritectic temperature, depending on the cooling rate.

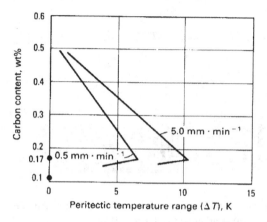

Figure 10.12. Temperature range of peritectic reaction in Fe-C alloys as a function of carbon content and solidification velocity at a temperature gradient of 60 K cm^{-1} (Fredriksson and Stjerndahl, 1982).

For multicomponent alloys it is necessary to use numerical models that depend on microsegregation models that describe the multiple solutal fluxes (Thuinet *et al.*, 2003).

10.3.3   Growth of banded (layered) peritectic structure

The formation of layered structures has been observed in several peritectic systems including Sn-Cd and Zn-Cu (Boettinger, 1974), Sn-Sb (Titchener and Spittle, 1975), Ag-Zn (Ostrowski and Langer, 1977) and Pb-Bi (Barker and Hellawell 1974, Tokieda *et al.* 1999). Layered structures have been observed in both hypo-peritectic and hyper-peritectic systems, but always at high G/V ratios. In principle the banded structure results when the second phase nucleates ahead of the planar primary phase. If the lateral growth of the secondary phase is higher than the normal growth of the primary phase, a planar band of secondary phase will form ahead of the planar primary phase. If lateral growth is slower than normal growth incomplete bands will result. Some typical microstructure for layered growth are given in Figure 10.5 a,b,f and in Figure 10.13.

To determine conditions that control the volume fraction and spacing of bands Trivedi (1995) has developed a model based on the following assumptions: solute transport by liquid diffusion only; no liquid convection; negligible diffusion in solid; growth conditions are such as to produce planar S/L interface. The widths of the α and β layers were derived to be:

$$\lambda_i = \frac{D_L}{V\,k_i}\ln\Lambda_i \qquad (10.12)$$

where $i = \alpha$ or β, the concentrations and slopes are defined as in Figure 10.14, and the functions Λ_i are given by:

$$\Lambda_\alpha = \left[1 - \frac{C_{\alpha p}}{C_o}\left(1 - \frac{\Delta T_N^\alpha}{C_p\left(m_L^\beta - m_L^\alpha\right)}\right)\right]\Bigg/\left[1 - \frac{C_{\alpha p}}{C_o}\left(1 + \frac{\Delta T_N^\beta}{C_p\left(m_L^\beta - m_L^\alpha\right)}\right)\right] \quad \text{and}$$

$$\Lambda_\beta = \left[1 - \frac{C_{\beta p}}{C_o}\left(1 - \frac{\Delta T_N^\beta}{C_p\left(m_L^\beta - m_L^\alpha\right)}\right)\right]\Bigg/\left[1 - \frac{C_{\beta p}}{C_o}\left(1 + \frac{\Delta T_N^\alpha}{C_p\left(m_L^\beta - m_L^\alpha\right)}\right)\right] \qquad (10.13)$$

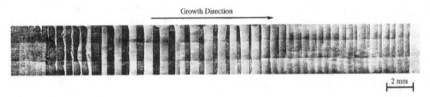

Growth Direction

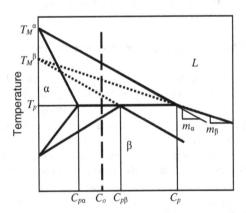

Figure 10.13. Banded structure in a Pb-33 at.%Bi alloy grown at $G = 2.7 \cdot 10^4$ K m^{-1} and $V = 0.56$ μm s^{-1} (black – α phase, white – β phase) (Barker and Hellawell, 1974).

Figure 10.14. Definition of quantities in Eq. (10.13).

The periodicity of the layers can then be written as:

$$\lambda = \lambda_\alpha + \lambda_\beta = \frac{D_L}{V}\left(\ln\Lambda_\alpha^{1/k_\alpha} + \ln\Lambda_\beta^{1/k_\beta}\right) \qquad (10.14)$$

From the analysis of the last two equations it is apparent that the thickness of the layers and their periodicity scales inversely with velocity for a given composition. It is also clear that the nucleation undercooling of the two phases plays a significant role in establishing the length scale of the layers.

However, in experiments by Tokieda *et al.* (1999) with hyperperitectic Pb-33at%Bi alloy, no dependence of the band spacing on velocity was observed. They explained the morphology of the banded structure based on the interface temperature. As the temperature gradient increases, the growth rate required for a planar interface also increases, decreasing macrosegregation in the growth direction. Therefore, not low growth rate but high temperature gradient is important for the formation of banded structures.

10.4 Applications

Application 10.1

Calculate the solute redistribution curve during the directional solidification of an *Ag*-10% *Sn* alloy. The phase diagram is given in Figure 10.15. Assume Scheil-type solidification.

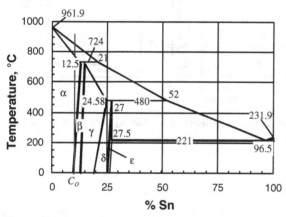

Figure 10.15. The Ag - Sn phase diagram.

Answer:
From the Scheil equation the evolution of solid fraction can be calculated as:

$$f_S = 1 - \left(C_S / kC_o\right)^{\frac{1}{k-1}}$$

For the α-phase the partition coefficient is: $k = 12.5/21 = 0.595$. Also $C_o = 10$ and $C_S = 12.5$. Then, the amount of α-phase will be: $f_\alpha = 0.84$.

For the γ-phase the partition coefficient is: $k = 24.58/52 = 0.47$. Also $C_o = 21$ and $C_S = 24.58$. Then, the amount of γ-phase will be:

$$f_\gamma = (1 - f_\alpha)f_S = (1 - 0.84)\left[1 - \left(\frac{24.58}{0.47 \cdot 21}\right)^{\frac{1}{0.47-1}}\right] = 0.13$$

For the ε-phase the partition coefficient is: $k = 27.5/96.5 = 0.28$. Also $C_o = 52$ and $C_S = 27.5$. Then, the amount of ε-phase will be:

$$f_\varepsilon = (1 - f_\alpha - f_\gamma)f_S = (1 - 0.84 - 0.13)\left[1 - \left(\frac{27.5}{0.28 \cdot 52}\right)^{\frac{1}{0.28-1}}\right] = 0.017$$

Finally, the amount of eutectic is simply $f_E = 1 - f_\alpha - f_\gamma - f_\varepsilon = 0.013$. The solid fraction evolution as a function of composition is given in Figure 10.16.

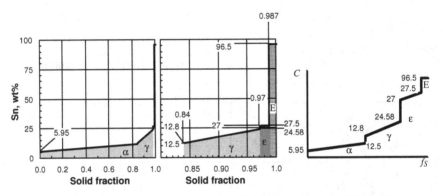

Figure 10.16. Compositional profile during the solidification of an Ag-10% Sn alloy assuming Scheil-type solidification.

References

Barker N.J.W. and Hellawell A., 1974, *Metal Sci.*, **8**:353
Boettinger W.J., 1974, *Metall. Trans.*, **5**:2023-31
Boettinger W.J., Coriell S.R., Greer A.L., Karma A., Kurz W., Rappaz M. and Trivedi R., 2000, *Acta mater.* **48**:43-70
Bosze W.P. and Trivedi R., 1974, *Metall. Trans*, **5**:511-12
Chuang Y.K., Reinisch D. and Schwendtfeger K., 1975, *Metall. Trans. A*, **6A**:235-38
Dhindaw B.K., Antonsson T., Tinoco J. and Fredriksson H., 2004, *Metall. Mater. Trans. A*, **35A**:2869-287
Flemings M.C., 1974, *Solidification Processing*, McGraw-Hill, New York
Fredriksson H., 1988, in: *Metals Handbook Ninth Edition, vol 15 Casting*, D.M. Stefanescu ed., ASM International, Ohio, p.125
Fredriksson H. and Nylen T., 1982, *Met. Sci.*, **16**:283-94
Fredriksson H. and Stjerndahl J., 1982, *Met. Sci.*, **16**:575-85
Hillert M., 1979, in *Solidification and Casting of Metals*, The Metals Society, p 81-87

Hunziker O., Vandyoussefi M., and Kurz W., 1998, *Acta Mater.*, **46**:6325-633

Izumi T. and Shiohara Y., 2005, *J. of Physics and Chemistry of Solids* **66**:535-545

Jin L. and Xiao F., 2004, *Thermochimica Acta*, **424**:1-5

Kerr H.W., Cissé J., and Bolling G.F., 1974, *Acta Metall.* **22**:667

Löser W., Leonhardt M., Lindenkreuz H.G. and Arnold B., 2004, *Mat. Sci. Eng. A*, **375-377**: 534-539

McDonald N.J. and Sridhar S., 2003, *Metall. and Mat. Trans. A*, **34A**:1931-1940

Ostrowski A. and Langer E.W., 1977, in *Int. Conf. on Solidification and Casting*, Sheffield, UK, Inst. of Metals, London, **1**:139-4

Shibata H., Arai Y. and Emi T., 2000, *Metall. and Mat. Trans. B*, **31B**:981-991

Stefanescu D.M., 2006, *ISIJ International*, **46**:786-794

Thuinet L., Lesoult G., and Combeau H., 2003, in *Modeling of Casting, Welding and Advanced Solidification Processes X*, D.M. Stefanescu et al. eds., TMS, p 237-244

Titchener A.P. and Spittle J.A., 1975, *Acta Metall.*, **23**:497-502

Tokieda K., Yasuda H. and Ohnaka I., 1999, *Mat. Sci. Eng.*, **A262**:238-45

Trivedi R., 1995, *Metall. Trans. A*, **26A**:1583

Ueshima Y., Mizoguchi S., Matsumiya T. and Kajioka H., 1986, *Metall. Trans. B*, **17B**:845-59

Umeda T., Okane T. and Kurz W., 1996, *Acta Mater.* **44**:4209-4216

Vandyoussefi M., Kerr H.W., and Kurz W., 2000, *Acta Mater.* **48**:2297-2306

Yasuda H., Ohnaka I., Tokieda K. and Notake N., 2003, in *Solidification and Casting*, B. Cantor and K. O'Reilly eds., Inst. of Physics Publishing, Bristol, p 160-74

11

MONOTECTIC SOLIDIFICATION

Monotectic solidification occurs in alloys where the liquid separates into two distinct liquid phases of different composition during cooling. While these alloys have limited commercial applications, one could mention Pb containing copper alloys, and the attempt to fabricate thin microfilters, as possible applications. In this latter case, Grugel and Hellawell (1981) have pursued directional solidification of monotectic alloys followed by the selective etching of the fibers.

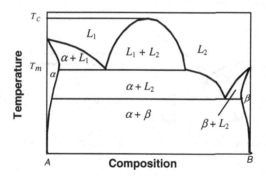

Figure 11.1. Phase diagram showing a monotectic reaction at temperature T_m.

A dome-shaped region within which the two liquids mix and coexist is seen on the phase diagram (Figure 11.1). The maximum temperature of this dome, T_c, is called the critical (or consolute) temperature. At the monotectic temperature, T_m, the monotectic reaction occurs:

$$L_1 = \alpha + L_2$$

11.1 Classification of monotectics

Monotectic alloys can be classified based on the difference $T_c - T_m$ as follows:

- high dome alloys (high $T_c - T_m$); such alloys include Al-In (206 °C) and Al-Bi (600 °C)
- low dome alloys (low $T_c - T_m$); typical examples include Cu-Pb (35 °C) and Cd-Ga (13 °C) alloys.

They can also be classified based on the T_m/T_c ratio.

11.2 Mechanism of monotectic solidification

The morphology of the microstructure produced during directional solidification is a function of the density difference between the two liquids, $\rho_{L_1} - \rho_{L_2}$, and of the wetting between L_2 and α. The role of contact angle is explained in Figure 11.2. It is seen that when L_1 wets the α phase the contact angle is 180°. The relationship between the three interface energies is governed by Young's equation:

$$\gamma_{\alpha L_1} = \gamma_{\alpha L_2} + \gamma_{L_1 L_2} \cos\theta \tag{11.1}$$

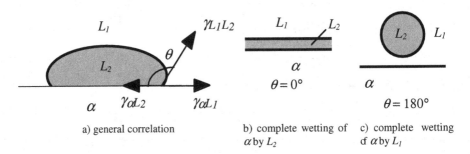

a) general correlation   b) complete wetting of   c) complete wetting
                              α by L_2 d α by L_1

Figure 11.2. Possible scenarios for interaction between L_1, L_2, and α.

Assuming negligible density difference between the two liquid phases, microstructure evolution is controlled by the interaction between L_1 and α (Cahn 1979, Elliot 1983). At the consolute temperature, T_c, there is total mixing between the two liquids. Thus, since there is no free surface, $\gamma_{L_1 L_2} = 0$. This is shown on Figure 11.3, assuming that $\gamma_{L_1 L_2}$ decreases with temperature.

From Young's equation, for $\gamma_{L_1 L_2} = 0$, it follows that at T_c we have $\gamma_{\alpha L_1} = \gamma_{\alpha L_2}$. This is also shown in Figure 11.3, assuming again some variation of the two surface energies with temperature.

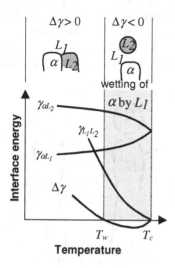

Figure 11.3. Correlation between surface energies at various temperatures (up) and profile of monotectic growth front (down) (after Cahn, 1979). With kind permission of Springer Science and Business Media.

From the examination of the phase diagram it is concluded that, during the monotectic reaction, the interface energies are correlated by $\gamma_{\alpha L_1} + \gamma_{L_1 L_2} \rightarrow \gamma_{\alpha L_2}$. Thus, the difference in free energy during the reaction is:

$$\Delta \gamma = \left(\gamma_{\alpha L_1} + \gamma_{L_1 L_2} \right) - \gamma_{\alpha L_2} \tag{11.2}$$

$\Delta \gamma$ can be zero at two temperatures as shown in Figure 11.3. The first case is when the temperature is T_c, where $\gamma_{\alpha L_1} = \gamma_{\alpha L_2}$ and $\gamma_{L_1 L_2} = 0$. It can also be zero when complete wetting of α by L_1 occurs. Then the contact angle in Figure 11.2 is 180°, and Young's equation gives $\gamma_{\alpha L_1} + \gamma_{L_1 L_2} = \gamma_{\alpha L_2}$. This temperature is the highest temperature above which wetting can occur and is defined as the wetting temperature, T_w.

Based on experimental work it was assumed (Cahn, 1979) that for low-dome alloys the monotectic temperature is higher than the wetting temperature, $T_w < T_m < T_c$. Consequently, at T_m, $\Delta \gamma < 0$ and thus:

$$\gamma_{\alpha L_2} > \gamma_{\alpha L_1} + \gamma_{L_1 L_2}$$

The system will choose the configuration with the lowest free energy, *i.e.*, where $\gamma_{\alpha L_1}$ and $\gamma_{L_1 L_2}$ exist, that is when αL_1 and $L_1 L_2$ interfaces exist. The phases α and L_2 are separated by L_1 as shown on Figure 11.3. At low growth rate, L_2 particles are pushed by the S/L interface (Figure 11.4a). If the solidification velocity increases above a critical velocity, V_{cr}, L_2 is incorporated with formation of an irregular fibrous composite (Figure 11.4b). A detailed discussion on particle engulfment and pushing by solidifying interfaces is given in section 13.1. This mechanism has been observed experimentally in a number of transparent organic alloys (*e.g.*, succinonitrile – 20% ethanol by Grugel *et al.* 1984, Figure 11.5) as well as metallic alloys. For a Bi-50% Ga alloy, Dhindaw *et al.* (1988) report that even at very high solidification velocities only a short fibrous structure was obtained (Figure 11.6). The same authors found that, for a hyper-monotectic Cu-Pb alloy, the critical solidification velocity above which the lead droplets are engulfed decreases as the size of the droplets increases.

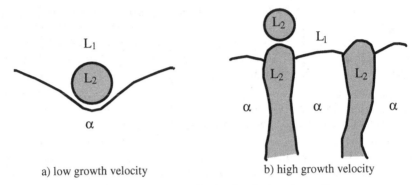

a) low growth velocity   b) high growth velocity

Figure 11.4. Monotectic solidification for low dome alloys.

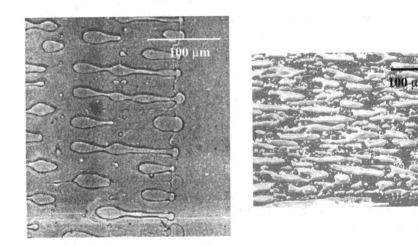

Figure 11.5. Growth front in succinonitrile – 20% ethanol, showing incorporation of ethanol droplets. $V = 0.27\mu m\ s^{-1}$, $G = 4.8K$ mm^{-1} (Grugel *et al.* 1984). With kind permission of Springer Science and Business Media.

Figure 11.6. Microstructure of a Cu-70% Pb alloy solidified at $V = 778\ \mu m\ s^{-1}$ and $G = 12$ K mm^{-1}. Solidification direction right-to-left (Dhindaw *et al.*, 1988). With kind permission of Springer Science and Business Media.

The range of existence of the fibrous composite is limited by the constitutional undercooling on one side and by the critical velocity of pushing-to-engulfment transition on the other (Figure 11.7). When the solidification velocity is smaller than V_{cr} a banded structure may result. An example of such a structure is provided in Figure 11.8. It is suggested the L_2 phase, which precipitates at the solid/liquid interface, piles up and covers the S/L interface. This produces a Pb-rich layer and increases the undercooling of the L_1/L_2 interface with respect to the monotectic temperature. Then, nucleation of the α-Cu phase occurs on the Pb-rich layer. The temperature at the growth front is also returned to the monotectic temperature. The repetition of this process will result in the banded structure.

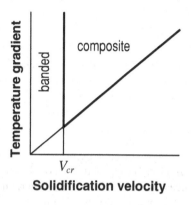

Figure 11.7. Restriction on composite growth of low-dome alloys imposed by the critical velocity for the pushing-engulfment transition and by constitutional undercooling.

Figure 11.8. Microstructure of upward DS of a Cu-37.7 wt% Pb alloy in longitudinal section. V = 4.4 µm/s (Aoi *et al.*, 2001). Reprinted with permission from Elsevier.

For high dome alloys it is assumed that the monotectic temperature is below the wetting temperature, $T_m < T_w < T_c$ and $\Delta\gamma > 0$. Thus, $\gamma_{\alpha L_2} < \gamma_{\alpha L_1} + \gamma_{L_1 L_2}$. In this case, the lowest energy exist when an α/L_2 interface exists. Consequently, α and L_2 will grow together (L_2 wets α) resulting in a regular (uniform) fibrous composite (Figure 11.9). An assessment of the validity of the theory can be made by studying the experimental data in Table 11.1. It is seen that the high dome ($T_m/T_c < 0.9$) alloys exhibit a fibrous structure at all velocities, while the low dome alloys ($T_m/T_c > 0.9$) have irregular fibrous structures only at high velocities, as predicted by the theory.

Table 11.1. Dome size, interphase spacing, and structure in selected mono-tectic alloys (after Grugel and Hellawell, 1984)

System	$T_c - T_m$, °C	T_m/T_c, K/K	$\lambda^2 V$, m³ s⁻¹ 10⁻¹⁶	Structure
Cu-Pb	35	0.97	284	irregular fibers at high V
Cd-Ga	13	0.98	120	irregular fibers at high V
Al-In	206	0.75	4.5	regular fibers
Al-Bi	600	0.59	250	regular fibers
Sb-Sb$_2$S$_3$		~0.5	4.8	regular fibers

Further analysis of the data in Table 11.1 reveals that the $\lambda^2 \cdot V$ relationship is about two orders of magnitude larger for irregular than for regular composites (with the exception of the Al-Bi alloy), and about one order of magnitude higher for regular monotectic composites than for regular eutectics. Indeed, for example, a Sn-Pb eutectic has $\lambda^2 \cdot V = 0.25$. This difference comes from the controlling mechanism. For irregular fibrous eutectics, this is the pushing-engulfment transition, which is a function of solidification velocity and surface energy. For regular fibrous monotectics, the spacing is controlled by surface energy, while for eutectics by diffusion.

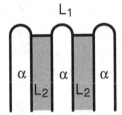

L_1

α α α

L_2 L_2

Figure 11.9. Monotectic solidification for high dome alloys.

A theoretical model based on Jackson-Hunt's analysis of eutectic growth has been proposed by Arikawa et al. (1994). Coriell *et al.* (1997) extended the model to treat immiscible alloys with large density differences between phases. However, they concluded that, for reasons that they could not explain, there is a significant discrepancy between theory and experiment.

Ratke (2003) pointed out that an additional mode of mass transfer must be considered: the thermocapillary effect causes Marangoni convection at the interface between the liquid L_2 phase and the molten L_1 matrix. The resulting flow affects solute transport and thus constitutional undercooling at the S/L interface. Following the Jackson-Hunt approach with ΔT_c modified by the thermocapillary convection, He derived analytical equations that extend the classic $V \cdot r^2$ = const. relationship by the term $V \cdot G_T \cdot r^2$. This implies that the inter-rod distance decreases with higher gradient. It should be noted that the dependence of lamellar spacing on the temperature gradient has been derived for eutectics from purely thermal considerations (see section 9.2.2).

References

Aoi I., Ishino M., Yoshida M., Fukunaga H., Nakae H., 2001, *J. Crystal Growth* **222**:806-815

Arikawa Y., Andrews J.B., S.R. Coriell, and Mitchell W.F., 1994, in: *Experimental Methods for Microgravity Mat. Science*, R.A. Schiffman and J.B. Andrews eds., TMS, p.137

Cahn, J.W., 1979, *Metall. Trans.* **10A**:119

Coriell S.R., Mitchell W.F., Murray B.T., Andrews J.B, and Arikawa Y., 1997, *J. Crystal Growth* 179:647-57

Dhindaw B.K., Stefanescu D.M., Singh A.K., and Curreri P.A., 1988, *Metall. Trans.* **19A**: 2839

Grugel R.N. and Hellawell A., 1981, *Metall. Trans.* **12A**:669

Grugel R.N., Lagrasso T.A., and Hellawell A., 1984, *Metall. Trans.* **15A**:1003

Ratke L., *Metall. Trans.* 34A:449-57

MICROSTRUCTURES OBTAINED THROUGH RAPID SOLIDIFICATION

As discussed in some detail in section 2.8, an increase in the rate of heat extraction (cooling rate) results in gradual departure from equilibrium up to global and inter-face non-equilibrium. As the rate of heat extraction increases the microstructure length scale of solidified alloys decreases. Eventually solidification without crys-tallization may occur.

Crystallization must occur via a process of nucleation and growth. If nucleation is suppressed (for example through the imposition of high undercooling on the system), the liquid will solidify without crystallization as an amorphous material also termed glass. Thus, condensed matter can be classified into three categories, as listed in Table 12.1. Liquids are stable above the fusion temperature, T_f, crystal-line solids are stable under the solidification temperature T_S, and glasses are stable at a temperature lower than the glass transition temperature, T_g.

Table 12.1. Classification of condensed matter

Condensed matter	Temperature range	Thermal condition	Atomic configu-ration
Liquid	$T > T_f$	stable	disordered
Solid - Crystal	$T < T_S$	stable	ordered
Solid - Amorphous (glass)	$T < T_g$	metastable	disordered

The glass transition temperature is defined as the temperature at which the material exhibits a sudden change in the derivative thermodynamic properties, such as heat capacity and expansion coefficient, from crystal-like to liquid-like values. This is illustrated in Figure 12.1. When the cooling rate is low phase transformation occurs at T_f with formation of a crystalline solid. When the cooling rate is high, the prop-erty changes continuously and the extrapolation method (broken lines) is used to determine T_g.

Thus, depending on the solidification velocity and the degree of interface non-equilibrium, rapidly solidified materials may exhibit a crystalline or amorphous microstructure.

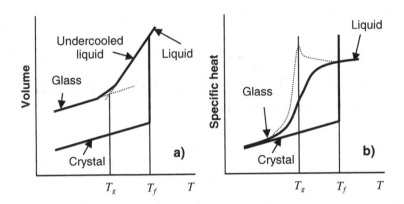

Figure 12.1. Schematic illustration of the change in volume (a) and specific heat (b) as the liquid is cooled. Solid lines are experimental results. Broken lines illustrate extrapolations for determining T_g (Yonezawa, 1991).

12.1 Rapidly solidified crystalline alloys

The term rapid solidification is normally applied to casting processes in which the liquid cooling rate exceeds 100K/s (Boettinger, 1974). This definition may be outdated in light of recent work on bulk metallic glasses. While different alloys respond differently to high cooling rates, some microstructures observed in rapidly solidified alloys can be achieved by slow cooling when large liquid undercooling is achieved prior to nucleation.

The following techniques are used to produce rapidly solidified alloys:

- melt spinning, planar flow casting, or melt extraction, which produce thin (~25 to 100μm) ribbon, tape, sheet, or fiber;
- atomization, which produces powder (~10 to 200μm);
- surface melting and re-solidification, which produce thin surface layers

Some examples of such techniques are presented in Figure 12.2. These methods may be considered casting techniques where at least one physical dimension of the final product is small. Consolidation is used to yield large products from rapidly solidified alloys. This consolidation often alters the solidification microstructure of the final products. Yet, many features of the solidification structure can remain in the final product.

At "normal' rates of cooling the tip radius of the dendrite decreases as the solidification velocity increases (Figure 12.3). However, as the cooling rate increases in the rapid solidification range the tip radius increases. This is accompanied by a decrease in branching. The equiaxed dendrite becomes globular/cellular. Typical examples of the evolution of the microstructure as a function of the solidification velocity are given in Figure 12.4. Figure 12.4a shows a transverse section of a fine

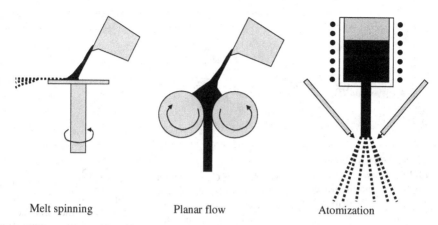

Melt spinning   Planar flow   Atomization

Figure 12.2. Examples of techniques for producing rapidly solidifying alloys.

cellular structure of the Ag-rich phase in Ag-15%Cu alloy. Most of the intercellular regions are filled with the Cu-rich phase, not the Ag-Cu eutectic as the Gulliver-Scheil equation would predict. Figure 12.4b shows a longitudinal view of a cellular solidification structure. In Figure 12.4c, the alloy has solidified with a planar interface to produce a microsegregation-free alloy. The fine particles are the result of a solid-state precipitation.

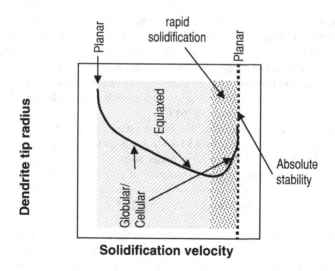

Figure 12.3. Schematic correlation between solidification velocity and dendrite tip radius.

A common occurrence in some rapidly solidified alloys is a change in the identity of the primary solidification phase from that observed for slow solidification. Many examples are found in hypereutectic Al alloys containing transition elements

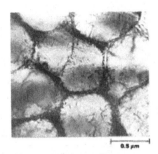

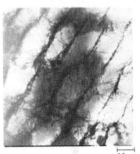

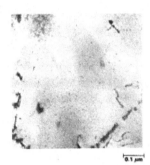

a) Ag-15%Cu alloy re-solidified at 0.025m/s. Globular microsegregation pattern. Magnification: 32000 X

b) Ag-15%Cu alloy re-solidified at 0.3m/s. Cellular microsegregation pattern revealed by dislocation networks along cell walls. Magnification: 18000 X.

c) Ag-15%Cu alloy re-solidified at approximately 0.6m/s. No cellular structure. The solid produced is uniform in composition except for fine Cu precipitates formed during solid-state cooling. Magnification: 87000 X.

Figure 12.4. Microstructures of Ag-Cu alloys electron beam melted and re-solidified at different velocities. Thin foil transmission electron micrograph prepared by ion milling (Boettinger *et al.*, 1984). With kind permission of Springer Science and Business Media.

such as Fe, Mn, or Cr. If the alloy is hypereutectic, slowly cooled castings will contain intermetallics such as Al_3Fe or Al_6Mn as the primary (or first) phase to solidify. However, under rapid solidification conditions the primary phase in these alloys is the Al solid solution, usually found in a cellular structure with an intermetallic in the intercellular regions. This transition from an intermetallic to an Al solid solution as the primary phase can be understood by a careful examination of the kinetics of the competitive nucleation and growth of the intermetallic and α-Al solid solution (Hughes and Jones, 1976). □

In some cases, an intermetallic that is not given on the equilibrium phase diagram may compete with α-Al. In Al-Fe alloys, a metastable phase, Al_6Fe, rather than the stable phase, Al_3Fe, can form under some rapid solidification conditions. This situation is analogous to the appearance of cementite rather than graphite in some cast irons. The use of metastable phase diagrams to assist in the interpretation of rapidly solidified microstructures is described by Perepezko and Boettinger (1983).

As explained in section 2.8, other rapidly solidified alloys have microsegregation-free structures formed by a liquid-solid transformation similar to a massive solid-solid transformation (partitionless or diffusionless transformation). The liquid transforms to solid without a change in composition. The ratio of the solid composition at the interface to the liquid composition is $C_S/C_L = 1$, rather than the equilibrium partition coefficient. Velocities required to produce partitionless solidification must exceed 5m/s. The phase diagrams do not apply in this situation.

Rapidly solidified alloy powders exhibit a broad spectrum of solidification structures, depending on alloy composition and solidification conditions. Figure 12.5 shows single powder particles of stainless steel with dendritic or cellular structure. The size of the particles is less than 25 μm. Different degrees of under-cooling prior to nucleation for particles of almost the same size determine the type of structure. The dendritic structure radiates from a point on the surface where nucleation has occurred. The scale of the structure is relatively uniform across the powder particle.

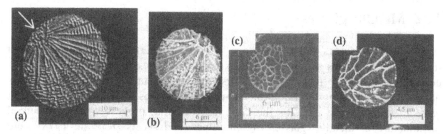

Figure 12.5. SEM micrographs of atomized droplets of martensitic stainless steel. (a) and (b) – dendritic structures; (c) and (d) – cellular structures (Pryds and Pedersen, 2002). With kind permission of Springer Science and Business Media.

Other rapidly solidified powders often show significant microstructural variations across individual powder particles. Initial growth of the solid may occur very rapidly in a partitionless manner. The interface velocity decreases as the S/L interface crosses the particle, because of the release of the latent heat of fusion and warming of the powder particle, and the solidification front becomes cellular.

The growth velocity of rapidly solidified dendrites can be calculated with Eqs. (8.19) and (8.21). For the eutectic solidification the simplifications used in the Jackson-Hunt model (low Péclet number, interface composition in the liquid close to the eutectic composition, constant partition coefficient) cannot be used anymore. The constant in the $\lambda^2 V = ct. = \mu_\lambda$ relationship becomes a function of the Péclet

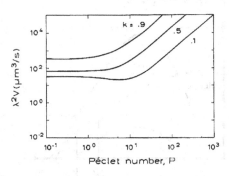

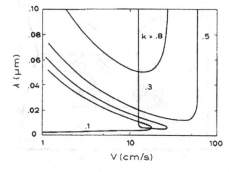

Figure 12.6. Eutectic growth at high Péclet numbers (Kurz and Fisher, 1998).

Figure 12.7. Eutectic spacing–growth velocity relationship (Kurz and Fisher, 1998).

number at $Pe > \sim 10$, as shown in Figure 12.6. The partition coefficient has a significant effect on both the growth parameter μ_λ and the velocity at which absolute stability is reached. Indeed, as seen in Figure 12.7, for relatively high k, λ decreases with V before increasing again upon approaching the limit of stability. However, for small k-values, the behavior is more complicated because of the increase in eutectic undercooling, which decreases diffusivity and bends the curve back before increasing it again.

12.2 Metallic glasses

While liquid polymers and silicates can be easily converted into amorphous solid forms at cooling rates as low as 1-10K/s, metallic glasses are relatively new products of the scientific ingenuity. The first reported metallic glass, $Au_{75}Si_{25}$, was obtained by quenching the liquid metal at very high rates of 10^5-10^6 K/s (Klement et al., 1960). By the late 70s the continuous casting processes for commercial manufacture of metallic glasses ribbons, lines, and sheets (Kavesh, 1978) were developed.

Turnbull and Fisher (1949) predicted that a ratio between the glass transition temperature, T_g, and the melting or liquidus temperature of an alloy, T_f, referred to as the reduced glass transition temperature $T_{rg} = T_g/T_f$, can be used as a criterion for determining the glass-forming ability (GFA) of an alloy. According to Turnbull's criterion (Turnbull, 1969), a liquid with $T_g/T_f = 2/3$ can only crystallize within a very narrow temperature range. Such liquids can thus be easily undercooled at relatively low cooling rates to solidify as glasses.

The concepts of eutectic coupled zone introduced in section 9.4 are also useful in understanding glass formation. For the symmetric coupled zone (regular) eutectics, glass will form preferentially to eutectic when the growth velocity is higher than the critical growth velocity for glass formation, V_a (Figure 12.8). For eutectic

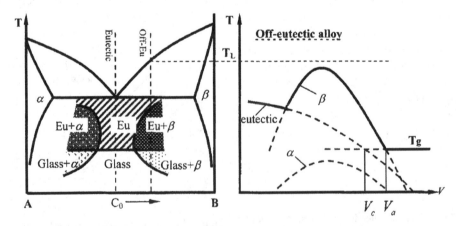

Figure 12.8. Schematic phase diagrams for symmetric coupled zone eutectics (Li, 2005).

composition fully amorphous phase will form, while for off-eutectic compositions glass + β dendrites campsites will form at velocities between V_a and V_c, where V_c is the critical growth velocity for composite formation. Experimental work seems to indicate that the complete glass region is rather narrow.

For the asymmetric coupled zone (irregular) eutectics, eutectic will be replaced with glass in a larger region once the growth velocity is above V_a'. (Figure 12.9). Thus it appears that the best glass forming alloy should be at an off-eutectic composition, on the faceted size of the eutectic.

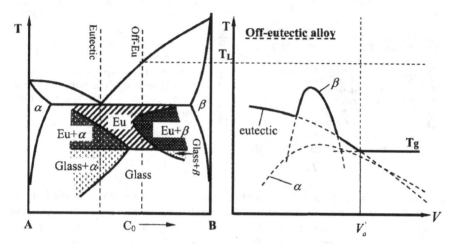

Figure 12.9. Schematic phase diagrams for asymmetric coupled zone eutectics (Li, 2005).

For all practical purposes, it is not necessary to completely eliminate nucleation in order to produce a glass. If the crystallized volume fraction is below the detection limit, the material is practically amorphous. The cooling rate necessary to achieve this depends on the nucleation and growth rates and their temperature dependence. In the absence of convection or some adiabatic cooling mechanism, the cooling rate is limited by thermal conduction in the liquid, which scales with the square of the smallest sample dimension (Peker and Johnson, 1993). As a result, a critical cooling rate for glass formation corresponds to a critical thickness for glass formation (see for example Figure 12.10). This critical thickness for glass formation can be used to quantify the concept of glass-forming ability (GFA). It is directly related to the critical cooling rate for glass formation, but easier to determine experimentally.

The Turnbull criterion for the suppression of crystallization in undercooled melts has played a key role in the development of various metallic glasses including the new materials termed bulk metallic glasses (BMGs). If, as shown on Figure 12.10, one arbitrarily defines the millimeter scale as "bulk" (Wang et al., 2004), the first bulk metallic glass was the ternary Pd–Cu–Si alloy prepared in the form of millimeter-diameter rods through a suction-casting method in 1974 (Chen, 1974). The cooling rate of 10^3K/s was significantly lower than reported before for the production of metallic glasses. Recently, amorphous Fe-based alloys with high carbon content termed by the researchers "structural amorphous steels" were cast in rods of up to 12mm diameter. A typical composition is for example $Fe_{44.3}Cr_5Co_5Mo_{12.8}Mn_{11.2}C_{15.8}B_{5.9}$ (Lu *et al.*, 2004). BMGs can now be produced at cooling rates as low as 0.1K/s which brings them in the range of castable materials.

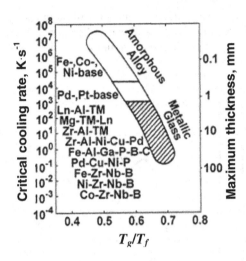

Figure 12.10. Relationship between the critical cooling rate for glass formation, maximum sample thickness for glass formation, and reduced glass transition temperature (T_g/T_f) for bulk amorphous alloys. The data of the ordinary amorphous alloys, which require high cooling rates for glass formation, are also shown for comparison (Inoue 2000).

According to Inoue (2000) there are three empirical rules for BMG formation: (1) multi-component systems consisting of more than three elements; (2) significant difference in atomic size ratios above ~12% among the three main constituent elements; and (3) negative heats of mixing among the three main constituent elements.

In general, the GFA in BMGs increases as more components are added to the alloy (the "confusion principle") (Greer, 1993). This is construed to mean that larger number of components in an alloy system destabilizes competing crystalline phases that may form during cooling. Thus the crystallization tendency of the melt increases. It is claimed that the alloys satisfying the three empirical rules have atomic configurations in the liquid state that are significantly different from those of the corresponding crystalline phases and that favor glass formation in terms of thermodynamics (free energy), kinetics (atom mobility) and microstructure evolution.

If the liquid-to-glass transformation is assumed to be the result of crystallization suppression within the supercooled liquid because of lack of nucleation, the steady state nucleation equation can be used to identify both the thermodynamic and the kinetic controlling parameters. The nucleation rate can be written as:

$$I = AD_{ef} \exp\left(-\frac{16\pi\gamma^3}{3kT\,\Delta G_{LS}^2}\right) \tag{12.1}$$

where A is a constant, k_B is the Boltzmann's constant, T is the absolute temperature, D_{ef} is the effective diffusivity, γ is the solid/liquid interface energy, and ΔG_{LS} is the energy difference between the liquid state and the crystalline state. The derivation of this equation is discussed in Section 15. ΔG_{LS} is the driving force for crystallization. Note that diffusivity can be related to viscosity through the Stokes-Einstein equation, $D = k_B T / 6\pi\eta r$ where r is the atomic radius. Based on the above considerations, the driving force (thermodynamic factor), diffusivity or viscosity (kinetic factor) and configuration (structural factor) are crucial parameters for understanding the glass formation in multicomponent alloys.

Further expanding on thermodynamic considerations, from Eq.12.1 it follows that high γ and small ΔG_{LS} are conducive to low nucleation rates and thus favors high GFA. In turn, ΔG_{LS} can be calculated as:

$$\Delta G_{LS}(T) = \Delta H_f - T_f \Delta S_f - \int_T^{T_f} \Delta c_p^{LS}(T) dT + \int_T^{T_f} \frac{\Delta c_p^{LS}(T)}{T} dT \qquad (12.2)$$

where Δc_p^{LS} is the specific heat difference between the liquid and solid states. Low ΔH_f and high ΔS_f will thus decrease the nucleation rate. As ΔS_f is proportional to the number of microscopic states (Inoue, 1995), a large ΔS_f is expected to be associated with multicomponent alloys. Therefore, a higher number of alloy components leads to the increase in ΔS_f and causes the increase in the degree of dense random packing in the liquid state. This is favorable to the decrease in ΔH_f and the S/L interfacial energy. The concept is consistent with the "confusion principle" and Inoue's first empirical rule (Wang et al., 2004).

The strong liquid behavior implies high viscosity and sluggish kinetics in the supercooled liquid state. The nucleation and growth of the thermodynamically favored phases is inhibited by the poor atom mobility resulting in high GFA and thermal stability of the supercooled liquid, as illustrated in Figure 12.11.

The second empirical criterion for BMG formation requires large difference in

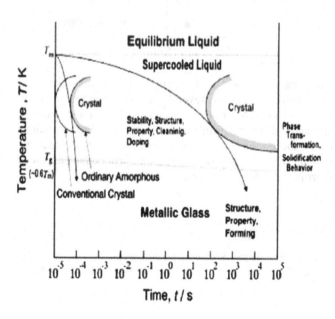

Figure 12.11. Schematic diagram showing the high stability of the BMG forming supercooled liquid for long periods reaching several thousands of second (Inoue and Takeuchi, 2002).

the size of the component atoms. This is thought to lead to complex structures that experience difficulties in crystallization. Density measurements show that the density difference between BMG and fully crystallized state is in the range 0.3–1.0% (Inoue *et al.* 1998, Wang *et al.* 2000), which is much smaller than the previously reported range of about 2% for ordinary amorphous alloys. Such small differences in values indicate that the BMGs have higher dense randomly packed atomic configurations.

Because the novel BMG-forming liquids can be studied on significantly larger time and temperature scales, the opportunities for studying nucleation and growth in undercooled liquids and the glass transition have been highly improved. It is now possible to produce time–temperature-transformation (TTT) diagrams that describe the competition between the increased driving force for crystallization produced by increasing undercooling and the deceleration of atom movement (effective diffusivity). The TTT diagram for Vitalloy 1 in Figure 12.12, shows the typical "C" curve and a minimum crystallization time of 60s at 895K and a critical cooling rate of about 1K/s. For older glass-forming alloys, the times were of the order of 10^3s.

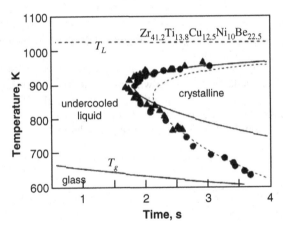

Figure 12.12. A time–temperature-transformation diagram for the primary crystallization of Vitalloy 1. Data obtained by electrostatic levitation (●) and processing in high-purity carbon crucibles (▲) are included (Busch, 2000).

In summary, the BMG-forming liquids are typically dense liquids with small free volumes and high viscosities, which are several orders of magnitude higher than those of previously known amorphous metal and alloys. Their microstructure configurations are significantly different from those of conventional amorphous metals. Thermodynamically, these melts are energetically closer to the crystalline state than other metallic melts. They have high packing density in conjunction with a tendency to develop short-range order (Wang *et al.*, 2004).

References

Boettinger W.J., 1974, *Metall. and Mat. Trans.* **5**:2026
Boettinger W.J., D. Shechtman, R.J. Schaefer, and F.S. Biancaniello, 1984, *Met. Trans. A* **15**:55
Busch R., 2000, *JOM* **52**:39.

Chen H.S., 1974, Acta Metall. **22**:1505.

Greer A.L., 1993, *Nature* **366**:303

Hughes I.R. and H. Jones, 1976, *J. Mater. Sci.* **11**:1781

Inoue A., 1995, *Mater. Trans. JIM* **36**:866

Inoue A., 2000, *Acta Mater.* **48**:279

Inoue A., Negishi T., Kimura H.M., Zhang T., Yavari A.R., 1998, *Mater. Trans. JIM* **39**:318

Inoue A., Takeuchi A., 2002, *Mater. Trans. JIM* **43**:1892

Kavesh S., 1978, in: *Metallic Glasses*, J.J. Gillman, H.L. Leamy, eds.), ASM Int., Metals Park, OH, (Chapter 2).

Klement W., Willens R.H., Duwez P., 1960, *Nature* **187**:869

Kurz W., Fisher D.J., 1998, *Fundamentals of Solidification*, 4th edition, Trans tech Publ., Switzerland

Li Y., 2005, *JOM* **March**:60-63

Lu Z.P., Liu C.T., Thompson J.R., Porter W.D., 2004, *Phys. Rev. Lett.* **92**:245503

Peker A., Johnson W.L., 1993, *Appl. Phys. Lett.*, **63**:2342

Perepezko J.H. and W.J. Boettinger, 1983, in *Mat. Res. Soc. Symp. Proc.* **19**:223

Pryds N.H. and A.S. Pedersen, 2002, *Metall. and Mat. Trans. A* **33A**:3755-3761

Turnbull D., 1969, *Contemp. Phys.* **10**:437

Turnbull D., Fisher J.C., 1949, *J. Chem. Phys.* **17**:71

Wang W.H., Wang R.J., Zhao D.Q., Pan M.X., Yao Y.S., 2000, *Phys. Rev. B* **62**:11292

Wang D., Li Y., Sun B.B., Sui M.L., Lu K., Ma E., 2004, *Appl. Phys. Lett.* **84**:4029

Yonezawa F., 1991, in: *Solid State Physics - Advances in Research and Applications*, H. Ehrenreich and D. Turnbull, eds., Academic Press, Boston p.179-254

SOLIDIFICATION IN THE PRESENCE OF A THIRD PHASE

In the mathematical treatment of microstructure evolution presented so far, only two phases were considered, solid and liquid. However, casting alloys often exhibit inclusions that degrade their mechanical properties. The solid, liquid or gaseous inclusions, that constitute a third phase, can considerably affect solidification morphology. In turn, solidification conditions influence the size and distributions of these inclusions. This chapter will describe two of the most significant phenomena related to the interaction between inclusions and the solidification front, interaction of solid particles with the S/L interface and micro-shrinkage formation.

13.1 Interaction of solid inclusions with the solid/liquid interface

The phenomenon of interaction of particles with solid-liquid interfaces has been studied since mid 1960's. While the original interest stemmed from geology applications (frost heaving in soil), researchers soon realized that understanding particle behavior at solidifying interfaces might yield practical benefits in other fields, including metallurgy. The issue is the location of particles with respect to grain boundaries at the end of solidification. Considerable amount of experimental and theoretical research was lately focused on applications to metal matrix composites produced by casting (Stefanescu et al. 1988, Kennedy and Clyne 1991) or spray forming techniques (Wu et al. 1994, Lawrynowicz et al. 1997). In the most common cast metal matrix composites, Al-Si alloy with SiC particles, the particles are normally distributed at the grain boundaries (Figure 13.1). This results in decreased plastic properties. Similar issues are pertinent to inclusion management in steel (Shibata et al., 1998). The particle S/L interface interaction was also found to play an important role in the solidification of ternary eutectics (Hecht and Rex, 2001) as well as in the formation of microporosity during solidification (Mohanty et al., 1995).

Another application of particle – S/L interface interaction is in the growing of $Y_1Ba_2Cu_3O_{7-\delta}$ (123) superconductor crystals from an undercooled liquid (Endo et

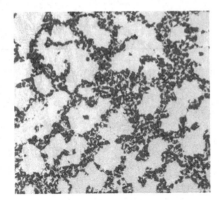

Figure 13.1. Microstructure of investment cast A356 Al-Si alloy with 20% SiC particles (Kennedy, 1991).

al. 1996, Shiohara *et al.* 1997). The oxide melt contains $Y_2Ba_1Cu_1O_5$ (211) precipitates, which act as flux pinning sites. Other applications include phagocytosis (literally "cell-eating", i.e. large particles are enveloped by the cell membrane of a larger cell and internalized to form a food vacuole) (Torza and Mason, 1969), and particle chromatography (separation of various solids) (Kuo and Wilcox, 1973).

For directional solidification (DS), the large body of experimental data on various organic and inorganic systems (e.g., Uhlmann *et al.* 1964, Zubko *et al.* 1973, Omenyi and Neumann 1976, Körber *et al.* 1985, Stefanescu *et al.* 1988, Shibata *et al.* 1998), demonstrates that there exist a critical velocity of the planar solid-liquid interface below which particles are pushed ahead of the advancing interface, and above which particle engulfment occurs. As shown on Figure 13.2a under certain conditions the particles are engulfed by the columnar front. On Figure 13.2b particles are pushed during directional solidification (DS) and then engulfed as the particle volume fraction builds up.

 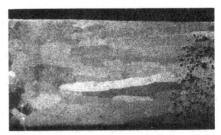

a) particles engulfed at $G_T = 117$ K/s   b) particles pushed at $G_T = 74$ K/s, then engulfed

Figure 13.2. Pushing and engulfment of 10-150 μm SiC particles during DS (from left to right) of an Al-2%Mg alloy. $V_{SL} = 8$ μm/s. The transition from the equiaxed zone at the left to the columnar zone marks the beginning of DS (Stefanescu *et al.* 1988). With kind permission of Springer Science and Business Media.

However, the problem is more complicated because in most commercial alloys dendritic interfaces must be considered. Indeed, most data available on metallic alloys, as summarized by Juretzko *et al.* (1998), are on dendritic structures. At high solidification velocity (V_{SL}) or low temperature gradient (G_T) cellular/ dendritic or even equiaxed interfaces will develop. The tips of cells or dendrites may *engulf* the particles. Alternatively, solute trapping in the particle/ interface gap decreases interface curvature to the point that it changes sign, resulting in tip splitting followed by engulfment. However, because convection will move the particles in the interdendritic regions, *entrapment* between the dendrites arms will be more common. In general, for cellular/dendritic and equiaxed interfaces the probability of engulfment is much smaller than that of entrapment. Thus, most particles will be distributed at the grain boundaries (Stefanescu *et al.*, 2000). The physics of these two phenomena, engulfment and entrapment, is quite different.

13.1.1 Particle interaction with a planar interface

In liquids solidifying with planar interfaces it has been observed that the interface can instantaneously engulf the particles, the interface may continuously push the particle ahead of it, or the interface may push the particle up to a certain distance before engulfing it (Omenyi and Neumann, 1976). Most experimental findings point out to the existence of a critical velocity, V_{cr}, above which particles are engulfed (the pushing - engulfment transition - PET), and to an inverse relationship between V_{cr} and particle radius, r_P. An example is given in Figure 13.3 for water doped with nylon particles (Azouni *et al.*, 1990). The V_{cr} - r_P relationship can be described by a power function:

$$V_{cr} = M \cdot r_P^{-m} \tag{13.1}$$

where M is a material constant and m is an exponent that can have different values. Some examples of the experimental values of these parameters for a number of systems are given in Table 13.1.

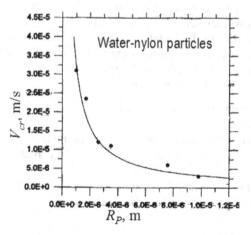

Figure 13.3. Correlation between the critical velocity for PET and particle radius (Azouni *et al.*, 1990). Reprinted with permission from Elsevier.

As the particle approaches the S/L interface, the difference in thermal conductivity between the particle, k_P, and the matrix (liquid), k_L, will impose an interface curvature (Figure 13.4). If the thermal conductivity ratio $k^* = k_P/k_L$ is larger than unity, the interface will form a trough. In the opposite case, a bump will grow on the interface. For engulfment to occur, however, a trough should form on the bump. These effects have been demonstrated on transparent organic metal analog materials (TOMA) as well as on aluminum matrices through X-ray transmission microscopy (Omenyi and Neumann 1976, Uhlmann *et al.* 1964, Sen *et al.* 1997).

Table 13.1. Values of parameters in Eq. (13.1) obtained from experimental data on particle pushing by a planar interface.

Matrix/particle system	m	$M, \mu m^2/s$	Reference
Biphenyl/acetal	0.90	1132	Omenyi and Neumann (1976)
Biphenyl/nylon	0.64	199	Omenyi and Neumann (1976)
Naphthalene/ acetal	0.30	195	Omenyi and Neumann (1976)
Naphthalene/nylon	0.46	142	Omenyi and Neumann (1976)
Succinonitrile/polystyrene	1.0	12.1	Pang *et al.* (1993)
Steel/silica-alumina (liquid)	1.0	24	Shibata *et al.* (1997)
Aluminum/zirconia	1.0	250	Juretzko *et al.* (1998)

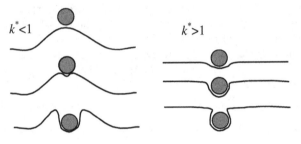

Figure 13.4. Interface shape during particle engulfment.

However, as demonstrated by Hadji (1999) the interface shape in pure substances is not determined solely by the thermal conductivity ratio, but also by the local pressure. In alloys, the solutal field will also affect interface shape.

The PET is also affected by the flow of the liquid at the S/L interface. As summarized in Figure 13.5, two distinct liquid flow patterns exist: flow into the interface generated by solidification, and flow parallel to the interface at velocity V_L induced by natural convection. Four main forces are identified:

- the interaction force between the particle and the S/L interface F_I, (typically a repulsive force),
- the drag force exercised by the solidification induced liquid flow around the particle into the interface F_D, (which pushes the particle into the interface)
- the lift force produced by the liquid flow parallel to the interface, F_L, (which pushes the particle away from the interface)
- the gravity force, F_g.

There is agreement on the fact that at low convection regime in the melt particle-interface interaction governs the PET, while under conditions of high convection there is no particle-interface interaction and particles are continuously pushed.

Considering the two velocities, V_{SL} and V_L, it can be rationalized that the behavior of the particle is governed by either the interaction with the interface, or by the fluid flow velocity between the particle and the interface, as summarized in Table 13.2.

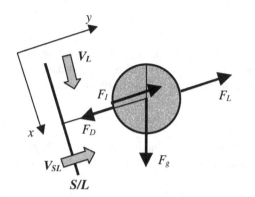

Figure 13.5. Schematic representation of the forces acting on a particle in the vicinity of the S/L interface.

Table 13.2. Regimes of particle-interface interaction.

Melt convection	Interface velocity	Governing phenomenon	Outcome
zero or small	low	interface interaction	pushing
zero or small	high	interface interaction	engulfment
large	any	fluid flow velocity	no particle-interface interaction (pushing)

To explain the experimental findings several theories have been proposed. They fall largely into two categories: i) theories that attribute the PET to select material properties such as surface energy or thermal properties, and ii) theories that acknowledge the role of solidification and particle kinetics. Numerous models have been suggested based on these approaches.

13.1.2   Material properties models

The net change in free energy of a single spherical particle that moves from the liquid to the solid is:

$$\Delta \gamma_o = \gamma_{PS} - \gamma_{PL} < 0 \qquad (13.2)$$

where γ_{PS} and γ_{PL} are the particle-solid and particle-liquid interface energy, respectively. Omenyi and Neumann (1976) have postulated that if $\Delta \gamma_o < 0$, engulfment is to be expected; while for $\Delta \gamma_o > 0$, pushing should result. $\square$

Since the thermal properties of the particle and the liquid affect the shape of the interface, some researchers assumed that the ratio between these properties determines particle behavior. Zubko *et al.* (1973) suggested that the value of k^* determines the outcome. If $\Delta \gamma_o > 1$, the trough that forms at the S/L interface will engulf the particle. On the contrary, if $\Delta\gamma_o < 1$, a bump forms and the particle will be continuously pushed. Following a similar line of thinking, Surappa and Rohatgi (1981) proposed an empirical heat diffusivity criterion; engulfment is postulated when $(k_P c_P \rho_P / k_L c_L \rho_L)^{1/2} > 1$, where c and ρ are the specific heat and density, respectively.

13.1.3   Kinetic models

Kinetic models are complex attempts at describing the physics of PET and include energy and mass transport (diffusive or convective), particle motion, and the interaction force between the particle and the S/L interface. This problem does not yet have a complete solution.

13.1.3.1   Forces acting on the particle

The physics of PET is best described in terms of force balance as shown in Figure 13.5. The governing equation is the equation of particle motion (Catalina *et al.*, 2000):

$$F_I + F_L - F_D - F_g - F_o = m\frac{dV_P}{dt} \tag{13.3}$$

where F_o is the force required to accelerate the fluid that adheres to the particle, m is the particle mass, and dV_P/dt is the particle acceleration. An additional force, the thermal force was suggested by Hadji (2001). Formulating the various forces *dynamic models* can be developed. Assuming steady-state, the equation simplifies to:

$$F_I + F_L - F_D - F_g = 0 \tag{13.4}$$

Further simplification is possible by ignoring the gravitational acceleration (microgravity experiments):

$$F_I - F_D = 0 \tag{13.5}$$

These last two equations have been used in developing *steady-state models* that calculate an *equilibrium velocity*.

Let us now address the problem of formulating the various forces in the governing equation.

The interface force

The formulation of the interface force depends on the source considered for this force. In the earliest model proposed for PET, Uhlmann *et al.* (1964) assumed that the interfacial repulsive force results from the variation of the surface free energy with the distance d_{gap} from the interface given as:

$$\Delta\gamma = \Delta\gamma_o \cdot (d_o/d_{gap})^n \quad \text{where} \quad \Delta\gamma_o = \gamma_{PS} - \gamma_{PL} - \gamma_{LS} \tag{13.6}$$

Here, γ is the surface tension, d_o is the minimum separation distance between particle and solid, and the subscripts P, S, L stand for particle, solid and liquid, respectively. Note that as $d_{gap} \to \infty$, $\Delta \gamma \to \Delta \gamma_o$. The value of the exponent n was assumed to be between 4 and 5. Further assuming $n = 2$ and non-retarded Van der Waals interaction the interaction force was derived (Stefanescu et al. 1988, Pötschke and Rogge 1989, Shangguan et al. 1992) to be:

$$F_I = 2\pi \Delta \gamma_o \, d_o^2 \, \frac{r_P}{d_{gap}^2} \, \xi \tag{13.7}$$

where ξ is a correction factor for the curved interface. For example, in the SAS model (Shangguan et al. 1992), for $d_{gap} \ll r_P$, the repulsive interface force was derived as:

$$F_I = 2\pi \, r_P \, \Delta \gamma_o \left(\frac{a_o}{a_o + d} \right)^2 \frac{r_I}{r_I - r_P} = 2\pi \, r_P \, \Delta \gamma_o \left(\frac{a_o}{a_o + d} \right)^2 k^* \tag{13.8}$$

where r_I is the radius of the S/L interface, a_o is the atomic diameter of the matrix material, and d_{gap} is the equilibrium distance between the interface and the particle.

The main problem with this formulation is the calculation of $\Delta \gamma_o$. In spite of extensive efforts and sometimes bitter debate (Kaptay 1999, 2000) it is clear that because of the uncertainties in the evaluation of the various interface energies, $\Delta \gamma_o$ is at best a fitting parameter.

Attempting to avoid the complexities of solid-solid interface energy, Ode et al. (1999) assumed a sinusoidal interface and proposed the equation:

$$F_I = \pi^2 \gamma_{SL} \sqrt{(r_P + d_{gap})l} \left(\frac{d_0}{d_0 + d_{gap}} \right)^2 \tag{13.9}$$

where l is the amplitude of the sin function.

Another approach to the estimation of the interface force was proposed by Chernov et al. (1976, 1977). They assumed that as the interface approaches the particles, the difference in chemical potential between the bulk liquid, μ_L^∞, and that in the liquid film between the particle and the interface, μ_L, produces a disjoining pressure given by:

$$\Pi = \frac{\mu_L^\infty - \mu_L}{v_o} = \frac{B_n}{d^n} \tag{13.10}$$

where v_o is the molecular volume of the liquid and B_n is a constant. The disjoining pressure (which is in fact a volumic energy expressed in J/m³) is assumed positive if the films thickens, and negative in the opposite case. The disjoining pressure introduces a local undercooling, ΔT_P. Further assuming that the disjoining pressure is the source of the interface force, the following equation was derived:

$$F_I = \pi B_3 \frac{r_P}{d_{gap}^2} \xi \qquad (13.11)$$

where B_3 is a constant suggested to be 10^{-21} J. However, just as $\Delta \gamma_o$, B_3 becomes a fitting parameter. The correlation between B_3, the Hammaker constant and surface energy is discussed by Stefanescu (2002) and Asthana and Tewari (1993).

Experimental measurements of the interface (repulsive) force were attempted by Smith et al. (1993). By observing particles being pushed up an incline by an advancing S/L interface, they measured the repulsive force between the particle and the front in three systems where the energy of particle adhesion to the solidification front was known. The equation of their force of adhesion is identical with that for $\Delta \gamma_o$ given in Eq. (13.6). A linear relationship was found between $\Delta \gamma_o$ and F_I, as predicted by the theory. A summary of their experimental data is given in Table 13.3. Note that the minimum separation distance calculated from these experiments is constant at about 20 nm.

Table 13.3. The interface (repulsive) force between acetal particles in three matrix materials (Smith et al., 1993).

Matrix	$2 \cdot r_P$, μm	F_I, nN	d_o, nm
Salol	20	0.0010	22
Benzophenone	20	0.0051	18
Biphenyl	20	0.0145	19

The drag force

To maintain a stable liquid film between the particle and the S/L interface during particle pushing, liquid must continue to flow into the gap. The pressure gradient behind the particle is the driving force for the fluid flow. It induces the drag force on the particle. In its most general form the drag force on a particle of radius r_P can be expressed as:

$$F_D = \frac{1}{2} C_D \rho_L V_P^2 \pi r_P^2 \qquad (13.12)$$

where C_D is the drag coefficient. The simplest expression for the drag force was derived by Stokes for a spherical particle in an unbounded fluid, $F_D = 6\pi \eta V_P r_P$, where η is the dynamic viscosity. It is only valid for Reynolds numbers (Re $= V_P r_P \rho/\eta$) smaller than 0.5.

When the particle is very close to the interface, the assumption of unbounded flow does not hold anymore. The drag force can be calculated from the lubrication theory assuming the process is controlled by fluid flow in the particle-interface gap. For a gap of width d much smaller than the particle radius, the first approximation of the drag force as calculated by Leal (1992) is:

for a circle: $F_D = 6\pi\eta V_P \dfrac{r_P^2}{d}$ for a cylinder: $F_D = 3\sqrt{2}\pi\eta V_P \left(\dfrac{r_P}{d}\right)^{1.5}$ (13.13)

where η is the viscosity. Note that these equations are approximations obtained by integrating the pressure distribution only within the gap between the particle and the interface, and not all around the particle.

This equation has been modified by different investigators to account for the interface curvature (e.g., Pötschke and Rogge 1989, Shangguan et al. 1992). For locally deformed interface, it was derived that the drag force becomes:

$$F_D = 6\pi\eta V_P \frac{r_P^2}{d}\left(\frac{r_I}{r_I - r_P}\right)^2 = 6\pi\eta V_P \frac{r_P^2}{d}k^{*2}$$ (13.14)

Because of the limitations introduced by the assumptions used in the derivation, this equation is valid for trough formation in the interface ($k^* < 1$). For $k^* \geq 1$, calculated results match better the experimental data when k^* is assumed equal to one in this equation as well as in Eq.(13.8).

Using numerical modeling (Catalina et al., 2000) it was demonstrated that the drag force is higher than given by the preceding equations. For a cylinder it is:

$$F_D = \sqrt{3}\pi\eta V_P (r_P/d)^{1.92}$$ (13.15)

However, recent numerical work by Garvin and Udaykumar (2004) seems to confirm Leal's equation. Indeed, they calculated that:

$$F_D = 10\eta V_P (r_P/d)^{1.53}$$ (13.16)

Regardless of which equation is more accurate, the main problem in the calculation of the drag force rests with the value used for viscosity. As the gap width is of the atomic distance order, it is to be expected that the value of the viscosity in the gap will significantly differ from that in the bulk liquid.

The gravity force
The net gravity force acting on the particle is:

$$F_g = \frac{4}{3}\pi r_P^3 (\rho_L - \rho_P)g$$ (13.17)

Lift forces
The general expression for the lift forces is Eq. (13.12) where the drag coefficient is replaced by the lift coefficient, C_L. Using the method of matched asymptotic expansions, Saffman (1965) derived the following equation for the lift force on a

rigid sphere translating parallel the streamlines of a unidirectional linear shear flow field:

$$F_S = 6.46\eta V_{rel}\, r_P^2 \sqrt{\frac{\rho_L}{\eta}\left(\frac{dV_{Lx}}{dy}\right)_{avg}}$$
(13.18)

where V_{rel} is the velocity of the particle relative to the liquid, and $(dV_{Lx}/dy)_{avg}$ is the average liquid velocity gradient over the particle diameter. This equation holds only for small Reynolds numbers.

An additional lift force comes from particle rotation – the Magnus force, F_M. For a particle rotating with an angular velocity, ω, and translating with a velocity, V_{rel}, relative to the liquid, the lift coefficient for the Magnus force was calculated (Rubinow and Keller, 1961)) to be:

$$C_{LM} = 2r_P\, \omega/V_{rel}$$

13.1.3.2   Steady-state models

A large number of analytical steady-state models for PET were developed over the years. The first kinetic steady-state model was proposed by Uhlmann et al. (1964). It assumed that mass transport in the particle-solid gap is by mass diffusion alone. In most cases, this model underestimates the critical velocity.

Numerous kinetic models were developed based on Eq. (13.5). By equating the repulsive and the drag force, these models calculate an equilibrium velocity at which the particle is continuously pushed. The first such model was published in by Stefanescu et al. (1988). Assuming that the equilibrium velocity is the critical velocity they suggested that:

$$V_{cr} = \frac{\Delta\gamma_o\, d_o}{6\eta\, r_P}\left(2 - \frac{k_P}{k_L}\right)$$
(13.19)

One year later, Pötschke and Rogge (1989) obtained an equation for V_{cr} by solving numerically an analytical solution that included the effect of the solutal field. Many other models followed. In most of these models the equation is of the form $V_e \propto 1/d_e$ and states that at any interface velocity the particle will find an interface distance (equilibrium distance) at which steady-state exists. An additional criterion must be imposed to obtain the critical velocity. The criteria imposed by various researchers ranged from complex mathematical exercises (e.g., Chernov et al. 1976 - CTM model, Bolling and Cissé 1971) to additional hypotheses on the mechanism of engulfment including maximization of equilibrium velocity (Pötschke and Rogge 1989, Shangguan et al. 1992) or on gap thickness (Sen et al. 1997, Stefanescu et al. 1998).

For example, in the Shangguan/Ahuja/Stefanescu (1992) (SAS) model, when equating Eqs. (13.8) and (13.14) the following equilibrium velocity is obtained:

$$V_e = \frac{\Delta\gamma_o}{3\eta\,k^*}\frac{d}{r_P}\left(\frac{a_o}{a_o+d}\right)^2$$

Maximizing this equation with respect to d (i.e., $dV_e/dd = 0$) gives the minimum separation distance (critical distance): $d_{cr} = a_o$. Substituting the critical distance in the equilibrium distance gives the critical velocity:

$$V_{cr} = \frac{a_o\Delta\gamma_o}{12\eta\,k^*\,r_P} \tag{13.20}$$

In the CTM model the equilibrium velocity is obtained by equating Eqs. (13.11) and (13.14): $V_e = B_3/(6\eta d\,r_P)$. It becomes the critical velocity if d is substituted by the minimum separation distance d_o. However, in this case, both the velocity and the distance are unknown. Additional manipulations are required to derive the critical velocity without imposing a critical distance.

Pötschke and Roge (1989) (PR model) assumed that the interaction force could be treated as the Van der Waals force between two spheres. Their final analytical solution for the critical velocity was solved numerically, to yield:

$$V_{cr} = \frac{1.3\Delta\gamma_o}{\mu}\left[16\left(\frac{r_P}{a_o}\right)^2 k^*\left(15k^* + x\right) + x^2\right]^{-1/2} \quad \text{with} \quad x = \frac{C_o|m_L|\Delta\gamma_o}{k\,G_L\eta\,D_L} \tag{13.21}$$

For pure metals $x = 0$, and the above equation becomes:

$$V_{cr} = \frac{0.084 a_o\Delta\gamma_o}{\mu\,k^*\,r_P} \tag{13.22}$$

Note that this equation is identical to Eq. (13.20). Both the SAS and PR model overestimate the critical velocity because they use a critical distance $d_{cr} = a_o$, which is too small.

Rempel and Worster (2001) derived a similar equation to the CTM equation for $r_P < 500$ μm. They elaborated on the role of intermolecular forces other than those dominated by non-retarded van der Waals interactions, such as long-range electrical interactions and retarded van der Waals interaction. Using linear stability analysis, Hadji (2003) demonstrated that presence of a particle in the melt modifies the threshold value of the thermal gradient for the inset of morphological instability. With the exception of the UCJ model derived on the assumption that the controlling mechanism is mass diffusion in the P-S gap, all others assume flow in the P-S gap as the controlling mechanism.

The equations for most of the models have been summarized in some references including Stefanescu (2002) and Youssef et al. (2005). A summary of the steady-state equations for critical velocity derived by different researchers is provided in Table 13.4.

Table 13.4. Selected steady-state models for PET

Model/Reference	Critical velocity, $V_{cr} =$	Exponent m in Eq. (13.1)	Assumption for d_{cr}
Uhlmann/Chalmers/ Jackson (UCJ)	$\dfrac{n+1}{2}\left(\dfrac{\Delta H_f a_o v_a D}{k_B T r_P^2}\right)$	2	
Bolling/Cissé (BC)	$\left(\dfrac{1.36 k_B T a_o \gamma_{SL}}{\pi}\right)^{1/2} \dfrac{1}{3\eta r_P^{3/2}}$	3/2	maximum force
Chernov/Temkin/ Melnikova (CTM)	$r_P > 500\ \mu m\quad \dfrac{B^{3/4}\left(\Delta S_f G_T\right)^{1/4}}{24\eta r_P \left(k^*\right)^{3/4}}$	1	
	$r_P < 500\ \mu m\quad \dfrac{0.14 B^{2/3} \gamma_{SL}^{1/3}}{\eta r_P^{4/3}}$	4/3	
Pötschke/Rogge (PR)	$\dfrac{0.084 a_o \Delta\gamma_o}{\eta k^* r_P}$	1	$dV_c/dd = 0$
Sen et al. (1997) (modified SAS)	$\dfrac{\Delta\gamma_o a_o}{156\eta k^* r_P}$	1	$d_{cr} = 50 a_o$
Stefanescu et al. (1998) (SC)	$\left(\dfrac{\Delta\gamma_o a_o^2}{3\eta k^* r_P}\right)^{1/2}$	1/2	$dV_c/dd = -1$
Kim/Rohatgi (1998) (KR)	$\dfrac{a_o \Delta\gamma_o}{18\eta}\left[\dfrac{G a_o}{\Gamma}\left(\dfrac{k^*-1}{3}+\dfrac{1}{r_P}\right)\right]$	1	$d_{cr} = a_o$
Hadgi (1999a)	$\|A\|G_L\left[\dfrac{36\eta\left[\Delta T_L+(1-k^*)\Delta T_P\right]}{\pi r_P+6\pi\eta(1-k^*)}\dfrac{12(3\pi)^{1/3}\gamma_{SL}v_a}{\|A\|^{1/3}\Delta H_f}r_P^{2/3}\right]^{-1}$ A: Hammker ct., ΔT_L, ΔT_P: temperature change across the liquid and particle		limit of vanishing disjoining pressure

A comparison between calculation with some of these models for the systems biphenyl/glass particles and succinonitrile/polystyrene particles on one hand and experimental results on the other is presented in Figure 13.6. The BC, CTM and modified SAS models calculate critical velocities in the same range. The UCJ model underestimates V_{cr}, while PR overestimates it by several orders of magnitude.

Interesting results were obtained when theoretical predictions with the SC model were compared with the experimental results of Shibata et al. (1998) on inclusions in steel. The experimental data are given in Figure 13.7 for alumina agglomerates and for complex globular inclusions in steel. The lower limit for the

PET of alumina clusters is fitted to the experimental data as the $V = 1.9 \cdot 10^{-11}/r_P$ curve. Most clusters that have been pushed are located under this lower limit. For the globular inclusions, a fitted PET critical velocity is superimposed on the experimental points. The curve corresponds to the equation $V = 2.3 \cdot 10^{-11}/r_P$, as suggested by Shibata *et al.* It is seen that in both cases predictions with the SC model are reasonable.

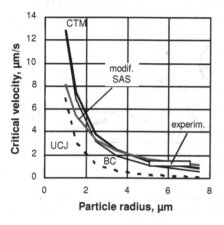

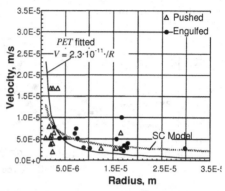

Figure 13.6. Experimental and calculated critical velocities for the biphenyl /glass system (Sen *et al.*, 1997a). Reprinted with permission from Elsevier.

Figure 13.7. Experimental (Shibata *et al.*, 1998) and calculated (Stefanescu and Catalina, 1998) correlation between the critical velocity for the pushing/engulfment transition of globular inclusions in 0.01 wt% C steel.

13.1.3.3 Interface shape models

With the development of numerical techniques, it became possible to attempt to calculate the change in interface shape at least in two dimensions. In most cases, the particle was assumed to be at rest, while the S/L interface advanced toward the particle. The first work on this subject seems to be by Sasikumar *et al.* (1989, 1991) who developed a steady-state heat flow numerical model that considered curvature and pressure undercooling and accounted for the role of solutal field. Fluid flow around the particle was ignored. The critical velocity for PET was calculated from the maximization of the equation of particle velocity. While they demonstrated for the first time that for $k^* < 1$ a trough will occur on the bump under certain conditions, they were unable to produce interface shape plots because of the complications of their numerical method.

Casses and Azouni (1993) used numerical simulation, including disjoining pressure and curvature effects, to demonstrate the change in the shape of a pure substance S/L interface approaching a particle that has a different thermal conductivity than that of the liquid (bump and trough formation). Kim and Rohatgi (1998) studied the effect of diffusion field in the gap, but did not present any calculated interfaces.

A more complete analysis was performed by Catalina and Stefanescu (1999). They developed a diffusive model for interface shape based on a 2D numerical model for interface tracking. The interface temperature is controlled by solute concentration and curvature. Fluid flow around the particle was ignored. Some typical results are shown in Figure 13.8. Note the higher concentration in the gap. Later (Catalina et al., 2004), the model was validated for the system Au-H_2 pores, and used to explain the mechanism of comet tail-shape segregation observed behind the particles in many experiments (Figure 13.9).

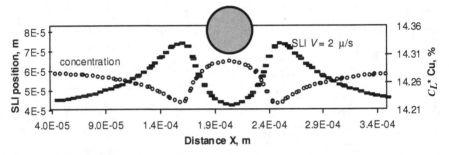

Figure 13.8. S/L interface shape and liquid interface Cu concentration (*Al*-2 wt% *Cu* alloy, ZrO_2 particle $r_P = 22.5$ μm, $V_{SL} = 2$ μm/s) (Catalina and Stefanescu, 1999).

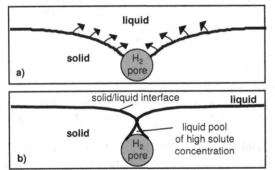

Figure 13.9. The mechanism of comet-tail-shaped segregation region behind an H_2 pore: a) solute diffusion toward the sample centerline before complete engulfment; b) solute trapping behind the pore at the time of complete engulfment (Catalina et al., 2004). With kind permission of Springer Science and Business Media.

This model demonstrated elegantly the formation of a trough-in-bump for alloys, because of solute accumulation in the gap. Thus, engulfment of particles for the case of $k^* < 1$ was explained. However, it remained to unambiguously prove that a trough-in-bump can form even for the case of pure matter. Otherwise, engulfment in pure matter for $k^* < 1$ would not be possible.

After Hadji (1999) used perturbation analysis to demonstrate that the interface shape is affected not only by the thermal conductivity, but also by the disjoining pressure, Catalina et al. (2003) developed a semi-analytical model for the interface shape that accounts for the Gibbs-Thomson and disjoining pressure effects. Calculations were performed for the SCN/polystyrene system for which $k_P < k_L$. It was found that the shape of the S/L interface is fundamentally different from the situation when the Gibbs-Thomson and disjoining pressure effects are neglected (Figure

13.10). For systems characterized by $k_P < k_L$ the disjoining pressure causes the sign change of the interface curvature near the particle. The increase of the temperature gradient in the liquid diminishes the effect of the disjoining pressure.

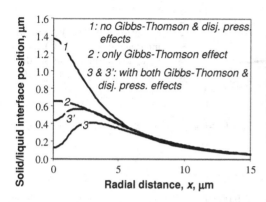

Figure 13.10. The influence of the Gibbs-Thomson and disjoining pressure effects on the shape of the solid/liquid interface (SCN/ polystyrene system, r_P=5 μm, d_{gap} = 11·a_o, G_T =6 K/mm except for curve 3' for which G_T =8 K/mm) (Catalina et al., 2003).

13.1.3.4 Dynamic models

Dynamic models consider the non-steady-state nature of PET, meaning that a particle, initially at rest, must have an accelerated motion in order to reach the steady-state velocity, which is the solidification velocity. The governing equation is the equation of particle motion, introduced earlier as Eq. (13.3).

The first dynamic models were presented at the same conference by Schvezov (1999) and by Catalina and Stefanescu (1999). In the latter model, Eq. (13.7) was used for the interface force with $\Delta\gamma_o$ given by Eq. (13.2). The drag force for a sphere was obtained from manipulations of Eqs. (13.13) and (13.15). Lift and gravity forces were ignored. The force required to accelerate the fluid that adheres to the particle was included in the formulation. Model calculations results for a constant solidification velocity smaller than V_{cr} are shown in Figure 13.11a. It is seen that the particle velocity, V_P, increases from zero to the velocity of the interface as the particle-interface gap, d, decreases. The velocity of the interface under the particle, called tip velocity, V_t, behaves in an oscillatory manner, and so does the particle-interface distance. Eventually, when steady-state is reached, $V_P = V_t$.

The evolution of V_P and V_t for the case when the solidification velocity is equal to the critical value is presented in Figure 13.11b. As solidification proceeds, both V_P and V_t are continuously increasing, as in the previous case. In the second stage, when V_t decreases, it is seen that it only decreases to a value close to V_{SL}, without going below this value. At the same time, V_P increases close to V_{SL} but without exceeding it. At the critical velocity, the drag force becomes higher than the pushing force and when the particle comes close enough to the S/L interface, it is eventually caught by its tip and then engulfed into the solid. Clearly this model describes well the observed experiments. It demonstrates that the interaction is essentially non-steady-state and that steady-state eventually occurs only when solidification is conducted at sub-critical velocities.

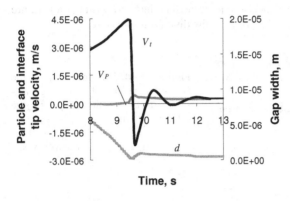

a) Particle velocity (V_p) and S/L interface tip velocity (V_t) versus time, for a subcritical solidification velocity $V_{SL} = 0.3$ μm/s

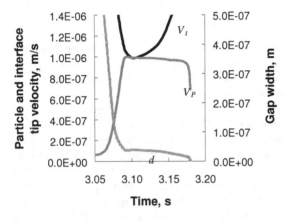

b) Time evolution of particle velocity and S/L interface tip velocity for $V_{SL} = V_{cr} = 1$ μm/s

Figure 13.11. Calculations for the Al-ZrO$_2$ particle system (Al viscosity was taken as 2 mPa s, and $r_P = 250$ μm) (Catalina and Stefanescu 1999, Catalina *et al.* 2000). With kind permission of Springer Science and Business Media.

13.1.3.5   Models considering fluid flow

In all the models discussed so far fluid flow parallel to the S/L interface has been ignored. Han and Hunt (1995) introduced in their analysis the force acting on a particle because of the difference in flow velocity in the region between the particle and the interface and in the region on the opposite side of the particle ($F_L \neq 0$). They also proposed a different mechanism for engulfment, as follows: the particle near the solidification front rolls/slides on the interface and is captured by the front because of its roughness. However, because of the unavailability of data regarding fluid velocity and friction conditions, this model is impractical.

Mukherjee and Stefanescu (2000, 2004) further developed the Catalina/Stefanescu (1999) model to include all terms in Eq. (13.3). The governing equation, which is the equation for particle acceleration, was written in its complete form as:

$$\frac{4\pi}{3}\rho_P r_P^{\,3}\frac{dV_P}{dt} = 2\pi r_P\Delta\gamma_o\left(\frac{d_o}{d}\right)^2 k^* - 6\pi\eta V_P\frac{r_P^{\,2}}{d}\left(\frac{r_P}{d}\right)^{0.423}k^{*2} -$$

$$C_A\frac{4\pi}{3}\rho_L r_P^{\,3}\frac{dV_P}{dt} + 6.46\eta V_{rel}r_P^{\,2}\sqrt{\frac{1}{\nu}\left(\frac{dV_{Lx}}{dy}\right)_{avg}} + r_P^{\,3}\rho_L\omega V_{rel}$$

(13.23)

where the LHT is the force on the particle, the first RHT is the interface force, the second RHT is the drag force, the third RHT is the force required to accelerate the virtual "added" mass, the fourth RHT is the Safmann force, and the fifth RHT is the Magnus force. ω is the angular velocity and V_{rel} is the velocity of the particle relative to the fluid. Then, the particle acceleration was expressed in terms of the Reynolds and Weber numbers. The effect of particle radius and the Reynolds number of flow, modified by changing the gravity level, on the critical velocity of engulfment is shown in Figure 13.12. It is seen that the critical velocity increases as the flow Reynolds number increases. No effect of the level of convection on the critical velocity of engulfment was observed in the low-convection regime. In the moderate-convection regime (0.1<Re<0.6), the critical velocity is significantly increased with the increase in the Reynolds number. When Re>0.6, *i.e.* when a high-convection regime is established, no interaction occurs between the interface and the particle. For this regime, the convection velocities are so large as to cause the particles to be swept away from the interface. This was calculated for **g** = 1. In a subsequent paper, Mukherjee *et al.* (2004) verified the validity of the analytical equations proposed for the drag and lift forces through a 2D numerical calculation.

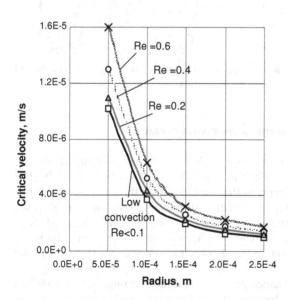

Figure 13.12. Effect of Reynolds number of flow on the critical velocity for engulfment as a function of particle radius for the aluminum-zirconia particle system (Mukherjee and Stefanescu, 2004). With kind permission of Springer Science and Business Media.

13.1.3.6   Microstructure visualization models

A phase filed model by Ode *et al.* (2000) applied to a Fe-C alloy with an alumina particle was able to reproduce the interface movement during particle pushing and engulfment and to estimate the critical velocity for PET. Phase field simulation was also used to describe the growth of an L_2 droplet in front of a solid planar front in a monotectic system (Nestler *et al.*, 2000). The computed pictures compared well with the results of experimental work on monotectics with transparent organic metal analogues.

Shelton and Dunand (1996) carried out 2-D cellular automaton computer simulations to model the geometric interactions between mobile, equiaxed particles and growing matrix grains. The model allows the study of particle pushing by growing grains, resulting in particle accumulation and clustering at the grain boundaries. It was found that certain parameters such as particle area fraction, particle settling speed, particle cluster mobility and grain nucleation rate strongly affect the spatial distribution of particles. An example of computed microstructures is given in Figure 13.13.

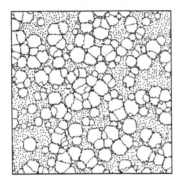 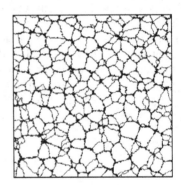

Figure 13.13. Particles segregation at grain boundaries after 13 time-steps (left) and after 24 time-steps (right) (Shelton and Dunand, 1996). Reprinted with permission from Elsevier.

13.1.4   Mechanism of engulfment (planar S/L interface)

Most steady-state models do not provide a clear mechanism for engulfment. In principle, the majority of them attempt to calculate the breakdown of steady state and consider that engulfment occurs at that time. Some of them suggest that when the particle–interface gap becomes so small as to prevent mass transport by either diffusion or fluid flow, the particle "sticks" to the interface and is engulfed. Note that only in two-dimensions the particle-interface contact will prevent mass transport. In three dimensions, a point contact between a sphere and a surface will not be sufficient to stop mass transport.

Stefanescu *et al.* (1995) explained particle engulfment through the "sinking" of the particle in the S/L interface because of the combined effect of the thermal and solutal field. As shown in Figure 13.4, local perturbation of the interface allows the solid to grow around the particle and engulf it. In other words, there is no requirement of particle "sticking" to the interface for engulfment to occur.

There is reliable experimental evidence to support such a mechanism for alloys. An example is provided from work on transparent organic materials in Figure 13.14. It is seen that a SiC particle initiates destabilization of the interface and is eventually engulfed by the solid. As the interface approaches the particle at a distance smaller than D/V, the particle obstructs solute diffusion. The composition gradient at the interface decreases and the interface under the particle is decelerated, while the rest of the interface moves at the velocity imposed by the thermal field. A trough is formed on the interface. This trough appears in the bump of systems with $k^*>1$, or increases the trough of systems with $k^*<1$. This mechanism has been confirmed through numerical calculations.

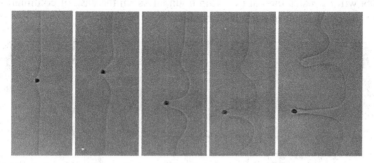

Figure 13.14. Engulfment of a silicon carbide particle by the S/L interface in succinonitrile. Velocity was gradually increased from 9.2 to 21 μm/s (Stefanescu *et al.*, 1995).

In the case of pure materials trough formation in systems with $k^*<1$ explains engulfment. A trough can also occur for systems where $k^*>1$ (Sasikumar *et al.*, 1989), because of the undercooling produced by the disjoining pressure term. This effect is demonstrated in Figure 13.10.

13.1.5   Particle interaction with a cellular/dendritic interface

The information on particles interaction with dendritic or cellular interfaces is scarce. Although some information for metallic alloys has been published (*e.g.*, Stefanescu *et al.* 1988, Premkumar and Chu 1993, Yaohui *et al.* 1993, Kennedy and Clyne 1991, Fasoyinu and Schvezov 1990, Hecht and Rex 1997) the details of the interaction have not been clearly described because the analysis has to rely on quenched samples. A better understanding can be obtained at this stage from experiments with transparent organic materials. When conducting directional solidification experiments with succinonitrile (99.5% SCN, balance water) and a variety of particles (nickel, alumina, cobalt), Sekhar and Trivedi (1990) observed that engulfment occurred in cells and dendrites, as for planar interfaces. Particle engulfment into the dendrite tip caused tip splitting. Particle entrapment was observed between the secondary dendrites.

More recently, Stefanescu *et al.* (2000) observed a variety of particle behaviors during microgravity experiments with succinonitrile - polystyrene particles systems solidifying with a cellular/ dendritic interface. In some instances, particles were engulfed as for planar interfaces with local deformation of the interface (Figure

13.15a). In some other cases, solute trapping in the particle/interface gap decreased interface curvature to the point that it changed sign. This means that the tip of the cell has split (Figure 13.15b). Dendrite/cell tip splitting results in engulfment. Thus, in dendritic solidification engulfment is still possible. However, because in a 1-**g** environment convection will move the particles, the residency time of a particle on a particular dendrite tip is short. The particle will most likely be moved in the interdendritic regions where it will be entrapped. In μ**g** the tip of the cell/dendrite may push the particle, which is then entrapped in the intercellular space (Figure 13.15c). Consequently, for cellular/dendritic interfaces the probability of engulfment is much smaller than that of entrapment. Thus, a large number of particles will be distributed at the grain boundaries. In the case of columnar solidification, this will result in particle alignment.

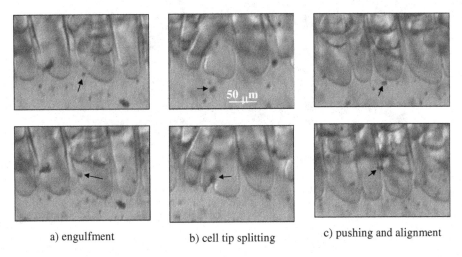

a) engulfment   b) cell tip splitting   c) pushing and alignment

Figure 13.15. Particle interface interaction during cellular/dendritic solidification at 10 μm/s of SCN-polystyrene particles systems in microgravity (Stefanescu *et al.*, 2000).

Perturbation of the solutal field by the particle changes the dendrite tip radius and its temperature. When the particle approaches the interface the solute gradient will decrease and the tip radius will increase since the tip radius, r_t, depends on the liquidus temperature gradient, G_L, produced by the solute. This effect is obvious from Eqs. (7.26) and (8.13) written here as $r_t = 2\pi\sqrt{\Gamma/(G_L - G_T)}$. Thus, a dendritic to cellular transition may occur. This effect is stronger as the number of particles ahead of the interface increases (Sekhar and Trivedi, 1990).

Dutta and Surappa (1998) have attempted a theoretical analysis of particle interaction with a dendritic S/L interface assuming no convection. They concluded that a higher growth velocity is required to engulf a particle during dendritic solidification than during planar solidification. They claim however, that if the growth velocity is sufficiently small, entrapment may result. Clearly, the convection level in the liquid will play particularly significant role in the PET of dendritic interface since it may prevent any significant interaction with the dendrite tips. Thus, the outcome cannot be predicted only on the basis energy and solute transport.

For a chart of the influence of experimental conditions on the outcome of particle-interface interaction for dendritic interface the reader is referred to Stefanescu (2002) p.240.

13.2 Shrinkage porosity

In most cases if the macroshrinkage is not limited to the riser the casting is rejected. On the other hand, while not always a reason for scrapping the casting, porosity impacts negatively on mechanical properties such as ductility (Uram *et al.* 1958, Samuel and Samuel 1999), dynamic properties (Skallerud *et al.*, 1993) and fatigue life. Confirming earlier work (Major, 1997), Boileau and Allison (2003) have shown that fatigue life decreases in direct relation to the increase in the maximum pore size (Figure 13.16) because of the effect of porosity on crack initiation. In turn, pore size increased as the local solidification time and the secondary dendrite arm spacing increased. Elimination of porosity through hot isostatic pressing yielded significant increase in fatigue life.

13.2.1 The physics of shrinkage porosity formation

During liquid cooling and solidification, a significant amount of the dissolved gas is rejected by the liquid and, if a critical pressure is overcome a gas bubble forms thereby initiating porosity. If the gas bubble is formed in the liquid it floats and failing to find an open liquid surface it interacts with the solid/liquid (S/L) interface eventually forming *gas porosity*. This is not a shrinkage defect. If the gas bubble forms in the mushy zone in the later stages of solidification, after dendrite coherency, it will be entrapped in the dendritic network and nucleate small local shrinkage cavities termed *microporosity* or *microshrinkage*.

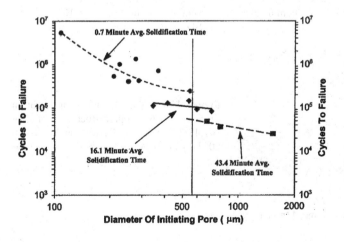

Figure 13.16. The relationship between pore size and fatigue life for aluminum alloy W319-T7 (alternating stress of 96.5 MPa) (Boileau and Allison, 2003). With kind permission of Springer Science and Business Media.

13.2.1.1   Gas porosity

The interaction of a gaseous inclusion with the S/L interface is rather complex. Recent work (Catalina *et al.*, 2004) on aluminum and Al-25% Au has shown that at a certain interface velocity a planar S/L interface may engulf the gas pore. The pore continues to grow during its engulfment. The real time measurements of pore growth in pure Al have revealed that when the pore is relatively far from the S/L interface the mechanism of pore growth is the hydrogen diffusion through the liquid phase. A sudden increase of the pore growth rate occurs when the solutal (*i.e.*, hydrogen) field ahead of the S/L interface begins interacting with the pore.

If the interface is nonplanar the pore is entrapped by the solid. Its size may be of the order of μm to mm.

13.2.1.2   Microporosity (microshrinkage)

In many instances the appearance of the pore is not spherical but follows the dendrites shape (Figure 13.18c,d and Figure 13.19). This defect is termed microporosity or microshrinkage and can range in size from μm to hundreds of μm.

Both the size and the shape of the microshrinkage are important. As shown in Figure 13.17, as the solidification time increases the size of the porosity increases together with the SDAS. In addition, the fatigue strength also decreases (Boileau and Allison, 2003).

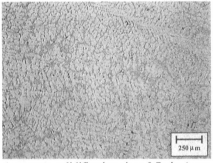

a) average solidification time 0.7min (average SDAS 23 μm)

a) average solidification time 16min (average SDAS 70 μm)

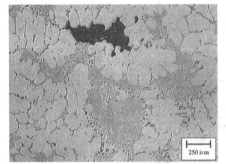

a) average solidification time 43min (average SDAS 100 μm)

Figure 13.17. The effect of solidification time on the microstructure of an Al7.4Si3.3Cu alloy (Boileau and Allison, 2003). Copyright 2003 American Foundry Soc., used with permission.

In well-degassed melts, microshrinkage takes the shape of the interdendritic liquid that remains just before eutectic solidification (Figure 13.18e), and the stress concentration factors resulting from these shapes are much higher than for spherical pores (Berry, 1995).

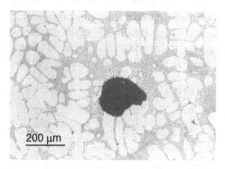

a) optical micrograph; Na-modified Al-Si alloy (Fuoco *et al.*, 1994). Copyright 1994 American Foundry Soc., used with permission.

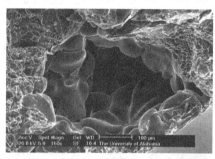

b) SEM image; unmodified Al-Si alloy (Piwonka, 2000)

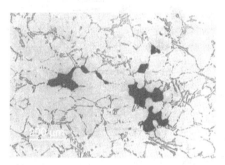

c) optical micrograph of interdendritic micro- shrinkage in Sb refined Al-Si alloy (Fuoco *et al.*, 1994). Copyright 1994 American Foundry Soc., used with permission.

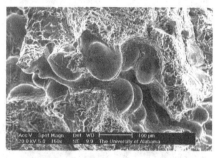

d) SEM image of interdendritic micro- shrinkage in Al-Si alloy (Piwonka, 2000)

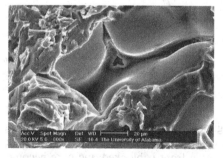

e) SEM image of microshrinkage between the eutectic grains (Piwonka, 2000)

Figure 13.18. Microporosity in aluminum alloys.

The present understanding of microporosity formation is that metal flows toward the region where shrinkage is occurring until flow is blocked, either by solid metal or by a solid or gaseous inclusion. The early theory by Walther *et al.* (1956) assumed that feeding ceases simply because the cross sectional area of the feeding channel continuously decreases during solidification. When the section of the channel has decreased too much the pressure drop ruptures the liquid in the channel, forming a pore. However, pure liquids have high tensile strengths capable of collapsing the surrounding solid, and preventing the fracture of the liquid (Campbell, 1991). Thus, in the absence of gas pressure the tensile stress in the liquid will prevent any discontinuity formation.

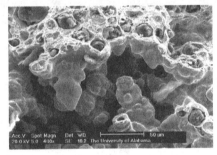

Figure 13.19. Microshrinkage in ductile iron. The figure on the bottom is an enlargement of the one on top (Ruxanda *et al.*, 2001).

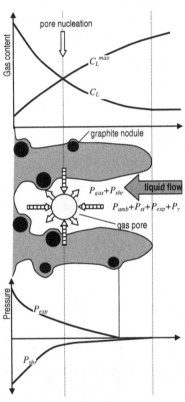

Figure 13.20. Pressure and gas content along the mushy zone of ductile iron (Stefanescu, 2005). Copyright 2005 Maney Publishing., used with permission. www.maney.co.uk/journals/ijcmr.

More recent models assume that when a gas pore appears in the mushy zone during late solidification, after dendrite coherency, it is entrapped in the dendritic network. When the metal flow toward the solidification front is blocked, the pore becomes the starting point of microshrinkage. Thus, microshrinkage formation depends on the nucleation and growth of micro-pores. This concept can be understood from the analysis of the local pressure summarized in Figure 13.20 for the case of ductile iron. Mathematically this is expressed through a pressure balance equation stating that the pressure exerted by gas evolution, P_G, must be higher than the sum of the local pressure in the mushy zone, P_{mush}, and the pressure induced by the surface tension on the pore, P_γ:

$$P_G > P_{mush} + P_\gamma \quad \text{where} \quad P_{mush} = P_{appl} + P_{st} + P_{exp} - P_{shr} \tag{13.24}$$

where P_{appl} is the applied pressure on the mold (e.g., atmospheric pressure), P_{st} is the metallostatic pressure, P_{exp} is the expansion pressure because of phase transformation, and P_{shr} is the negative pressure from resistance to shrinkage induced flow through the fixed dendrite network.

The equation can be rearranged to highlight the driving force on the left hand side:

$$P_{gas} + P_{shr} > P_{appl} + P_{st} + P_{exp} + P_\gamma \tag{13.25}$$

This equation shows that the driving forces for pore occurrence are the gas and shrinkage pressure. If the metal is completely degassed the shrinkage pressure must reach the level of the shear stress of liquid metal for a vacuum pore to occur.

Pore nucleation can occur when the gas dissolved in the liquid, C_L, exceeds the maximum solubility in the liquid, C_L^{max} (Figure 13.20.). The stability of such a pore is controlled by the surface energy pressure on the gas pore given by:

$$P_\gamma = 2\gamma_{LG}/r \tag{13.26}$$

where γ_{LG} is the gas-liquid surface energy and r is the pore radius. This equation suggests that homogeneous nucleation of a pore in the liquid is unlikely since P_γ is immense at the very small initial radius of the pore. Experimental evidence suggests that gas pores nucleate heterogeneously on inclusions that are present in the melt (e.g., Rooy 1993, Roy et al., 1996, Mohanty et al. 1995). This could explain why filtering molten aluminum alloys reduces porosity in castings.

In the case of ductile iron, eutectic graphite produces a positive expansion pressure, P_{exp}, which counteracts the shrinkage and gas pressure. If the expansion pressure equals or exceeds the shrinkage pressure, porosity is completely avoided. However, this requires a completely rigid mold which is rarely the case in practice.

Microshrinkage is a major defect in aluminum-silicon alloys because of the hydrogen dissolved in the liquid. In general, but not always, the amount of microshrinkage decreases with the amount of hydrogen dissolved in the melt. As shown in Figure 13.21, work on small tapered plates and end-chilled plates of Al 7Si0.4Mg found that at short solidification times (small DAS) the pore density rose with increasing hydrogen content. However, at long solidification times the pore density fell as hydrogen content increased (Tynelius et al. 1993).

According to Campbell (2003), this apparently confusing effect can be explained if it is accepted that pore nucleation occurs inside oxide films as thin as 20 nm formed in liquid aluminum during mold filling. Oxide films may fold and produce bifilms (Figure 13.22). If the liquid is assumed full of bifilms (i.e. cracks) the liquid has the potential to initiate pores with negligible difficulty. The bifilms simply open by the separation of their unbonded halves.

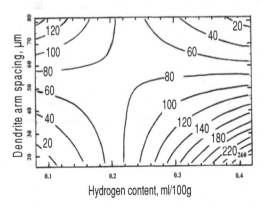

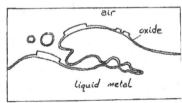

Figure 13.21. The effect of gas content and DAS and on the number of pores per square cm (Tynelius *et al*. 1993). Copyright 1993 American Foundry Soc., used with permission.

Figure 13.22. Schematic view of surface turbulence, acting to fold in an oxide film and bubbles (Campbell, 2003).

Such a pore initiation mechanism can explain the data in Figure 13.21 as follows. At low gas contents and short solidification times the bifilms remain folded and porosity is minimal. As the solidification time (DAS) increases, at the same gas level, the bifilms start opening and gas porosity increases. This is also true for high gas contents and low solidification times. At high gas contents and solidification times, large gas pores are formed, which decreases the area density of pores.

The shrinkage defects at different length scale (macro and micro) interact with one another and must be understood together. A certain correlation seems to exist between the total amount of porosity, open shrinkage cavity and caved surface. Awano and Morimoto (2004) who investigated the shrinkage behavior of Al-Si alloys with various silicon and gas (vacuum degassed, non-treated, and gas enriched) contents, concluded that the total amount of shrinkage is constant at the same silicon level, but varies with the amount of gas in the melt. As summarized in Figure 13.23, the total shrinkage depends little on the amount of gas, but decreases with higher silicon content. Microporosity, on the other hand is a direct function of the gas content, as further evident from Figure 13.24.

Awano and Morimoto further found that for alloys with large solidification interval (mushy solidification) the amount of pipe is constant in the low porosity region (low to moderate gas content), but decreases with increasing porosity in the high porosity region (high gas content). For alloys that solidify with small solidification interval (skin solidification) the amount of caved surface is constant in all porosity regions, while the amount of pipe decreases with increased porosity as the pore generation during the early stage of solidification compensates shrinkage.

Quenching experiments on a Al-4% Si alloy demonstrated that most of the pores are formed immediately after the start of primary solidification (Figure 13.25).

This analysis reveals the complexity of the problem. Computational modeling of microshrinkage formation must describe phenomena such as porosity nucleation and growth, elastic and plastic deformation of the solidification shell, and interdendritic flow.

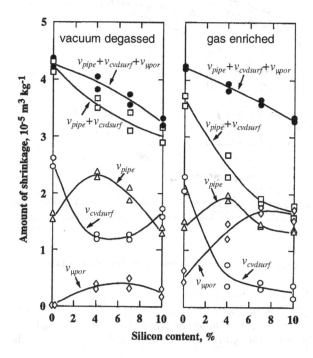

Figure 13.23. Effect of silicon content on the amount of shrinkage defects in Al-Si alloys. v_{pipe} is the volume of pipe, $v_{cvdsurf}$ is the volume of caved surface and $v_{\mu por}$ is the volume of microporosity (Awano and Morimoto, 2004). Copyright 2004 Maney Publishing., used with permission. www.maney.co.uk/journals/ijcmr.

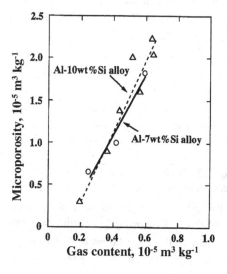

Figure 13.24. Correlation between microporosity and hydrogen content in two Al-Si alloys (Awano and Morimoto, 2004). Copyright 2004 Maney Publishing., used with permission. www.maney.co.uk/journals/ijcmr.

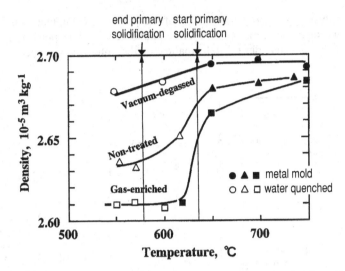

Figure 13.25. Effect of quenching temperature on the density of Al-4Si alloys (Awano and Morimoto, 2004). Copyright 2004 Maney Publishing., used with permission. www.maney.co.uk/journals/ijcmr.

13.2.2   Analytical models including nucleation and growth of gas pores

These models include in the analysis the contribution of rejected gas to microporosity formation. The full pressure balance Eq. (13.25) is used. It was also recognized that straight channel flow is a poor approximation of the flow through the tortuous interdendritic channels for alloys with large solidification intervals (mushy solidification).

Piwonka and Flemings (1966) were the first to consider the role of the dissolved gas pressure. Eq. (13.25) was written as $P_{gas} > P_{st} - P_{shr} + P_\gamma$. The gas pressure term for a diatomic gas such as hydrogen was formulated as:

$$P_{gas} = \left[\frac{v_i}{K_S(1 - f_L) + K_L f_L} \right]^2 \tag{13.27}$$

where v_i is the initial gas volume in the melt, and K_S and K_L are the equilibrium constants of the gas dissolution reaction for the solid and liquid, respectively.

Derivation of the gas pressure term in Eq. (13.27) (Piwonka and Flemings, 1966). A diatomic gas, G_2, dissolves in liquid metal according to the reaction: $(1/2)G_{2\,gas} \leftrightarrow \underline{G}$. Assuming that the gas behaves ideally, so that the activity, a_G, is proportional to the partial pressure, P_{G2}, the equilibrium constant of this reaction is $K_L = a_G / \sqrt{P_{G2}}$. Further assuming that the activity of the liquid follows Raoult's law (activity coefficient = 1), the weight percent gas in the liquid is given by $wt\%\underline{G} = K_L \sqrt{P_{G2}}$. This is Sievert's law. It can be written in terms of volumes of gas for both liquid and solid, as follows:

$$v_L = K_L \sqrt{P_G} \quad \text{and} \quad v_S = K_S \sqrt{P_G}$$

Mass balance dictates that $v_S(1 - f_L) + v_L f_L = v_i$, where v_i is the initial volume of gas in the liquid. Combining these equations, we obtain Eq. (13.27).

Piwonka and Flemings were also the first to use Darcy's law for the flow in a porous medium to describe the liquid flow through the coherent network of dendrites. They used a tortuosity factor, $\xi \geq 1$, to account for the fact that the liquid flow channels are not straight and smooth. After some manipulations (Stefanescu, 2005) an equation that shows dependency on the primary dendrite arm spacing (DAS), λ_I, is obtained:

$$P_{st} - P_{shr} = \frac{32\pi\mu\beta\varsigma^2 L^2}{(1-\beta)} \left(\frac{\xi\lambda_I}{r^2\pi r_{feed}^2} \right)^2 \tag{13.28}$$

where β is the shrinkage ratio, r_{feed} is the radius of the cylindrical feeding channel, L is the length of the feed (mushy) zone, r is the radius of the feeding capillary, and ς is the thermal function in Eq. (6.27). In general, $\xi^2 / \left(\pi r_{feed}^2 \, n \right) \ll 1$. Thus, by comparison with Eq. (6.27) pores in mushy freezing alloys will be much finer than those at the centerline of pure metals.

To verify the square dependency of the pressure drop on the DAS, they conducted some experiments and demonstrated that this is true at $f_L < 0.3$. The surface tension pressure on the gas pore was calculated with Eq. (13.26). It was assumed that there was no barrier to the nucleation of pores. This seems to be a good assumption for metal casting, where pores nucleate on inclusions and second phases.

However, the surface tension pressure, P_γ, is extremely high at the beginning of pore formation when $r \sim 0$. This requires some further assumption regarding pore nucleation. Kubo and Pehlke (1985) argued that the pores nucleate at the root of the secondary dendrite arms. The free energy change of pore formation is:

$$\Delta G = v(P_G - P_{mush}) + A_{SG} \cdot \gamma_{SG} + A_{LG} \cdot \gamma_{LG} - A_{SG} \cdot \gamma_{SL}$$

where v is the volume of the pore, and A is the area between the phases interfaces. If homogeneous nucleation is assumed, the first right-hand-term, which represents the free energy change during the liquid-to-gas phase transformation, must have a very large negative value to overcome the effect of P_γ. However, in the Kubo and Pehlke model, since the effect of surface energy is reduced by $A_{SG} \cdot \gamma_{SL}$ a large negative pressure is not required. The initial pore radius was thus assumed half of the secondary arm spacing.

A more complete mathematical description of microporosity formation on gas pores or on oxide films requires numerical treatment of the problem. It will be discussed is more detail in Chapter 13.

13.2.3   Analysis of shrinkage porosity models and defect prevention

A number of criteria-based models for shrinkage defect formation in the last solid to form were presented in section 6.5. Although the concepts used are of the macro-scale variety, the real intent of these models is to predict microporosity formation. Because microshrinkage requires the presence of gas and inclusions to nucleate pores and purely thermal analyses ignore both, criterion functions have not been overwhelmingly successful. Laurent and Rigaut (1992) concluded that they could account for only about 75% of the porosity that exists in castings. The effect of gas content on the use of criteria functions can be dealt with in several ways. The first way is to simply redefine that the proper use of criteria functions depends on having gas and inclusion-free melts (difficult to attain in practice). Alternatively, one can develop a customized function for a particular foundry that includes the density of a reduced pressure test sample, determined empirically (Chiesa et al. 1998, Chiesa and Mammen 1999). Yet another approach is to use statistic models that include not only the amount of gas but also the interaction between the amount of gas and some solidification parameters such as solidus velocity and final solidification time (Tynelius et al., 1993).

While criterion functions are not reliable in predicting microshrinkage in aluminum alloys, they are quite acceptable for ferrous castings (with the exception of ductile iron), where oxide films do not form, and gas evolution is not a problem, and for some copper based alloys. Indeed, the classical feeding distance rules for iron (Wallace and Evans, 1958), steel (Bishop et al. 1955, Caine 1950), and copper-base alloys (Weins et al. 1964) are essentially thermal criterion functions recast in geometrical terms.

Numerical models are more complete in the scope of the underlying physics, but have to resort to numerous simplifying assumptions because of lack of data or mathematical complexity. They have many fitting parameters. Typically, only one experimental fact is used in validation – final percentage of porosity. While the influence of alloy composition on hydrogen solubility is computed in some models, in most cases its effect on microstructure is not considered. However, there are indications that, because of the change in the phase morphology, variations in composition can affect the permeability values. There is also experimental evidence that minor additions such as carbon and hafnium in superalloys (Chen et al., 2004) reduce porosity.

Concerning pore nucleation, Poirier (1998) has shown that it is improbable for simple oxides to act as nuclei for pores. Nevertheless, oxide clusters or folded oxide films probably do.

From the previous discussion it appears that, unless the gas content of the liquid is reduced to a value below the solid solubility, the only way to prevent microporosity from forming is to apply pressure during solidification. Indeed, for magnesium alloys it was found that application of pressure greater than 40 MPa completely suppressed microporosity (Yong and Clegg, 2004). Possible methods to reduce the size and number of pores include melt degassing, filtering of the metal to remove nuclei and oxide films, and counter-gravity casting to decrease turbulence and to carry the remaining inclusions out of the casting cavity. A postmortem solution is the use of hot isostatic pressing, which may completely eliminate microshrinkage (Lei *et al.* 1998, Boileau and Allison 2003, Bor *et al.* 2004).

Finally, all models predict that microporosity occurs unless gas contents are below the solid solubility limit in the liquid. Since the most advanced models rely on pore nucleation and growth to calculate microporosity, it follows that in the absence of dissolved gas in the melt there will be no porosity. It may be argued that when the feeding channel is totally closed, the pressure drop may fracture the liquid and produce voids. On the other hand, burst feeding may then release the pressure, resulting in surface shrinkage but no microshrinkage. Unfortunately, the author could not find any report on experiments attempting to verify the theory that, in the absence of gas, there is no microshrinkage.

References

Asthana R. and S.N. Tewari, 1993, *J.Mater. Sci.* **28**:5414

Azouni, M.A., W. Kalita and M. Yemmou, 1990, *J. Crystal Growth* **99**:201

Awano Y. and K. Morimoto, 2004, *Inter. J. of Cast Metals Res.* **17**:107

Berry J.T., 1995, *AFS Trans.* **103**:837

Bishop H.F., E.T. Myskowski, and W.S. Pellini, 1955, *AFS Trans.* **63**:271

Boileau J.M.and J.E. Allison, 2003, *Met. Mater. Trans.* **34A**:1807

Bolling G.F. and J. Cissé. 1971, *J. Crystal Growth* **10**:56

Bor H.Y., C. Hsu and C.N. Wei, 2004, *Materials Chemistry and Physics* **84**:284

Caine J.B., 1950, *AFS Trans.* **58**:261

Campbell J., 1991, *Casting*, Butterworth Heinemann, Oxford, UK

Campbell J., 2003, in *Modeling of Casting, Welding and Adv. Solidif. Processes X*, eds. D.M. Stefanescu, J. Warren, M. Jolly and M. Krane, TMS, Warrendale PA, p.209

Casses P. and M.A. Azouni, 1993, *J. Crystal Growth* **130**:13

Catalina A.V. and D.M. Stefanescu, 1999, in: *Proc. of Modeling of Casting and Solidification Processes IV*, CP. Hong, J.K Choi and D.H. Kim eds., Center for Computer-Aided Mat. Proc., Seoul, p.3

Catalina A.V. and D.M. Stefanescu, in *Solidification 99*, 1999, W.H. Hofmeister, J.R. Rogers, N.B. Singh, S.P. Marsh and P.W. Vorhees eds., TMS, Warrendale, PA p.273

Catalina A.V., S. Mukherjee, and D.M. Stefanescu, 2000, *Metall. and Mater. Trans.* **31A**:2559

Catalina A.V., D.M. Stefanescu and S. Sen, 2003, in: *Modeling of Casting, Welding and Advanced Solidification Processes X*, D.M. Stefanescu et al. eds., TMS, Warrendale PA p. 125

Catalina A.V., D.M. Stefanescu, S. Sen and W. Kaukler, 2004, *Metall. and Mater. Trans.* **35A**:1525

Chen Q.Z., Y.H. Kong, C.N. Jones and D.M. Knowles, 2004, *Scripta Materialia* **51**:155

Chernov A.A., D.E. Temkin, and A.M. Mel'nikova, 1976, *Sov. Phys. Crystallogr.* **21**:369

Chernov A.A., D.E. Temkin, and A.M. Mel'nikova, 1977, *Sov. Phys. Crystallogr.* **22**:656

Chiesa F., M. Mammen and L.E. Smiley, 1998, *AFS Trans.* **106**:149

Chiesa F. and J. Mammen, 1999, *AFS Trans.* **107**:103

Dutta B. and M.K. Surappa, 1998, *Metall. and Mater. Trans.* **29A**:1319

Endo A., H.S. Chauhan, T. Egi, Y. Shiohara, 1996, *J. Mater. Res.* **11**:795

Fasoyinu Y. and C.E. Schvezov, 1990, in: *Proceedings F. Weinberg Intl. Symposium on Solidification Processing*, Ontario, Pergamon Press p.243

Fuoco R., H. Goldenstein, and J.E. Gruzleski, 1994, *AFS Trans.* **102**:297

Garvin J.W. and H.S. Udaykumar, 2004, *J. Crystal Growth*, **267**:
Hadji L., 1999, in: *Solidification 99*, W.H. Hofmeister, J.R. Rogers, N.B. Singh, S.P. Marsh and P.W. Vorhees, TMS, Warrendale, PA p.26
Hadji L., 1999a, *Phys. Rev. E* **60**:6180
Hadji L., 2001, *Physical Rev. E*, **64**:051502
Hadji L., 2003, *Scripta Materialia* **48**:665
Han Q. and J.D. Hunt, 1995, *ISIJ International*. **35**:693
Hecht U. and S. Rex, 2001, in: *The Sci. of Casting and Solidification*, D.M. Stefanescu, R. Ruxanda, M. Tierean, and C. Serban eds., Editura Lux Libris, Brasov, Romania p.53
Hecht U. and S. Rex, 1997, *Met. Trans.* **28A**:867
Juretzko F.R, B.K. Dhindaw, D.M. Stefanescu, S. Sen, and P. Curreri, 1998, *Metall. and Mater. Trans.* **29A**:1691
Kaptay G., 1999, *Metall. and Mater. Trans.* **30A**:1887
Kaptay G., 2000, *Metall. and Mater. Trans.* **31A**:1695
Kennedy D.O., 1991, *Advanced Mater. & Proc.* **6**:42
Kennedy A.R. and Clyne T.W., 1991, *Cast Metals* **4**,3:160
Kim J.K. and P.K. Rohatgi, 1998, *Metall. and Mater. Trans.* **29A**:351
Körber C., G. Rau, M.D. Cosman, and E.G. Cravalho, 1985, *J. Crystal Growth* **72**:649
Kubo K. and R.D. Pehlke, 1985, *Met. Trans.* **16B**:359
Kuo V.H.S. and Wilcox W.R., 1973, *Sep. Sci.* **8**:375
Laurent V. and C. Rigaut, 1992, *AFS Trans.* **100**:399
Lawrynowicz D.E., B. Li and J. Lavernia, 1997, *Met. Trans.* **28B**:877
Leal L.G., 1992, *Laminar Flow and Convective Transport Processes: Scaling Principles and Asymptotic Analysis*, Butterworth-Heinemann
Lei C.S., E.W. Lee and W.E. Frazier, 1998, in *Advances in Aluminum Casting Technology*, eds. M. Tiryakioglu and J. Campbell, ASM, Materials Park, OH p.113
Major J.F., 1997, *AFS Trans.* **105**:901
Mohanty P.S., F.H. Samuel and J.E. Gruzleski, 1995, *AFS Trans.* **103**:555
Mukherjee S. and D.M. Stefanescu, 2000, in *State of the Art in Cast Metal Matrix Composites in the Next Millennium*, P.K. Rohatgi ed., The Minerals, Metals & Materials Society, Warrendale PA p.89
Mukherjee S. and D.M. Stefanescu, 2004, *Metall. and Mater. Trans.* **35A**:613
Mukherjee S., M.A.R. Sharif and D.M. Stefanescu, 2004, *Metall. and Mater. Trans.* **35A**:623
Nestler B., A.A. Wheeler, L. Ratke, C. Stöcker, 2000, *Physica D* **141**:133
Ode M., J.S. Lee, S. G. Kim, W.T. Kim, and T. Suzuki, 2000, *ISIJ International* **40**:153
Omenyi S.N. and A.W. Neumann, 1976, *J. Applied Physics* **47**:3956
Pang H., D.M. Stefanescu and B.K. Dhindaw, 1993, in: *Proceedings of the 2nd International Conference on Cast Metal Matrix Composites*, D. M. Stefanescu and S. Sen editors, American Foundrymen's Soc. p.57
Piwonka T.S., 2000, in: *Proceedings of the Merton C. Flemings Symposium on Solidification and Materials Processing*, R. Abbaschian, H. Brody, and A. Mortensen eds., TMS, Warrendale Pa. p.363
Piwonka T.S. and M.C. Flemings, 1966, *Trans. AIME* **236**:1157
Poirier D.R., 1998, in: *Modeling of Casting, Welding and Advanced Solidification Processes VII*, B.G. Thomas and C. Beckermann, eds., TMS, Warrendale, PA p.837
Pötschke J. and V. Rogge, 1989, *J. Crystal Growth* **94**:726
Premkumar M.K., M. G. Chu, 1993, *Met. Trans.* **24A**:2358
Rempel A.W. and M.G. Worster, 2001, *J. Crystal Growth* **223**:420
Rooy E.L., 1993, *AFS Trans.* **101**:961
Roy N., A.M. Samuel and F.H. Samuel, 1996, *Met. Mater. Trans.* **27A**:415
Rubinow S.I. and J. B. Keller, 1961, *J. Fluid Mechanics.* **11**:447

Ruxanda R., L. Beltran-Sanchez, J. Massone, and D.M. Stefanescu, 2001, *Trans. AFS* **109**: 1037

Saffman P.G., 1965, *J. Fluid Mechanics*. **22**:385

Samuel A.M. and F.H. Samuel, 1999, *Met. Mater. Trans.* **26A**:2359

Sasikumar R., T.R. Ramamohan and B.C. Pai, 1989, *Acta Metall.* **37**:2085

Sasikumar R. and T.R. Ramamohan, 1991, *Acta Metall.* **39**:517

Schvezov C., in *Solidification 99*, 1999, W.H. Hofmeister, J.R. Rogers, N.B. Singh, S.P. Marsh and P.W. Vorhees eds., TMS, Warrendale, PA p.251

Sekhar J.A. and R. Trivedi, 1990, in: *Solidification of Metal Matrix Composites*, P. Rohatgi, ed., TMS, Warrendale, PA p.39

Sen S., W.F. Kaukler, P. Curreri, and D.M. Stefanescu, 1997, *Metall. and Mater. Trans* **28A**:2129

Sen S., Dhindaw B.K., Stefanescu D.M., Catalina A.V. and Curreri P., 1997a, *J. Crystal Growth* **173**:574

Shangguan D.K., S. Ahuja, D.M. Stefanescu, 1992, *Metall. Trans.* **23A**:669

Shelton R.K. and D.C. Dunand, 1996, *Acta mater.* **44**:4571

Shibata H., H. Yin, S. Yoshinaga, T. Emi and M. Suzuki, 1998, *ISIJ Int.* **38**:149

Shiohara Y., A. Endo, Y. Watanabe, H. Nomoto, and T. Umeda, 1997, in: *Solidification Processing 97*, J. Beech and H. Jones, eds., Univ. of Sheffield, UK p.456

Skallerud B., T. Iveland and G. Harkegard, 1993, *Eng. Fract. Mech.* **44**:857

Smith R.P., D. Li, W. Francis. J. Chappuis and A.W. Neumann, 1993, *J. Colloid Interf. Sci.* **157**:478

Stefanescu D.M., 2002, *Science and Engineering of Casting Solidification*, Kluwer Academic/Plenum Publishers, New York

Stefanescu D.M., 2005, *Int. J. of Cast Metals Res.* **18**(3):129-143

Stefanescu D.M., Dhindaw B.K., Kacar S.A., and Moitra A., 1988, *Met. Trans.* **19A**:2847

Stefanescu D.M., R.V. Phalnikar, H. Pang, S. Ahuja and B.K. Dhindaw, 1995, *ISIJ International* **35**:700

Stefanescu D.M. and A.V. Catalina, 1998, *ISIJ International* **38**:503

Stefanescu D.M., Juretzko F.R, Dhindaw B.K., Catalina A., Sen S., and Curreri P.A., 1998, *Metall. and Mater. Trans.* **29A**:1697

Stefanescu D.M., A.V. Catalina, F.R. Juretzko, S. Mukherjee, S. Sen and B.K. Dhindaw, 2000, in *Microgravity Research and Applications in Physical Sciences and Biotechnology*, SP-454 vol. 1 Sorrento, European Space Agency p. 621

Surappa M.K. and P.K. Rohatgi, 1981, *J. Mater. Sci. Lett.* **16**:765

Tynelius K., J.F. Major and D. Apelian, 1993, *Trans. AFS* **101**:401

Torza S. and Mason S.G., 1969, *Science* **162**:813

Uhlmann D.R., B. Chalmers and K.A. Jackson, 1964, *J. Appl. Phys.* **35**:2986

Uram S.Z., M.C. Flemings and H.F. Taylor, 1958, *AFS Trans.* **66**:129

Wallace J.F. and E.B. Evans, 1958, *AFS Trans.* **66**:49

Walther W.D., C.M. Adams and H.F. Taylor, 1956, *AFS Trans.* **64**:658

Weins M.J., J.L.S. Bottom and R.A. Flinn, 1964, *Trans. AFS* **72**:832

Wu Y., H. Liu and E.J. Lavernia, 1994, *Acta Metall. Mater.* **42**:825

Yaohui L., H. Zhenming, L. Shufan, Y. Zhanchao, 1993, *J. Mat. Sci.* **12**:254

Yong M.S. and A.J. Clegg, 2004, *J. of Materials Processing Technology* **145**:134

Youssef Y.M., R.J. Dashwood, P.D. Lee, 2005, *Composites* **A36**:747

Zubko A.M., V.G. Lobanov, and V.V. Nikonova, 1973, *Sov. Phys. Crystallogr.* **18**:239

NUMERICAL MICRO-MODELING OF SOLIDIFICATION

In the previous sections, the basic elements required to build a macro-solidification model were introduced. Until recently, casting properties were evaluated based on empirical correlations between the quantity of interest and some significant output parameter of the macro-model. Most commonly used are temperature and cooling rate, $\dot{T}$. However, the revolutionary development of numerical analysis and computational technology opened the door for prediction of microstructure evolution during solidification and subsequent cooling to room temperature through models that coupled macro-transport (MT) and transformation kinetics (TK) models. This in turn made possible prediction of mechanical properties based on direct micro-structure-properties correlations.

The age of computational modeling of microstructure evolution was started by the brilliancy of a scientist, W. Oldfield (1966), who developed a computer model that could calculate the cooling curves of lamellar graphite iron. To the best of our knowledge this is the first attempt to predict solidification microstructure through computational modeling, and the first attempt to validate such a model against cooling curves.

There are many different approaches to the description of MT, including multi-phase and solid movement, as well as to the description of TK. In principle, the methods used for the computational modeling of TK belong to one of the followings (see Figure 14.1):

- deterministic methods
- probabilistic/stochastic methods
- phase field methods

Deterministic modeling of TK is based on the solution of the continuum equations over some volume element. They are mostly concerned with the description of the solid fraction evolution over time.

Stochastic models are based on the on the local description of the material combined with some evolutionary rules, rather than solving the local integral and differential equations. They can output a graphical description of the microstructure (the computer becomes a dynamic microscope).

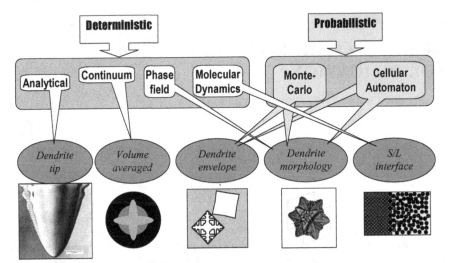

Figure 14.1. Classification of computational models for microstructure evolution.

The phase field method is an integrated simulation technique that typically solves two parabolic partial differential equations, where one of them governs the evolution of the phase field variable, which describes the type of phase, solid or liquid, present in the system. The solution of the phase field equation is affected by the solution of the second equation, either the solute or heat conservation equation, depending on the controlling process assumed for solidification.

14.1 Deterministic models

14.1.1 Problem formulation

A good description of the cooling rate throughout the casting can be obtained from the energy transport equation coupled with mass and momentum transport. The most important governing equation is derived from Eqs. 5.3 and 5.4:

$$\frac{\partial T}{\partial t} + \nabla \cdot (\mathbf{V}T) = \nabla \cdot (\alpha \nabla T) + \frac{\Delta H_f}{\rho c} \frac{\partial f_s}{\partial t} \tag{14.1}$$

As discussed previously, this equation can be solved if an appropriate description of f_S as a function of time or temperature is available. The heat evolution during solidification described in the source term depends strongly on both the macro transport (MT) of energy from the casting to the environment, and on the transformation kinetics (TK).

If it is assumed that TK does not influence MT, the two computations can be performed uncoupled. Typically, $\dot{T}$ is evaluated with an MT code, and then, the microstructure length scale that includes phase spacing, λ, and volumetric grain density, N, are calculated based on empirical equations as a function of $\dot{T}$. This methodology is presently used by classic macro-transport models that solve the mass, energy, momentum, and species macroscopic conservation equations, as discussed in the preceding chapter. They are inherently inaccurate in their attempt to predict microstructure because of the weakness of the uncoupled MT-TK assumption. Indeed, since, in effect, MT and TK are coupled during solidification, accurate prediction of microstructure evolution revolves around modeling both MT and TK, and then coupling them appropriately.

Consequently, the problem is to describe $f_S(x,t)$ in terms of transformation kinetics, and to select appropriate boundary conditions. The governing equation couples MT and TK through f_S, if f_S is calculated from TK.

While Oldfield's model for microstructural evolution included a MT computer model for heat flow across a cylinder similar to FDM, in some early models used analytical modeling for MT and time stepping procedures to calculate the cooling curve and the fraction solid evolution (*e.g.* Stefanescu and Trufinescu, 1974, Fras 1975, Aizawa 1978).

For complete numerical formulations the first step in building an MT-TK solidification model is to divide the computational space of the casting in macro-volume elements within which the temperature is assumed uniform (Figure 14.2).

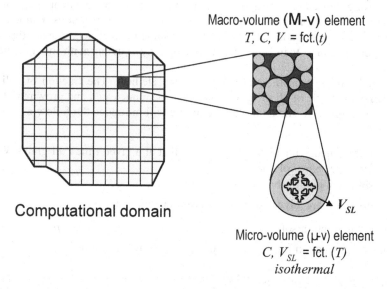

Macro-volume **(M-v)** element
T, C, V = fct.(t)

Computational domain

V_{SL}

Micro-volume (µ-v) element
C, V_{SL} = fct. (T)
isothermal

Figure 14.2. Division of the computational space for deterministic modeling of coupled macro-transport and transformation kinetics.

In the first numerical models (*e.g.*, Su *et al.* 1984, Stefanescu and Kanetkar 1986, Thévoz *et al.* 1989), the basic simplifying assumption was that the solid phase has zero velocity (*one velocity models*), meaning that once nucleated the grains remain in fixed positions. Grain coalescence and dissolution are ignored. The macro-volume element was assumed closed to mass and momentum transport but open to energy transport. Then, each of these elements was further subdivided in micro-volume elements, typically spherical, based on some nucleation law (Figure 14.2). Within each of these micro-volume elements only one spherical grain is growing at a velocity V, dictated by kinetic considerations. The isothermal micro-volume elements are considered open to species transport.

Assuming, for the time being, that the macro-volume element is closed to mass and momentum transport, and that the solid is fixed ($\mathbf{V} = 0$). The governing equation becomes:

$$\frac{\partial T}{\partial t} = \nabla(\alpha \nabla T) + \frac{\Delta H_f}{\rho c} \frac{\partial f_s}{\partial t} \qquad (14.2)$$

It must be noted that the assumption $V = 0$ is always valid once grain coherency is reached, which typically happens at about 0.2 to 0.4 fraction of solid. Thus, for many cases this is a reasonable approximation.

In later models (*e.g.*, Wang and Beckermann 1996) the macro-volume element is open to energy, mass and momentum transport. The solid may move freely with the liquid. Within each macro-volume element, solid grains are allowed to nucleate based on some empirical nucleation laws. Since the macro-volume element is open, solid grains can be transported in or out of the volume element. The volume averaging technique is used to manage the various phase quantities, temperature, species concentration, and velocity, as well as some microstructure quantities such as number of grains. $\mathbf{V}$ in Eq. (14.1) is obtained by coupling with the momentum equations.

Consider now an isothermal macro-volume element. The fraction of solid in this element is made of all the grains that have been nucleated and grown until that time, N, multiplied by the grain volume, v. We will introduce the following simplifying assumptions:

- instantaneous nucleation; this means that at a given time t all grains N will have the same volume;
- spherical grains, that is $v(t) = (4/3)\pi r(t)^3$;
- no grains are advected in the volume element, that is $\mathbf{V} = 0$;

Then, the fraction solid is $f_s = 4/3(N \pi r^3)$ and $df_s = 4N \pi r^2 dr$, which gives the following equation for the time evolution of the fraction solid:

$$df_s/dt = 4N \pi r^2 \, dr/dt = 4N \pi r^2 V \qquad (14.3)$$

where V is the solidification velocity of the grain, typically a function of under-cooling ($V = \mu \Delta T^2$). Toward the end of solidification the grains come in contact with one another and grain impingement occurs. To account for slowing of grain growth after grain impingement occurs a correction factor $(1 - f_S)$ is added to the equation:

$$df_s/dt = 4N \pi r^2 V(1 - f_s) \qquad (14.4)$$

This correction factor was originally derived by Kolmogorov (1937), Avrami (1939) and Johnson-Mehl (1939). A modified expression was proposed by Hillert (1955) as $(1 - f_S)^i$, where i is greater than unity if the transformed phase is aggregated and less than unity if it is random. Since equiaxed solid grains are typically random, Chang et al. (1992) used $(1 - f_S)^{f_s}$. Note that integration of Eq. (14.4) results in: $f_s = 1 - \exp(-(4/3)\pi N r^3)$.

In a general format, the evolution of the fraction of solid in time can be written as:

$$\frac{\partial f_s}{\partial t} = \frac{\partial}{\partial t}(N v) = \frac{\partial N}{\partial t} v + N \frac{\partial v}{\partial t} \qquad (14.5)$$

where N is the volumetric grain (nuclei) density and v is the grain volume. Here, the RHT1 describes the contribution from new nuclei (e.g., heterogeneous nucleation) and the RHT2 that of increased grains volume (grain growth).

The following conservation equation can be written to describe the contribution from new nuclei (Ni and Beckermann, 1991):

$$\frac{\partial N}{\partial t} = \dot{N} = \frac{dN}{dt} + \nabla \cdot (\mathbf{V_s} N) \qquad (14.6)$$

where $\dot{N}$ is the net nucleation rate accounting for the various mechanisms of nucleation and $dN(t)/dt$ is the local nucleation rate (rate of formation of grains per unit volume). The RH2 is the flux of grains due to a finite solid velocity.

The change in grain volume can be calculated as: $\partial v/\partial t = A V$, where V is the growth velocity of the grain, and A is its surface area.

In a simplified analysis, it can be assumed that the grains are of spherical shape. Then Eq. (14.5) describing the evolution of fraction of solid becomes (Stefanescu, 2001):

$$\frac{df_s}{dt} = 4\pi \left(\dot{N} \frac{r^3}{3} + N r^2 \frac{dr}{dt} \right)(1 - f_s) \qquad (14.7)$$

where $r(x,t)$ is the grain radius.

If instantaneous nucleation is assumed and if $\mathbf{V_S} = 0$, $\dot{N} = 0$ and the evolution of solid fraction simplifies to Eq. (14.4). The problem is then to formulate dN/dt and V (or $r(x,t)$). These calculations are based on specific nucleation kinetics and interface dynamics, to be discussed. If dN/dt is known, df_s/dt can be calculated and used in Eq. (14.2) to couple macro-transport with transformation kinetics. An enthalpy formulation can then be used.

14.1.2   Coupling of MT and TK codes

The coupling of the macro-a and micro-calculations can be done through the enthalpy method. However, much smaller time steps than obtained from the Fourier number are required to avoid massive calculated release of latent heat which will raise the temperature above the equilibrium temperature. This will result in an oscillatory behavior of the cooling curve around the equilibrium temperature.

To improve accuracy and/or reduce computational time, several coupling techniques between MT and TK have been developed. They include the Latent Heat Method (LHM) by Stefanescu and Kanetkar (1986) and Kanetkar *et al.* (1988), the Micro-Enthalpy Method (MEM) by Rappaz and Thévoz (1987) and the Micro Latent Heat Method (MLHM) by Nastac and Stefanescu (1992).

When the LHM is used, the governing equation is obtained by combining equations Eqs. (6.1) and (6.2):

$$\frac{\partial T}{\partial t} = \alpha \frac{\partial^2 T}{\partial x^2} + \frac{\Delta H_f}{c} \frac{\partial f_s}{\partial t}$$

Discretization of this equation gives:

$$T^{new} = T^{old} + Fo\left(T_E^{old} - 2T^{old} + T_W^{old}\right) + \frac{\Delta H_f}{c} \Delta f_s \qquad (14.8)$$

Here, Δf_S is calculated directly from nucleation and growth laws, *e.g.* Eq. (14.7). The LHM fully couples MT and TK and is the most accurate. However, because the time step increment necessary to solve the heat flow equation is limited by the microscopic phenomena, it has to be much smaller than the recalescence period in order to properly describe the microscopic solidification. Thus, much longer computational times are required as compared with codes that do not include TK (Table 14.1).

The MEM and the MLHM compute heat flow and microstructure evolution at two different scales (Nastac, 2004). At the macro level the heat flow is calculated through Eq. (5.1) or (5.3) without a source term using a large time step. Thus, the macro-enthalpy change, ΔH_{macro}, or the macro-temperature change, ΔT_{macro}, are obtained. Assuming constant rate of heat extraction during the L/S transformation, the micro-enthalpy (ΔH_{miacro}) or micro-temperature (ΔT_{miacro}) change is calculated using a much smaller time step by including the source term in Eq. (5.1) or (5.2). Thus the fraction solid evolution can be calculated from nucleation and growth kinetics. These two methods are partially coupled methods. While not as accurate as LHM, they substantially decrease the CPU time (Table 14.1). As for accuracy, in the case of eutectic gray iron, for cooling rates between 1 and 5 °C/s, a maximum error of 0.3% was calculated in the prediction of the recalescence and solidus temperatures with MLHM as compared to the LHM.

Table 14.1. Relative CPU-time for an explicit FDM scheme (Nastac and Stefanescu, 1992).

Coupling method	Enthalpy method (no TK)	Latent heat method	MEM or MLHM
CPU-time, %	100	510	106

For further details regarding the coupling schemes and stability criteria the reader is referred to Nastac (2004).

14.1.3   Models for dendritic microstructures

Two main techniques have been used to simulate the evolution of dendrites morphology during solidification: front tracking techniques and volume averaging techniques.

Front tracking models

Front tracking (FT) models (Juric and Tryggvason 1996, Jacot and Rappaz 2002, Zhao et al. 2003) attempt to simulate time-dependent dendritic growth. They follow the dynamics of the sharp SL interface by solving the heat and species conservation equations with appropriate interface conditions. The FT method explicitly provides the location and shape of the SL interface at all times by means of a set of extra marker nodes, independent of the mesh, that are defined at every time and move according to the interface conditions. FT models can handle the discontinuous properties at the interface, interfacial anisotropy and topology changes. However, the algorithm for explicitly tracking the position of the interface involves complex numerical calculations, particularly when 3D computations are considered (Udaykumar et al., 1999). Thus, this method seems inadequate to simulate multi-dendrite microstructures with well-developed side branches realistically, because of the solving difficulties caused by the explicit nature of the algorithm.

Volume averaged models

From the foregoing discussions, it must be obvious to the reader that at the present state of the art it is impossible to produce models that describe the complexities of dendrite growth, and are simple enough to have engineering usefulness. Accordingly, a number of simplified models, based on the volume averaging technique, have been proposed.

A summary of the geometry of the equiaxed grains assumed by some of the models that will be discussed is shown in Figure 14.3. The dendrite envelope is defined as the surface that touches all the tips of the primary and secondary dendrite arms. It includes the solid and the intradendritic liquid. It is the interface separating the intradendritic and extradendritic liquid phases. The equivalent dendrite envelope is a sphere that has the same volume as the dendrite envelope (r_E on Figure 14.3a). The equivalent dendrite volume is the sphere having the same volume as the solid dendrite.

Two main types of dendrites are seen: "condensed" and "extended" dendrites. The condensed dendrites have the equivalent dendrite volume of the same order with the volume of the dendrite envelope ($r_S \cong r_E$). They form typically at high undercooling. In the case of the extended dendrite the equivalent dendrite envelope is significantly higher than the equivalent dendrite volume ($r_E \gg r_S$). They are common at low undercooling, when primary arms develop fast with little secondary and higher order arms growth. The final radius of the grain, r_f, is in fact the micro-volume element. The main problem to solve is to formulate the radius of the equiaxed grain, or the grain growth velocity.

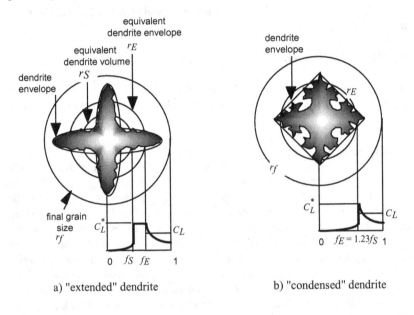

a) "extended" dendrite   b) "condensed" dendrite

Figure 14.3. Assumed morphologies and associated concentration profiles of solidifying equiaxed dendrites. With kind permission of Springer Science and Business Media.

To calculate the radius of the equiaxed grain it is necessary to develop a model that describes the growth of a simplified dendrite. From Figure 14.3 a number of solid fractions can be defined as follows:

$$f_S = v_S/v_f = \left(r_S/r_f\right)^3 \quad \text{fraction of solid in the volume element} \qquad (14.9a)$$

$$f_E = v_E/v_f = \left(r_E/r_f\right)^3 \quad \text{volume fraction of dendrite envelope} \qquad (14.9b)$$

$$f_i = f_S/f_E \qquad\qquad\qquad \text{internal fraction of solid} \qquad (14.9c)$$

These three relationships fully define the growth and morphology of the dendrite. Spherical envelopes must not necessarily be assumed. Stereological relationships such as shape factors and interface concentrations are used for non-spherical envelopes. The time derivative of these equations defines the growth velocity of the dendrite. Present volume average models assume that the movement of the dendrite envelope is governed by the growth model for the dendrite tip. Thus:

$$r_E = \int_0^t V_E \, dt \quad \text{until} \quad r_E = r_f = (4\pi N/3)^{-1/3} \tag{14.10}$$

Since r_f is considered known, one more equation is necessary to solve the system of equations (14.9). However, additional simplifications used for the condensed dendrite allow the implementation of simple models in the HT codes (see Applications 8.5 and 9.3)

A first model (Dustin and Kurz, 1986) was proposed for a compacted dendrite, in which the r_E sphere was not completely filled by the network-type dendrite. It was assumed, based on some experimental data, that the internal fraction of solid was $f_i = 0.3$. Both thermal and solutal undercooling were considered. The solutal field was calculated only at the tip of the dendrite, which implies the assumption of equilibrium solidification (complete diffusion in solid and in liquid), over the total volume of the grain.

Rappaz and Thévoz (1987) developed a 1D radial symmetry model for the case in Figure 14.3a. The solutal field ahead of an equivalent sphere having the same mass as the dendrite, r_E, growing within the micro-volume element r_f, was calculated assuming no diffusion in the solid. It was demonstrated that $f_i = \Omega \cdot f(P_c)$. Uniform composition in the mushy zone was assumed. Solutal balance over the elemental volume of the grain was used. The growth velocity of the sphere of radius r_E was calculated based on the growth velocity of the dendrite tip and a mass balance. Thermal undercooling was neglected. A modified form of the classic equation for the hemispherical dendrite tip was used:

$$V_E = \mu \Delta T_c^2$$

where the growth constant can be calculated from an analytical model for dendrite tip growth. ΔT_c is the constitutional undercooling taken as the temperature difference between the interface and the grain boundary, $T^* - T_{rf}$. Once the dendrite tip reached r_f, which for certain alloys can occur as early as when $f_S = 0.1$, the Scheil equation was used to calculate f_S for the remaining solidification path.

The numerical implementation of this model was done by assuming that the solid phase has zero velocity. This concept was discussed earlier in this text and is schematically described in Figure 14.2. The evolution of fraction of solid can be calculated with equations such as Eqs. (14.4) or (14.7).

To avoid the early use of the Scheil equation in calculations necessary because of the excessively high growth rate for large grains (e.g., $r_f = 100 \ \mu m$), Kanetkar and Stefanescu (1988) modified the Rappaz-Thévoz model. They used an average composition in the liquid $\langle C_L \rangle$ as the driving force for dendrite growth, rather than the composition at r_f. The interface undercooling (Eq. 2.29) becomes: $\Delta T = T_f + m \langle C_L \rangle - T_{bulk}$. This correction was later adopted also by Wang and Beckermann (1993a).

In the Nastac and Stefanescu (1996) model, it was assumed that nuclei grow as spheres until the radius of the sphere becomes larger than the radius of the minimum instability. Then, the sphere degenerates into a dendrite. Growth of the dendrite is also related to morphological stability, and is calculated as a function of the thermal and solutal undercooling of the melt, which is controlled by the bulk temperature and the average composition in the liquid. The complex shape of the equiaxed dendrite, which has several levels of instabilities (*i.e.*, primary, secondary, etc., arms), is converted into a sphere of equivalent volume (Figure 14.3a and b). This sphere has the same number of instabilities as the dendrite. Since the composition of the liquid, and thus the driving force for the growth of the dendritic instabilities is position and time dependent, the volume average of the liquid composition is used to compute the growth of instabilities on the solid sphere. The number of instabilities is given by the ratio between the surface area of the solid sphere and that of one instability. Thus, the evolution of the fraction of solid is directly related to the radius of the solid phase, the number of instabilities, and the tip growth velocity. Growth is calculated until the average composition becomes the eutectic composition, or until the fraction of solid becomes one.

The basic elements of the mathematical formulation are as follows. From Eq. (14.9) we have: $f_S = f_i \cdot f_E$. The time derivative of this equation is:

$$\frac{\partial f_S}{\partial t} = \frac{\partial f_E}{\partial t} f_i + \frac{\partial f_i}{\partial t} f_E \tag{14.11}$$

After calculation of shape factors, specific interfacial areas, and interfacial area concentrations, it was demonstrated that the temporal evolution of the solid fraction is:

$$\frac{\partial f_S}{\partial t} = \psi f_S \left(\frac{V_E}{r_E} \frac{1}{\chi_E} + \frac{V_S}{r_S^i} \frac{1}{\chi_S^i} \right)(1 - f_S) \tag{14.12}$$

with $\chi_E = 4\pi r_E^2 / A_E$ and $\chi_S^i = 4\pi (r_S^i)^2 / A_S^i$. Here, V_E and V_S are the average normal velocities of the dendrite envelope and the S/L interface, respectively, r_S^i is the radius of the instability, χ_E and χ_S^i are the shape factor of the envelope and of the instability, respectively, and A_E and A_S^i are the interfacial areas of the envelope and of the instability, respectively. ψ is a geometrical factor that is 3 for equiaxed grains and 2 for columnar grains. The factor $(1-f_S)$ on the RH side accounts for grain impingement. The first term in this equation involves calculations at the dendritic length scale, while the second term includes calculations at the instability length scale, describing both formation and coarsening of instabilities. Here coarsening is understood as a process in which larger instabilities grow at the expense of smaller ones through diffusion. An iterative method must be used to solve this equation, since the solid fraction is used implicitly.

Application of such a model to casting modeling is restricted by the lack of information regarding the shape factors and the interfacial areas. Some simplification can be introduced, as follows.

If the instabilities are assumed to be spherical, $\chi_S^i = 1$. Assuming further that the condensed dendrite has an envelope that preserves a cubic shape (Figure 14.3b), $\chi_E = \pi/6$. Thus, Eq. (14.12) simplifies to:

$$\frac{\partial f_s}{\partial t} = \psi f_s \left(\frac{6}{\pi} \frac{V_E}{r_E} + \frac{V_S}{r_S^i} \right) (1 - f_s) \tag{14.13}$$

Calculation of V_S and r_S^i is not trivial. However, if coarsening is neglected, since, as shown earlier, when the instability grows at the limit of stability, $r_S^i = 4\pi^2 \Gamma / \Delta T$:

$$\frac{V_S}{r_S^i} = \frac{1}{r_S^i} \frac{\partial r_S^i}{\partial t} = -\frac{1}{\Delta T} \frac{\partial \Delta T}{\partial t}$$

Note that from this relationship the internal fraction of solid can be calculated as: $\partial f_i / \partial t \cong f_i \dot{T} / \Delta T$.

If coarsening is considered, it was shown (Nastac - Stefanescu, 1996) that:

$$V_S = \frac{0.75}{\left(r_S^i \right)^2} \frac{D_L \Gamma}{m(k-1)C_d} \left(\frac{1}{f_i^{1/3}} - 1 \right)^{-1} \tag{14.14}$$

where C_d is the volume averaged intradendritic concentration. f_i and r_S^i are either obtained from the previous time step or calculated implicitly.

For the particular case of the condensed dendrite, $f_i = f_S / f_E = (\pi/6)^{1/3} = 0.806$ (or $f_i \approx 1$), $\partial f_i / \partial t = 0$, and the RHT2 in Eq. (14.11) disappears. Also, the shape factor is $\chi = 4\pi r_E^2 / \left[6(2r_E)^2 \right] = \pi/6$. Thus, under the assumption of constant f_i throughout solidification Eq. (14.13) simplifies to:

$$\frac{\partial f_s}{\partial t} = \psi \frac{6}{\pi} f_s \frac{V_E}{r_E} (1 - f_s) \tag{14.15}$$

In all these equations the movement of the dendrite envelope is directly related to dendrite tip velocity; V_E is calculated with some dendrite tip velocity model for either equiaxed (e.g. Eq. 8.22) or columnar dendrites. Nastac (2004) has used for the velocity of columnar dendrites the following equation:

$$V_{den} = \frac{D_L}{\pi^2 \Gamma k \Delta T_o} (\Delta T^*)^2 \quad \text{with} \quad \Delta T^* = d \frac{\dot{T}}{V}$$

where ΔT^* is the S/L interface undercooling, d is the mesh size, and $\dot{T}$ is the local cooling rate.

This model has the capability to calculate dendrite coherency. Coherency is reached when $f_E = 1$, that is when the equivalent dendrite envelope reaches the final grain radius. However, the calculated coherency fraction solid for a condensed dendrite, which is about 0.55, is significantly larger than the experimental one, which is typically 0.2-0.3 for most alloys.

In a series of papers, Wang and Beckermann (1993,1994,1996) proposed a multiphase solute diffusion model for dendritic alloy solidification. A control volume, containing columnar or equiaxed dendrites, is considered to consist of three phases: solid, interdendritic liquid (between the dendrite arms), and extradendritic liquid (between the dendrites) (Figure 14.4).

The two liquid phases are associated with different interfacial length scales and have different transport behaviors. Melt convection and solid transport were included in the last version of the model. Macroscopic conservation equations were derived for each phase, using a volume averaging technique. The model can incorporate coarsening, and was used to predict the columnar-to-equiaxed transition (CET).

The governing equations for the microscopic model were summarized as follows:

- dendrite envelope motion

$$\frac{\partial}{\partial t}(f_s + f_d) = \frac{A_E}{v_E} \frac{D_L m_L (k-1) C_E}{\pi^2 \Gamma} \left[I^{-1}(\Omega_c) \right]^2 \qquad (14.16a)$$

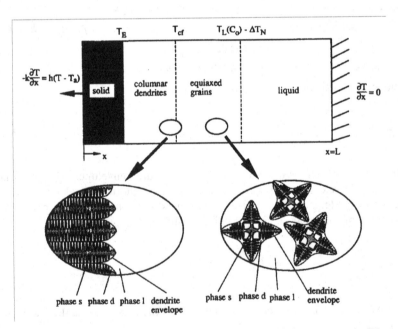

Figure 14.4. Illustration of the physical problem in the multiphase approach (Wang and Beckermann, 1994). With kind permission of Springer Science and Business Media.

- solute balance of the solid phase

$$\frac{\partial\left(f_s\langle C_s\rangle^s\right)}{\partial t} = \overline{C}_{Sd}\frac{\partial f_s}{\partial t} + \frac{A_S}{v_S}\frac{D_S}{l_{Sd}}\left(\overline{C}_{Sd} - \langle C_s\rangle^s\right) \qquad (14.16b)$$

- solute balance of the interdendritic liquid:

$$\frac{\partial\left(f_d\langle C_d\rangle^d\right)}{\partial t} = \left(C_E - \overline{C}_{dS}\right)\frac{\partial f_s}{\partial t} + C_E\frac{\partial f_d}{\partial t} + \frac{A_S}{v_S}\frac{D_d}{l_{dS}}\left(\overline{C}_{dS} - \langle C_d\rangle^d\right)$$
$$+ \frac{A_e}{v_e}\frac{D_d}{l_{dL}}\left(C_E - \langle C_d\rangle^d\right) \qquad (14.16c)$$

- solute balance of the liquid phase

$$\frac{\partial\left(f_L\langle C_L\rangle\right)}{\partial t} = C_E\frac{\partial f_L}{\partial t} + \frac{A_L}{v_L}\frac{D_L}{l_{Ld}}\left(C_E - \langle C_L\rangle^L\right) \qquad (14.16d)$$

- interfacial solute balance at the S-d interface

$$\left(\overline{C}_{dS} - \overline{C}_{Sd}\right)\frac{\partial f_s}{\partial t} = \frac{A_S}{v_S}\frac{D_S}{l_{dS}}\left(\overline{C}_{dS} - \langle C_d\rangle^d\right) + \frac{A_S}{v_S}\frac{D_S}{l_{Sd}}\left(\overline{C}_{Sd} - \langle C_s\rangle^s\right) \qquad (14.16e)$$

where the subscript d stands for the interdendritic liquid, double subscripts stand for interface quantity (e.g., dS means interdendritic S/L interface), l are the species diffusion lengths. $\Gamma^{-1}(\Omega_c)$ is the inverse Ivantsov function. The inverse Ivantsov function can be approximated with: $I^{-1}(\Omega_c) = 0.4567(\Omega/(1-\Omega))^{1.195}$ The definitions of the interphase concentrations are summarized in Figure 14.5. Other notations are as before. Note that the model for the dendrite tip velocity is essentially that used by Rappaz and Thévoz (1987).

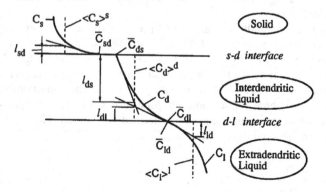

Figure 14.5. Definition of interface compositions (Wang and Beckermann, 1993). With kind permission of Springer Science and Business Media.

When supplying the expressions for the diffusion lengths, thermodynamic conditions, and geometrical relations, these five equations have five unknowns and thus, can be solved. They represent a complete model for solute diffusion, applicable to both equiaxed and columnar solidification. However, calculation of diffusion lengths and geometrical relations cannot be done without further simplifications. For example, assuming a 1D plate-like dendrite arm geometry it can be shown that $l_{dS} = d_S/6$, where d_S is the mean characteristic length or the diameter of the solid phase. A complete discussion is not possible within the constrains of the present text.

The main difference when applying this model to equiaxed (spherical grains) or columnar (cylindrical grains) solidification consists in the calculation of the interfacial area concentrations of the dendrite envelope. It was shown that they are:

for equiaxed grains: $\dfrac{A_E}{v_E} = \dfrac{1}{\chi_E}(36\pi)^{1/3} N^{1/3}(1-f_L)^{2/3}$

for columnar grains: $\dfrac{A_E}{v_E} = \dfrac{1}{\chi_E}(4\pi)^{1/2}\dfrac{1}{\lambda_1}(1-f_L)^{1/2}$

where $\lambda_1 = N^{-1/3}$ is the primary arm spacing of the columnar dendrites.

According to Ludwig and Wu (2002) the Wang-Beckermann model included a number of uncertainties such as detailed volumetric heat and mass transfer coefficients and stereological formulations for interfacial area concentrations. Addressing some of these issues they produced a two-phase (L, S) model for globular equiaxed solidification thus avoiding calculation of stereological equations. The exchange (source) terms take into account interactions between the melt and the solid, such as mass transfer (solidification and melting) friction and drag, solute redistribution, release of latent heat, and nucleation. In principle, they write the exchange of momentum, species, or enthalpy between the liquid and the solid as the sum of diffusional interaction (d) and phase change (p). For example, for species the source term is: $S_{C-LS} = S_{C-LS}^{d} + S_{C-LS}^{p}$. The detailed formulation of the source terms is discussed in the paper. The conservation equations were solved numerically by using the fully implicit, computational fluid dynamics (CFD) code FLUENT (by Ansys).

Wu and Ludwig (2006) extended their previous model to a three-phase model for mixed columnar-equiaxed solidification capable of predicting the CET. The three phases are the melt as the primary phase, and the solidifying columnar and equiaxed dendrites as two different secondary phases. The three phases are considered as spatially coupled and interpenetrating continua. The conservation equations are solved for all three phases, and an additional conservation equation for the number density of the equiaxed grains is defined and solved. The growth velocity of the globular equiaxed dendrites is calculated as:

$$V_{eq} = \frac{dr_{eq}}{dt} = \frac{D_L}{r_{eq}(1-k)}\left(1-\frac{C_L}{C_L^*}\right) \tag{14.17}$$

The columnar dendrites are approximated as growing cylinders. The growth velocity of the tip is calculated with the Lipton-Glicksman-Kurz model (Eqs. 8.14 to 8.17). The growth velocity in the radial direction of the cylindrical trunk was calculated as:

$$V_{col} = \frac{dr_{col}}{dt} = \frac{D_L}{r_{col}} \frac{C_L^* - C_L}{C_L^* - C_S^*} \ln^{-1}\left(\frac{r_f}{r_{col}}\right) \tag{14.18}$$

where r_f is half of the primary arm spacing. The columnar tip front tracking ignores the growth anisotropy determined by crystallographic orientation. While the model shows promise, the authors recognize that to obtain realistic predictions the model should consider more realistic approximations of the dendrite morphology, and should include the effect of melt convection on the diffusion field ahead of the S/L interface. Also, mechanical interactions between moving equiaxed and stationary columnar dendrites, as well as dendrite fragmentation are phenomena that should be included in more advanced numerical models.

When dendritic solidification is followed by eutectic solidification, the coupling of the two solidifications is done through the fraction evolution term as exemplified in Application 9.3.

More recently, Catalina et al. (2007) used a different approach for both calculation of the solid fraction and coupling of the macro- and micro-calculations. They calculated the evolution of f_s directly from the evolution of temperature and solutes fields (without the use of dendrite tip models) that are coupled at the solid/liquid interface through the temperature interface equation written as:

$$T^* = T_f + \sum_{i=1}^{n} m_i C_{Li} + \Gamma K \tag{14.19}$$

where T_f is the melting temperature of the base element, m_i is the liquidus slope of element i, C_{Li} is the liquid concentration of i at the S/L interface, n is the number of alloying elements, Γ is the Gibbs-Thomson coefficient, and K is the curvature of the S/L interface. For spherical grains $K = -2/r_g$, where r_g is the instantaneous grain radius.

In a solidifying volume element the average concentration, $\overline{C}_i$, of element i is given by:

$$\overline{C}_i = f_s \cdot \overline{C}_{Si} + (1 - f_s)\overline{C}_{Li} \tag{14.20}$$

where $\overline{C}_{Si}$ and $\overline{C}_{Li}$ are the average concentration of element i in the solid and liquid phase, respectively. The actual distribution of the concentration within each phase depends on the diffusion rate. For the particular case of steel it was assumed no diffusion in the solid for the substitutional elements (Mn, Si, Ni, Cr, Mo, P, and S) and complete mixing of the interstitial elements (H, N, C) (see also Sung et al. 2002). It was further assumed that complete mixing in the liquid phase, within a volume element, occurs for all the solutes, *i.e.*, $\overline{C}_{Li} = C_{Li}$. The effect of interstitial gases on solid fraction evolution was neglected. Consequently, $\overline{C}_{Si}$ can be expressed as:

$$\overline{C}_{Si} = \frac{1}{f_s} \int_0^{f_s} C_S(\xi) d\xi \tag{14.21}$$

where ξ is the coordinate variable. The change in time is given by:

$$\frac{d\overline{C}_{Si}}{dt} = \frac{C_{Si}^* - \overline{C}_{Si}}{f_s} \frac{df_s}{dt} \tag{14.22}$$

where C_{Si}^* is the solid concentration of i at the solid/liquid interface and is related to the liquid concentration through $C_{Si}^* = k_{pi} C_{Li}$, where k_{pi} is the partition coefficient of i. Then, using the previous equation, the change of the average concentration in the volume element, $\overline{C}_i$, can be calculated as:

$$\frac{d\overline{C}_i}{dt} = (1 - f_s) \frac{dC_{Li}}{dt} - (1 - k_{pi}) C_{Li} \frac{df_s}{dt} \tag{14.23}$$

Multiplying this equation by the liquidus slope, m_i, performing the summation over all the alloying elements, and then using Eq. (14.19) the following finite difference expression can be obtained:

$$\Delta f_s = \frac{1 - f_s^o}{\sum_{i=1}^{n} (1 - k_{pi}) m_i C_{Li}^o} \cdot T^n - \frac{(1 - f_s^o) \cdot \left(T_m + \Gamma K^n + \sum_{i=1}^{n} m_i C_{Li}^o \right) + \sum_{i=1}^{n} m_i (\overline{C}_i^n - \overline{C}_i^o)}{\sum_{i=1}^{n} (1 - k_{pi}) m_i C_{Li}^o} \tag{14.24}$$

where Δf_s is the change of the solid fraction during the time-step, and the superscripts n and o denote quantities at the end and beginning of the time step, respectively. After the energy equation is solved for T^n this equation is used to calculate Δf_s for the respective time-step. Note that $\overline{C}_i^n$ should first be calculated from the species conservation equation by means of an explicit procedure. The curvature, K^n, is taken from the previous time-step. Once Δf_s is known the solid and liquid concentrations can be updated by means of Eqs. (14.20) and (14.22). The concentrations of the fast diffusing interstitial elements should be updated by means of the relationship:

$$\overline{C}_i = \left[1 - (1 - k_{pi}) \cdot f_s \right] \cdot C_{Li} \tag{14.25}$$

This model was incorporated in Caterpillar's proprietary model, CAPS.

14.1.4   Microporosity models

Kubo and Pehlke (1985) developed a 2-D model, with constant thermal properties. The fraction liquid in the source term was calculated based on the equilibrium or the Scheil model. The continuity equation was used for each volume element, equating the shrinkage (first left-hand-term) to the interdendritic inflow of liquid metal (second and third left-hand-term) and the growth of the pore (third left-hand-term):

$$(\rho_S / \rho_L - 1)\partial f_L / \partial t - (\partial f_L / \partial x)V_x - (\partial f_L / \partial y)V_y + \partial f_G / \partial t = 0 \qquad (14.26)$$

The pressure drop through the mushy zone was calculated assuming Darcy type flow (Eq. 4.43). The permeability was formulated based on the Blake-Kozeny model, Eq. (4.47), with $C_2 = 1/180$, and using λ_{II} rather than λ_I.

As long as no porosity has formed P_{mush} is calculated from Eqs. (14.26) and (4.43). The gas pressure is then calculated from Eq. (13.24). The amount of porosity, f_G, is calculated from mass conservation assuming equilibrium, stating that the initial gas volume is:

$$v_i = v_S(1 - f_L) + v_L f_L + R_H P_G f_G / T$$

where $R_H = 27300/(f_L \rho_L + f_S \rho_S)$ is a units transformation parameter. The first, second and third right-hand-term are the amount of hydrogen in solid, liquid, and porosity, respectively. v_S and v_L is obtained from Sievert's law.

Their calculations were in good qualitative agreement with the experiment. They predicted that the amount of porosity increases with the amount of gas dissolved but decreases with higher cooling rate. The basic concepts introduced by Kubo and Pehlke have been used by many other researchers that improved on some details of the calculation as well as on the data-base.

Poirier et al. (1987) assumed nucleation of cylindrical pores (i.e., r_1 and $r_2 = \infty$) between the primary dendrite arms. They argued that less excess pressure would be required for the gas phase existing between the primary arms than between the secondary arms, since the former are larger. This resulted in a modified equation for the surface tension pressure, $P_\gamma = 4\gamma/(g_L \lambda_I)$, where λ_I is the primary dendrite arm spacing. This model was able to show that increasing the cooling rate or the thermal gradient (which decreases λ_I), also decreases the amount of segregation.

Zhu and Ohnaka (1991) assumed that nucleation occurs when the hydrogen content in the residual liquid exceeds the solubility limit. Selecting an arbitrary number for this supersaturation (in this case 0.1 $cm^3/100g$), a supersaturation pressure for porosity nucleation, ΔP_N, can be calculated. Accordingly, Eq. (13.24) was rewritten as:

$$P_G - P_{local} - \Delta P_N > P_\gamma \qquad (14.27)$$

They further assumed that a pore having a radius 1/2 DAS nucleates, and that the pore grows as a cylinder of constant radius in the interdendritic space. The porosity fraction can be calculated from the mass balance equation $f_G = R\,(C_L - C_L^{max})\,f_L \cdot \rho_L \cdot T/(100 M \cdot P_G)$, where R is the gas constant, C_L^{max} is the gas solubility limit in the liquid, and M is the molecular weight of the gas.

In their experimental results, porosity was found between secondary dendrite arms. Only in the relatively gassy heat (0.48 cm³/100g) was any porosity found between primary arms. The model correctly showed the effect of increasing ambient pressure, decreasing initial hydrogen content, and decreasing solidification time.

Zou and Doherty (1993) modeled porosity occurring in upward directional solidification assuming again Darcy-type flow. They relaxed the steady-state flow assumption of Ganesan and Poirier (1990). This yielded an interdendritic flow equation:

$$\nabla P + \rho_L \mathbf{g} - \frac{\mu}{K} \cdot \mathbf{V} = \rho_L \cdot \frac{\partial \mathbf{V}}{\partial t}$$

Assuming further that the pore volume is so small that it can be ignored in the mass balance, and constant solid and liquid densities, the continuity equation was simplified to:

$$\frac{\partial f_S}{\partial t}\left(\frac{\rho_S}{\rho_L}\right) - \frac{\partial f_G}{\partial f} + \nabla \mathbf{V} = 0$$

Solidification kinetics was also included in the model. The porosity versus distance from the chill results appears to have too little porosity, primarily because the authors did not include the effect of dissolved gas. However, in upward directional solidification it is entirely possible under some conditions that the gas pores be pushed ahead of the advancing solidification front.

Viswanathan *et al.* (1998) modified the equation for capillary pressure to $P_\gamma = (2\gamma/r)\sqrt{S(\theta)}$, where $S(\theta)$ is a factor that takes into account the effect of heterogeneous nucleation of a pore, and θ is the contact angle of the gas-melt interface with the pore. Knowing that the pore fraction must be greater than zero when pores form, and using the previous equation, they offered the following microporosity condition:

$$C_{H_2}^0 > (f_S k_H + f_L) K_L \sqrt{P_{mush} + \frac{2\gamma}{r}\sqrt{S(\theta)}} \qquad (14.28)$$

where $C_{H_2}^0$ is the initial gas concentration within the liquid, k_H is the partition coefficient of hydrogen between solid and liquid, and K_L is the equilibrium constant in Sievert's law.

Sabau and Viswanathan (2002) used these concepts in their 3-D model for porosity prediction in aluminum alloys. The model computes flow and pressure both in the liquid region and in the mushy zone and calculates shrinkage porosity when feeding flow is cut off. This resulted in numerical difficulties because the dynamic pressure in the liquid zone (typically <1 Pa) is much lower than the pressure drop in the mushy zone (typically 1-10 KPa). When feeding in a region is cut off, they no longer solve for pressure or velocity in the region, but rather compute porosity such that it compensates for all the shrinkage occurring in that region.

Péquet et al. (2002) coupled a microporosity model based on Darcy flow and gas microsegregation with a macroporosity and shrinkage pipe predictions model in 3-D. The governing equations of microporosity formation are solved in the mushy zone with boundary conditions imposed around this zone. The microporosity model is only applied in the mushy zone by superimposing a fine finite volume grid onto the coarser finite element mesh used for heat flow computations. Each liquid region of the casting is evaluated to find whether it is connected to a free surface, surrounded by solid or surrounded by a mushy zone. In the latter two cases, integral boundary conditions must be solved to determine the pressure boundary condition. Because of inaccuracies in this computation, the void fractions are adjusted to ensure global mass conservation.

Carlson et al. (2003) presented a one-domain multi-phase model that predicts melt pressure, feeding flow, and porosity in steel castings during solidification and calculates microshrinkage, closed and open shrinkage cavities throughout a shaped casting as it solidifies. The model assumes that each volume element in the casting is composed of some combination of solid metal (S), liquid metal (L), porosity (G), and air (a), such that:

$$g_S + g_L + g_G + g_a = 1 \qquad (14.29)$$

Mixture properties are obtained as the sum of the property values for each phase multiplied by their respective volume fractions. The mixture energy conservation assumed that the solid fraction is a known function of temperature and ignored the advection term due to shrinkage flow, which may be quite significant at riser necks.

By combining Darcy's law, which governs fluid flow in the mushy zone, with the equation for Stokes flow, which governs the motion of slow-flowing single-phase liquid, they derived a momentum equation that is valid everywhere in the solution domain:

$$\nabla^2 \mathbf{V} = \frac{g_L}{K} \mathbf{V} + \frac{g_L}{\mu_L} \nabla P - \frac{g_L}{\mu_L} \rho_{ref} \, \mathbf{g} \qquad (14.30)$$

where μ_L is the dynamic viscosity of the liquid (assumed to be constant), $\mathbf{g}$ is the gravity vector, and ρ_{ref} is the reference liquid density, taken as the melt density at the liquidus temperature. Note that this equation reduces to Stokes' equation in the single-phase liquid region, where K becomes very large. Also, in the mushy zone, the left-hand side of this equation becomes very small relative to the permeability term, and the equation then reduces to Darcy's law.

The mixture continuity equation does not include the air phase and assumes that the solid metal and the porosity are stationary. Then:

$$\frac{\partial}{\partial t}(g_S \rho_S + g_L \rho_L + g_G \rho_G) + \nabla \cdot (\rho_L \mathbf{V}) = 0 \tag{14.31}$$

where $\mathbf{V}$ denotes the superficial liquid velocity, $\mathbf{V} = g_L \mathbf{V}_L$. This equation shows that the divergence of the velocity field is a function of the solidification contraction, liquid density change, porosity evolution, and gradients in the liquid density.

By combining the momentum equation with the continuity equation a pressure equation was derived. The concentration of each gas species dissolved in the melt is obtained from the mixture species conservation equation. The partial pressure of the gas in the pores is calculated from Sievert's law.

The governing equations are solved to give the melt pressure, gas pressure in the pore, and feeding velocity throughout the casting. The criterion for porosity nucleation is Eq. (13.24). However, in this model the capillary pressure was set to zero. When porosity forms, the melt pressure at that location is forced to $P_{mush} = P_G - 2\gamma/r$. This allows the continuity equation (14.31), to be solved for the pore fraction, g_G. Once the pore fraction is known, the liquid fraction is updated using Eq. (14.29).

To simulate the formation of a shrinkage cavity open to the atmosphere the pressure is forced to atmospheric pressure in the volume elements that are emptying of liquid. Then, the continuity equation is solved for the liquid fraction, g_L, while keeping g_G constant (no pore formation when air is present). Finally, the air fraction is obtained from Eq. (14.29).

This multi-phase model has been implemented in a general-purpose casting simulation code. The predicted porosity distributions were compared to radiographs of steel castings produced in sand molds.

A predefined pore density number was used in the Péquet et al. (2002) and Carlson et al. (2003) models. Although porosity fractions calculated with these models matched well the experimental measurements, the size and number of pores were dependent on the assumed pore density number.

Catalina et al. (2007) developed a model for porosity prediction that accounts for mass conservation of the gas and local gas/melt equilibrium. The local amount of porosity as well as the number and size of the gas bubbles can be predicted. The gas concentration in the alloy (liquid or solid) was calculated from Sievert's law:

$$C_G = \frac{K_G}{g_G^a} \cdot \frac{\sqrt{P_G}}{\sqrt{101325}} \tag{14.32}$$

where C_G is the concentration (wt%) of the dissolved gas, g_g^a is the gas activity coefficient, P_G is the gas partial pressure in the gas phase (Pa), and K_G, a function of temperature, is the equilibrium coefficient.

Assuming a spherical shape of radius r_G for the gas bubbles, the relationship between P_G and local melt pressure, P_{local}, is given by $P_G = P_{local} + 2\gamma/r_G$. The local melt pressure is calculated from the momentum conservation equation:

$$g_L \rho_L \frac{\partial}{\partial t}\left(\frac{\mathbf{V}}{g_L}\right) + \rho_L \mathbf{V} \cdot \nabla\left(\frac{\mathbf{V}}{g_L}\right) = -g_L \nabla P_L + \eta \nabla^2 \mathbf{V} - g_L \frac{\eta}{K_{mush}} \mathbf{V} + g_L \rho_L \mathbf{g} \qquad (14.33)$$

and the continuity equation:

$$\frac{1-g_G}{\overline{\rho}} \frac{\partial \overline{\rho}}{\partial t} - \frac{\partial g_G}{\partial t} + \nabla \cdot \mathbf{V} = 0 \qquad (14.34)$$

where the first term in the LHS represents the volume fraction of metal shrinkage during the time dt, the second term represents the volume fraction of gas bubbles generated during dt, and the third is the volume fraction of liquid supplied to the volume element. g_S, g_L, and g_G are the volume fractions of solid, liquid, and gas bubbles, respectively ($g_S + g_L + g_G = 1$), $\mathbf{V}$ is the superficial velocity vector, $\mathbf{g}$ is the gravity vector, and K_{mush} is the permeability of the mushy zone that can be calculated with an equation such as (4.47). To calculate the gas bubbles evolution it is assumed that the gas obeys the ideal gas law:

$$P_G v_G^1 = N_m RT/n_G \qquad (14.35)$$

where v_G^1 is the volume of one gas bubble, n_G is the number of bubbles, N_m is the total number of moles of gas in the gas phase of a volume element, and R is the gas constant. At the end of a time-step N_m is composed of an initial number of moles, N_m^o, and the number of moles passing from solution into the gas phase, ΔN_m, during the time-step ($N_m = N_m^o + \Delta N_m$). With gas concentration expressed in wt% and assuming that only the gas dissolved in the liquid phase precipitates into bubbles, ΔN_m can be expressed as:

$$\Delta N_m = \frac{(1-g_S-g_G)\rho_L v_{elem}}{100 M_G}\left(C_{LG}^o - C_{LG}^n\right) \qquad (14.36)$$

where v_{elem} is the volume of the volume element, M_G is the molar mass of gas, and C_{LG}^o and C_{LG}^n are the gas concentrations in the liquid before and after precipitation, respectively. Assuming that the bubble/liquid equilibrium is reached at the end of the time-step:

$$C_{LG}^n = \left(\overline{K}_G/g_G^a\right)\sqrt{P_L + 2\gamma/r_G} = \left(\overline{K}_G/g_G^a\right)\sqrt{P_G} \qquad (14.37)$$

where the notation $\overline{K}_G = K_G/\sqrt{101325}$ has been used for simplicity. Further, by making use of Eqs. (14.35), (14.36), and (14.37) the following equation that relates r_G and n_G can be obtained:

$$\left(a_4 + a_3\sqrt{P_L + 2\gamma/r_G}\right)\cdot r_G^3 + a_2 r_G^2 + n_G^{-1}\left(a_1\sqrt{P_L + 2\gamma/r_G} + a_o\right) = 0 \qquad (14.38)$$

with the coefficients a_i ($i = 0$ to 4) given by:

$$a_o = -\left[\frac{(1-g_S)\rho_L v_{elem} C_{LG}^o}{100 M_G} + N_m^o\right]; \quad a_1 = \frac{(1-g_S)\rho_L v_{elem}\overline{K}_G}{100 M_G}; \quad a_2 = \frac{8\pi\gamma}{3RT}$$

$$a_3 = -\frac{4\pi}{3}\frac{\rho_L\overline{K}_G}{100 M_G}; \quad a_4 = \frac{4\pi}{3}\left(\frac{P_L}{RT} + \frac{\rho_L C_{LG}^o}{100 M_G}\right)$$

The Newton-Raphson method can be used to solve Eq. (14.38) by first assuming a value for n_G and then solving for r_G. First, a minimum radius of the bubble is calculated from the relationship $C_{LG}^o = \left(\overline{K}_G/g_G^o\right)\sqrt{P_L + 2\gamma/r_{min}}$ and then a maximum number of bubbles, n_{max}, of radius r_{min} that can fill the available volume $V_{available} = \left[\Delta t(1-g_G^o)\overline{\rho}^{-1}\partial\overline{\rho}/\partial t + g_G^o\right]v_{elem}$. Then n_G is decreased iteratively as long as Eq. (14.38) has a real and positive solution for r_G. In the end, the liquid gas concentration is updated based on Eq. (14.37) and then the average concentration is updated with the relationship given by Eq. (14.20).

In poorly fed regions it is possible that the computed liquid pressure drops below zero (cavitation). It was considered that gas porosity develops up to the point where pressure reaches the cavitation value after which the entire solidification shrinkage transforms into porosity. Application of this model to a Caterpillar's steel casting showed a good qualitative agreement with the actual defects (Catalina et al., 2007).

The concept of oxide entrapment during mold filling and heterogeneous nucleation on or in the entrapped oxide films suggested by Campbell (2003) was used by Sako et al. (2001) and Ohnaka (2003) to model microshrinkage formation in aluminum alloys. First, collision of free surfaces with each other and with the mold wall was evaluated. The free surface in a volume element was represented in Cartesian coordinates as shown in Figure 14.6. A criterion for collision based on the filling ratio of the element and the velocity on the surface of the element was formulated. Then marker particles were generated to symbolize entrapped oxides in the volume elements where the collision occurred (Figure 14.7). The markers contain information of oxide number (N_{of}) and average surface area (S_{of}^{av}), which are calculated from the relative collision velocity ($|V_1 - V_2|$) and collision area with the following equations:

$$N_{of} = \alpha_1|V_1 - V_2| + \alpha_2 \quad \text{and} \quad A_{of}^{av} = \sum A_{of}/N_{of}$$

where α_1 and α_2 are parameters depending on alloy composition determined experimentally, and A_{of}^{av} and A_{of} are the average and individual surface area of the oxide films. Then, assuming that the markers move with the same velocity as the melt their movement is calculated from the flow velocity of the melt. The final distribution of the oxide markers in an Al-7.5% Si casting is compared with the observed shrinkage porosity in Figure 14.8a,b.

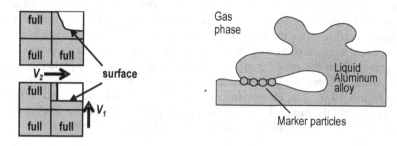

Figure 14.6. Modeling of free surface of melt (Sako *et al.*, 2001).

Figure 14.7. Generation of markers (Sako *et al.*, 2001).

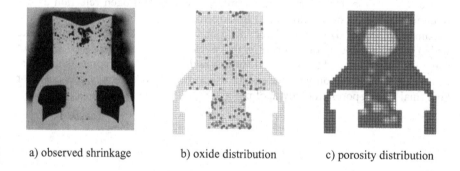

a) observed shrinkage   b) oxide distribution   c) porosity distribution

Figure 14.8. Comparison of simulated and observed results for an Al-7.5% Si casting (Ohnaka, 2003).

Recognizing that the above method cannot evaluate macroporosity, Ohnaka *et al.* (2002) developed another model in which the temperature, the solid volume change, and pore growth are calculated for each element at each time step. Before the beginning of solidification the change in pore volume because of hydrogen diffusion is calculated under the assumption of quasi-steady-state diffusion and ideal gas. After the beginning of solidification the pressure field in the mushy zone is calculated through mass conservation and Darcy's law. To avoid solving a non-linear equation the pore volume is estimated by assuming that all the hydrogen rejected from the solid moves to the pores and extrapolating the pore pressure from the previous pressure change. The error of this estimation is claimed to be less than 10%. Another complication arises when calculating the pressure field after a mushy or liquid region is enclosed by a solid layer. It requires solving simultaneously for Darcy flow, pore growth and solid layer deformation. In a simplified approach, the pressure change from a reference pressure and correction of the pore volume to satisfy mass balance are used. An example of calculation results are given in Figure 14.8c.

Based on in-situ experimental observations of the kinetic and pore nucleation and growth in Al-Cu alloys using an X-ray temperature gradient stage, Lee and Hunt (1997) concluded that, for small mushy zone, microporosity formation is controlled by gas diffusion. Accordingly, Atwood et al. (2000) developed a deterministic model to solve the hydrogen-diffusion-controlled growth. The pore grows in the liquid because of hydrogen diffusion which is enriched in the liquid because of rejection from the growing solid. Once the pore and the solid grain impinge, the pore grows between the grains assuming a regular closed packed spherical structure (Figure 14.9). The change in curvature occurring during growth affects the pressure within the pore. The region around the pore was approximated as a sphere of radius r_f within which diffusion was governed by Fick's second law:

$$\frac{\partial C_G}{\partial t} = D_G \frac{1}{r^2} \frac{\partial}{\partial r}\left(r^2 \frac{\partial C_G}{\partial r} \right) + S_t \qquad (14.39)$$

where S_t is a time-dependent source term representing hydrogen rejection as determined by the solidification rate. Impingement was calculated based on the maximum spherical size that could be incorporated in the octahedral interstitial site within a lattice made up of spherical grain envelopes. Although this model did not solve for fluid flow and thus ignored shrinkage drive growth of pores, the authors claim that the correlation with the in-situ experimental observations was good for both final size of pores and growth of pores.

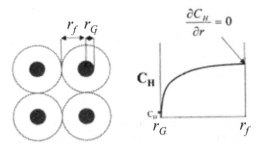

Figure 14.9. Periodic distribution of pores (dashed circles indicates impingement of hydrogen depletion layers) (left) and hydrogen profile (Atwood et al., 2000). Reprinted with permission from Elsevier.

Atwood and Lee (2003) and then Lee et al. (2004) also developed a stochastic model that includes grain growth and pore growth sub-models. The descriptions of the nucleation algorithm for solid and pores are unclear. Growth of the solid is calculated through standard cellular automaton techniques. The pore radius is calculated from the combination of the following three equations:

$$P_G \, v_G = N_m RT \qquad P_G = P_{st} + \frac{2\gamma}{r} \qquad v_G = \frac{4\pi}{3} r_G^3$$

The solubility of H_2 in Al is calculated as a function of temperature and the effect of dissolved alloying elements. The solubility of H_2 at equilibrium with the pore pressure was determined from Sievert's law. If v_G exceeded the total volume of the cell, the microshrinkage was allowed to expand into one of the neighboring cells as a complex agglomeration of spheres.

Statistical models based on regression analyses of empirical data have also been proposed. Tynelius *et al.* (1993) argued that the quantity and size of microporosity in a casting for which the thermal field of the casting/mold assembly is known can be predicted from a statistical model. Based on multiple regression analysis of experimental data on alloy A356 they expressed the percentage porosity as:

$$\%porosity = a \cdot [H]^2 + b \cdot [H] \, t_f + c \cdot [Sr][H] + d \cdot [H] \, V_S + e \cdot (m_{GR})$$

where [H] is the hydrogen content, [Sr] is the strontium content, (m_{GR}) is the amount of grain refiner, and a, b, c, d, e are fitted coefficients.

14.2 Stochastic models

Deterministic models for microstructure evolution have made tremendous progress rewarded by recognition and acceptance by industry. However, their main short-coming is their inability to provide a graphic description of the microstructure. Another problem is in the very mathematics that is used. Recent advances in electron microscopy make possible the examination of materials with near-atomic resolution. Such nano-scale structures cannot be described adequately through continuum approximations. As stated by Kirkaldy (1995), "Materials scientists who persist in ignoring the microscopic and mesoscopic physics in favor of mathematics of the continuum will ultimately be seen to have rejected an assured path towards a complete theoretical quantification of their discipline."

More recently, simulation of microstructure evolution has been approached through stochastic models based on the local description of the material combined with some evolutionary rules, rather than solving the differential and integral equations. Two basic techniques have been developed: the *Monte Carlo* (MC) technique, where the evolutionary rules are stochastic, and the *cellular automaton* (CA) technique, where algorithms or probabilistic rules control evolution.

Stochastic models approach solidification modeling at the meso-scale, micro-scale, nano-scale, or a combination of these scales. They may attempt to describe the growth of the dendrite grain envelope or of the eutectic grain (commonly referred to as *grain growth models*), or the development of the morphology of the dendrite grain (termed *dendrite growth* in this text). The two cases are described graphically in Figure 14.10.

Grain growth models are in principle meso-scale models that ignore the growth morphology at the scale of the grain. The graphical output of these models is limited to showing the grain boundaries, as they follow the dynamics of the grain envelope. They are useful for eutectic solidification or for solid-state grain growth. However, they do not give a description of the interface dynamics within the dendrite envelope. The primary and higher order dendrites arms, internal fraction of solid and dendrite coherency are not modeled. However, these dendrite features influence fluid flow and thus the outcome of solidification.

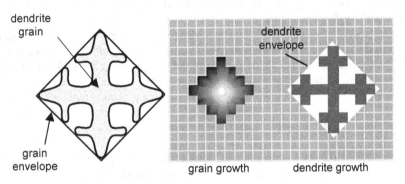

Figure 14.10. Features described by stochastic growth models (left) and discretization of the computational space (right) (Stefanescu, 2001).

14.2.1   Monte-Carlo models

The MC technique is based on the minimization of the interface energy of a grain assembly. It has been developed to study the kinetics of grain growth (Sahni *et al.* 1983, Anderson *et al.* 1984). As summarized by Stefanescu (2001), the microstructure was first mapped on a discrete lattice as shown as an example in Figure 14.11. Each lattice site was assigned an integer (grain index), I_j, from 1 to some number. Lattice sites having the same I_j belong to the same grain. The grain index indicates the local crystallographic orientation. A grain boundary segment lies between two sites of unlike orientation. The initial distribution of orientations is chosen at random and the system evolves to reduce the number of nearest neighbors pairs of unlike crystallographic orientation. This is equivalent to minimizing the interfacial energy. The grain boundary energy is specified by defining an interaction between nearest neighbors lattice sites. When a site changes its index from I_j to that of its

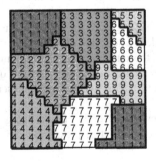

Figure 14.11. Example of a microstructure mapped on a rectangular lattice; the integers denote orientation and the lines represent grain boundaries.

neighbor, I_k, the variation in energy can be calculated from the Hamiltonian[1] describing the interaction between nearest neighbor lattice sites:

$$\Delta G = -\gamma \sum \left(\delta_{I_j I_k} - 1 \right) \tag{14.40}$$

where γ is the interface energy and $\delta_{I_j I_k}$ is the Kronecker delta. The sum is taken over all nearest neighbors. The Kronecker delta has the property that $\delta_{I_j I_k} = 0$ and $\delta_{I_j I_j} = 1$. Thus, unlike nearest pairs contribute γ to the system energy, while like pairs contribute zero. The change of the site index is then decided on the basis of the transition probability, which is:

$$W = \begin{cases} \exp\left(-\Delta G / k_B \cdot T \right) & \Delta G > 0 \\ 1 & \Delta G \le 0 \end{cases} \tag{14.41}$$

where k_B is the Boltzmann constant. A change of the site index corresponds to grain boundary migration. Therefore, a segment of grain boundary moves with a velocity given by:

$$V = C \left(1 - \exp\left(-\frac{\Delta G_i}{k_B T} \right) \right) \tag{14.42}$$

C is a boundary mobility reflecting the symmetry of the mapped lattice, and ΔG_i is the local chemical potential difference. Note that this equation is similar to that derived from classical reaction rate theory.

Dendrite envelope growth models - meso-scale

We use the term "dendrite envelope" rather than "dendrite grains" to emphasize that the object of this modeling is not to describe the morphology of the dendrite but rather the shape of the grain, equiaxed or columnar, at the end of solidification. Chronologically, Spittle and Brown (1989) seem to be the first to use the Monte Carlo approach to model solidification. Their meso-scale model was able to qualitatively predict the grain structure of small castings. However, their study was based on a hypothetical material and the correspondence between the MC time step used in the calculations and real time was not clear. Consequently, it was not possible to analyze quantitatively the effects of process variables and material parameters.

To correct this inadequacy of previous models, Zhu and Smith (1992) coupled the MC method with heat and solute transport equations. They accounted for heterogeneous nucleation by using Oldfield's nucleation model Eq. 7.3. The probability model of crystal growth was based on a lowest free energy change algorithm.

---

[1] A Hamiltonian cycle is a cycle that contains all the vertices of a graph.

ΔG in Eq. (14.41) was calculated as the difference between the free energy determined by the undercooling, ΔG_v, and the interface energy existing between liquid/solid and solid/solid grains, ΔG_γ:

$$\Delta G = \Delta G_v - \Delta G_\gamma = \frac{\Delta H_f \Delta T}{T_m} v_m + l d\left(n_{SL} \gamma_{SL} + n_{SL} \gamma_{SS}\right) \tag{14.43}$$

where ΔH_f is the latent heat of fusion, ΔT is the local undercooling, T_m is the equilibrium melting temperature, v_m is the volume of the cell associated with each lattice site, l and d are the lattice length and thickness, respectively, n_{SL} is the difference between the number of new L/S interfaces and overlapped L/S interfaces, n_{SS} is the number of S/S interfaces between grains with different crystallographic orientations, and γ_{LS} and γ_{SS} are the interface energies. The time elapsed (MC time step) was calculated using the macroscopic heat transfer method, where the time elapsed corresponds to the amount of liquid that has solidified under given cooling conditions.

Some typical computational results of this meso-scale model were presented earlier in this text in Figure 8.20. A clear transition from columnar to equiaxed solidification is seen when the superheating temperature, and therefore the undercooling, is increased.

Eutectic S/L interface growth models - micro-scale

A more fundamental approach has been used by Xiao et al. (1992). They simulated microscopic solidification morphologies of binary systems using a probabilistic MC model that accounts for bulk diffusion, attachment and detachment kinetics, and surface diffusion. An isothermal two-component system contained in a volume element subdivided by a square grid was considered. Initially, the region was filled by a liquid that consists of particles A and B that occupy each grid point according to a preset concentration ratio. Diffusion in the undercooled liquid was modeled by random walks on the grid. Through variation of interaction energies and undercooling, a broad range of microstructures was obtained, including eutectic systems and layered and ionic compounds. It was shown that, depending on the interaction energies between atoms, the microscopic growth structure could range from complete mixing to complete segregation. For the same interaction energy, as undercooling increases, the phase spacing of lamellar eutectics decreases (Figure 14.12). Thus, continuum derived laws, such as $\lambda \cdot \Delta T$ = const., were recovered through a combination between a probabilistic model and physical laws. Further, the composition in the liquid ahead of the interface is very similar to that predicted from the classic Jackson- Hunt model for eutectics, as shown in Figure 14.13. However, this model can only be used for qualitative predictions, because the length scale of the lattice, a, is several orders of magnitude higher than the atomic size. Thus, this is not a nano-scale model, but rather a compromise between a nano- and a micro-scale model. The model is limited to micro-scale level calculations, and cannot be coupled to the macro-scale because of computing time limitations. Nevertheless, the main merit of this work is that it has demonstrated that physical laws can be successfully combined with probabilistic calculation for transformation kinetic

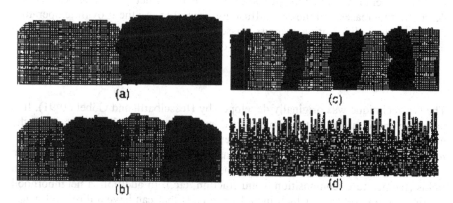

Figure 14.12. Role of increased undercooling on microscopic growth morphology: (a) $\Delta\mu/k_BT = 0.1$; (b) $\Delta\mu/k_BT = 0.5$; (c) $\Delta\mu/k_BT = 5$; (d) $\Delta\mu/k_BT = \infty$. $\Delta\mu$ is the chemical potential (Xiao *et al.*, 1992). Reprinted with permission from Phys. Rev. Copyright 1992 by the American Physical Soc. www.aps.org

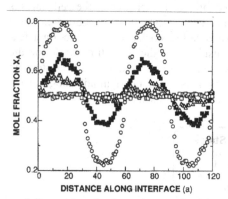

Figure 14.13. Concentration in mole fraction parallel to the liquid/solid interface at $2a$ (o), $10a$ (■), $20a$ (△) and $50a$ (□) into the liquid; a is the lattice constant (Xiao *et al.* 1992). Reprinted with permission from Phys. Rev. Copyright 1992 by the American Physical Soc. www.aps.org

modeling. Also, it prove that to obtain satisfactory results in terms of length scale of the microstructure, it is sufficient to model the diffusion of atoms aggregates, perhaps as large as 1000 atoms per aggregate. These aggregates represent the statistically averaged trajectories of the atoms contained within.

A similar approach was use by Das and Mittemeijer (2000). They simulated the isothermal solidification structure of binary alloys taking into account the simultaneous diffusion of all liquid atoms, instead of random walk of single liquid atoms, one at a time, as considered in earlier simulations. The thermodynamic driving force is composed of a volume energy change during solidification. It includes bond energies of the like and unlike nearest and next nearest neighbors and the S/L interface energy change determined by the difference in unsatisfied solid bonds at the interface, before and after solidification. The probability of solidification or remelting of a particular atom is controlled by the change in Gibbs energy during phase transition. At the microscopic length scale, the energy minimization of the system is achieved through a rearrangement process at the S/L interface.

The model predicts that the period of the lamellar structure decreases as the undercooling increases and the S/L diffusivity decreases. As the growth temperature decreases the solidification velocity passes through a maximum, because of the increased undercooling and the decreased diffusivity in the liquid.

14.2.2   Cellular automaton models

The CA technique was originally developed by Hesselbarth and Göbel (1991). It is based on the division of the simulation domain into cells, which contain all the necessary information to represent a given solidification process, as shown schematically in Figure 14.14. Each cell is assigned information regarding the state (solid, liquid, interface, grain orientation, etc.), and the value of the calculated fields (temperature, composition, solid fraction, etc.). In addition, a neighborhood configuration is selected, which includes the cells that can have a direct influence on a given cell. The definition of a neighbor can be modified by introducing a *weighted neighbor* (Dilthey *et al.*, 1997). This means that its influence on the other cell is taken with some weight, depending for example on the distance that separates them. The fields of the cells are calculated by analytical or numerical solutions of the transport and transformation equations. The change of the cell states is calculated through transition rules, which can be analytical or probabilistic. When these rules are probabilistic, the technique is called stochastic. The important feature of the method is that all cells are considered at the same time to define the state of the system in the following time step. Thus the computational time step can be directly related to the physical time step. This gives a certain advantage of the CA technique over the MC method where cells are chosen randomly.

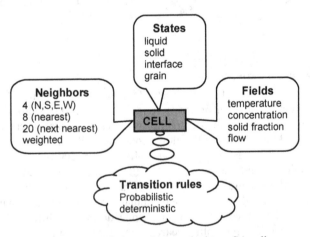

Figure 14.14. Information attached to a CA cell.

Cellular automaton grain growth models - meso-scale
The CA method has been perfected by Rappaz and Gandin (1993) and then Rappaz and co-workers (*e.g.* 1995, 1996). They coupled a probabilistic nucleation model

with a CA growth model and a Finite Element heat flow model. The result is a meso-scale model (CAFE). First, the explicit temperature and the enthalpy variation were interpolated at a cell location. Then, nucleation and growth of grains were simulated using the CA algorithm. This algorithm accounts for the heterogeneous nucleation in the bulk and at the mold /casting interface, for the preferential growth of dendrites in certain directions, and for microsegregation. Since in their models dendritic tip growth law was used and growth velocity was then averaged for a given micro-volume element, the final product of the simulation was dendritic grains, not dendritic crystals. In this respect, the method relies heavily on averaged quantities, just as continuum deterministic models do, but can display grain structure on the computer screen. However, the graphical output is limited at showing grain boundaries.

An example of the CA setup for a plate is provided in Figure 14.15. The volume is divided into a square lattice of regular cells. The Von-Neumann neighborhood configuration was adopted in calculation (*i.e.*, only the nearest neighbor of the cell are considered). The index that defines the state of the cell is zero when the cell is liquid and an integer when the cell is solid. The integer is associated with the crystallographic orientation of the grain and it is randomly generated.

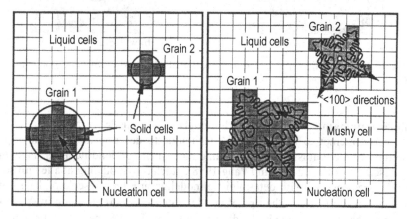

Figure 14.15. Computational space of a 2D cellular automaton for dendrite envelope grain growth (Rappaz *et al.*, 1995).

For eutectic grains, the velocity is normal to the interface, while for dendritic grains it corresponds to the preferential growth direction of the primary and secondary arms (Rappaz *et al.*, 1995). The extension of each grain is then calculated by integrating the growth rate over time. All the liquid cells captured during this process are given the same index as that of the parent nucleus. The computed structure can be either columnar or equiaxed.

One of the most interesting features of the CAFE models is their ability to predict the CET in the presence of fluid flow. As shown in Figure 14.16, without convection the predicted grain structure is fully equiaxed. It does not reproduce the sedimentation cone and the columnar grain structure observed experimentally. Even such details as the deflection of columnar grains because of fluid flow parallel to the solidification interface can be modeled with the CA technique (Lee *et al.*, 1998).

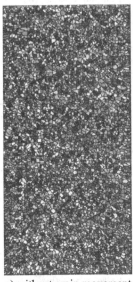

a) without grain movement   b) with grain movement

Figure 14.16. Simulation of CET in a conventionally cast Al-7%Si alloy (Gandin *et al*, 1998).

Cellular automaton dendrite growth models - micro-scale

In an early attempt to extend the CA technique to describe the crystallographic anisotropy of the grains and the branching mechanism of dendrite arms (Pang and Stefanescu 1996, Stefanescu and Pang 1998), stochastic modeling at a length scale of 10^{-6}m was coupled with deterministic modeling at a length scale of 10^{-4}m. Atoms attach much faster at the tips of the dendrite, which are growing in the $\langle 100 \rangle$ direction. The lowest rate is in the direction of $\theta = 45°$ with respect to the dendrite axes. Accordingly, the growth velocity of the dendrite liquid interface was expressed as $V_{dendr} = V_{max} \cdot f(\theta)$, where $f(\theta)$ is a function of the orientation of the cell with respect to the dendrite axes. This function must be 1 in the $\langle 100 \rangle$ directions and zero at $\theta = 45°$. V_{max} and arm thickening were calculated with deterministic laws derived from the dendrite tip velocity. However, the overall growth of dendrite arms was derived from CA probabilistic calculations. Branching of dendrites arm was allowed to occur based on the classic criterion for morphological instability. Thus the dendrite morphology, rather than the grain structure, was simulated.

Expanding on Dilthey *et al.* (1997) model, Nastac (1999) developed a stochastic model to simulate the evolution of dendritic crystals. The model includes time-dependent calculations for temperature distribution, solute redistribution in the liquid and solid phases, curvature, and growth anisotropy. Previously developed stochastic procedures for dendritic grains (Nastac and Stefanescu, 1997) were used to control the nucleation and growth of dendrites. A numerical algorithm based on a Eulerian-Lagrangian approach was developed to explicitly track the sharp S/L interface on a fixed Cartesian grid. 2D calculations at the dendrite tip length scale were performed to simulate the evolution of columnar and equiaxed dendritic morphologies including the occurrence of the columnar-to-equiaxed transition.

The classic transport equations in 2D and Cartesian coordinates were used, that is Eqs. (6.1) and (6.2) for energy transport, and Eq. (4.2) assuming $V = 0$ for solute transport. Eq. (4.2) was written for solid and then for liquid. The boundary conditions were:

- solute conservation at the interface ($\vec{n}$ is the interface normal vector):

$$V_n\left(C_L^* - C_S^*\right) = D\left(\frac{\partial^2 C}{\partial x^2} + \frac{\partial^2 C}{\partial y^2}\right) \cdot \vec{n}\Big|_-^+ \tag{14.44}$$

- local equilibrium at the interface: $C_S^* = kC_L^*$

Curvature was used in the formulation of both interface temperature and interface concentration according to the equations:

$$T^* = T_L + \left(C_L^* - C_0\right)m_L - \Gamma\,\bar{\kappa}\,f(\varphi, \theta) \tag{14.45}$$

$$C_L^* = C_O + \left(T^* - T_L^{EQ} + \Gamma\,\bar{\kappa}\,f(\varphi, \theta)\right)m_L^{-1} \tag{14.46}$$

where $\bar{\kappa}$ is the mean curvature of the S/L interface, $f(\varphi, \theta)$ is a coefficient used to account for growth anisotropy, θ is the growth angle (*i.e.*, the angle between the normal and the x-axis), and φ is the preferential crystallographic orientation angle.

The treatment of the interface curvature merits further discussion. The average interface curvature for a cell with the solid fraction f_S was calculated as:

$$\bar{\kappa} = \frac{1}{a}\left(1 - 2\left(f_S + \sum_{k=1}^{N} f_S(k)\right)(N+1)^{-1}\right) \tag{14.47}$$

where N is the number of neighboring cells taken as $N = 24$. This includes all the first and second order neighboring cells. Eq. (14.47), a modification of the method proposed by Sasikumar and Sreenivisan (1994), is a simple counting-cell technique that approximates the mean geometrical curvature, and not the local geometrical curvature. Thus, its applicability for this problem is limited.

The anisotropy of the surface tension was calculated as per Dilthey and Pavlik (1998):

$$f(\varphi, \theta) = 1 + \delta\cos(4(\varphi - \theta)) \quad \text{with} \quad \varphi = a\cos\left(\frac{V_x}{\left[(V_x)^2 + (V_y)^2\right]^{1/2}}\right) \tag{14.48}$$

where V_x and V_y are the growth velocities in the x and y directions, respectively, and δ accounts for the degree of anisotropy. For four-fold symmetry, $\delta = 0.04$.

Some typical examples of calculation results are given in Figure 14.17. Note that all dendrites are oriented in the x-y direction.

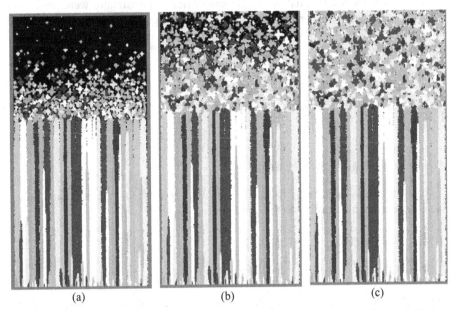

(a)   (b)   (c)

Figure 14.17. Simulated microstructure (columnar cellular/equiaxed dendritic morphologies and CET formation) in unidirectional solidification of IN718-5 wt.% Nb alloy (initial melt temperature is 1400°C, 10x20mm-domain, $a = 20\mu m$, $h = 10^4$ W m^{-2} K^{-1}) (Nastac, 1999). Reprinted with permission from Elsevier.

Building on an earlier meso-scale CA model, Zhu and Hong (2001) developed a modified micro-scale CA model by incorporating the effect of the constitutional and the curvature undercoolings on the equilibrium interface temperature. The model has been applied to simulate two-dimensional (2D) and three-dimensional (3D) single and multi-dendritic growth, non-dendritic and globular microstructure evolution in semi-solid process (Zhu *et al.*, 2001), dendritic growth in the presence of melt convection, and microstructure formation in regular and irregular eutectics alloys (Zhu and Hong, 2004). However, as the velocity of the SL interface is calculated from the analytical theories of dendrite growth or by introducing a kinetics coefficient, the model leads to mostly qualitative graphical outputs.

Recognizing the problems of mesh-induced anisotropy in crystallographic orientation Beltran-Sanchez and Stefanescu (BSS) (2003, 2004) developed a model for the simulation of solutal dendritic growth in the low Péclet number regime that does not use an analytical solution to determine the velocity of the S/L interface. The model adopted a methodology similar to that of Nastac (1999) by calculating growth kinetics from the complete solution of the solute and heat transport equations and by incorporating the boundary condition of solute conservation. The heat equation was solved using an FD method with an implicit scheme. For the case of the solute diffusion equation, a special scheme was used to overcome the problem of the discontinuity at the interface. The diffusion equation was solved in terms of a potential, which consist of an equivalent composition at every point in the domain. Thus, the whole domain is treated as a single phase for computational purposes. The potential is defined as:

$$P = C_L \qquad\qquad\qquad\qquad \text{for the liquid phase}$$

$$P = C_S/k \qquad\qquad\qquad\qquad \text{for the solid phase}$$

$$P = C_L^* \qquad\qquad\qquad\qquad \text{for the interface cells}$$

Using these definitions the solute conservation equation becomes:

$$\frac{\partial \overline{C}}{\partial t} = D\left(\frac{\partial^2 P}{\partial x^2}\right) + D\left(\frac{\partial^2 P}{\partial y^2}\right) \tag{14.49}$$

where $\overline{C} = f_S C_S + (1 - f_S)C_L$.

In the first model BSS (2003) the x and y components of the velocity at the interface were calculated from Eq. (14.44) using an FD scheme. The composition at the interface was calculated with Eq. (14.46). The fraction of solid was then evaluated with the equation (Dilthey et al., 1997):

$$\Delta f_s = \frac{\Delta t}{\Delta s}\left(V_x + V_y - V_x V_y \frac{\Delta t}{\Delta s}\right) \tag{14.50}$$

While the equations for V_x and V_y derived from the solute conservation at the interface Eq. (14.44) are correct from the mathematical point of view, they describe the motion of an interface that is planar in the two Cartesian directions, which cannot be true physically. The result is the evaluation of a solid fraction assuming that the interface moves in the x and y direction as a planar front. This produces an overestimation of the actual solid fraction. Since the solute balance is valid only at the limit of a vanishing SL interface thickness in the normal direction, the only physically realistic velocity that can be calculated from this balance is the normal velocity and not the components. Accordingly, in their second model, BSS (2004) derived the following equation to evaluate the normal SL interface velocity and solid fraction increments in 2D Cartesian coordinates:

$$V_n = \frac{1}{C_L^*(1 - k_o)}\left[-D_L(D_n C_L)\big|_{\text{interface}} + D_S(D_n C_S)\big|_{\text{interface}}\right]$$

where D_n is the directional derivative operator on the direction of the normal. The change in solid fraction was then computed as $\Delta f_s = V_n \Delta t / L_\varphi$, where L_φ is the length of the line along the normal passing trough the cell center. The length L_φ represents the distance to be covered by a point on the S/L interface so that it could be considered solid. Note that this distance is normalized with the direction of motion of the S/L interface so as to minimize the effect of the artificial mesh anisotropy on the rate of advance of the interface. Thus, while the model is based on the CA concepts, it is using virtual tracking of the sharp S/L interface. Combined with the trapping rules for new interface cells, the mesh dependency of calculations was eliminated allowing the model to simulate dendrites growing at any crystallographic direction.

As the calculation of local curvature with a counting cell method used in previous models is mesh dependent, BSS introduced another method that converges to a finite value when the mesh is refined. It is based on the variation of the unit vector normal to the S/L interface along the direction of the interface. The final equation in vectorial form is $K = D_{\hat{T}}\hat{n}\cdot\hat{T}$, where $D_{\hat{T}}$ is the directional derivative operator in the direction of the tangent $\hat{T}$. The tangent vector is perpendicular to the unit normal $\hat{n} = -\nabla f_S / \|\nabla f_S\|$. In 2D Cartesian coordinates the equation for curvature takes the following form:

$$K = \left[\left(\frac{\partial f_s}{\partial x}\right)^2 + \left(\frac{\partial f_s}{\partial y}\right)^2\right]^{-3/2}\left[2\frac{\partial f_s}{\partial x}\frac{\partial f_s}{\partial y}\frac{\partial^2 f_s}{\partial x \partial y} - \left(\frac{\partial f_s}{\partial x}\right)^2\frac{\partial^2 f_s}{\partial y^2} - \frac{\partial^2 f_s}{\partial x^2}\left(\frac{\partial f_s}{\partial y}\right)^2\right]$$

This last equation was originally proposed by Kothe et al. (1991) using a definition of curvature as:

$$K = -(\nabla\cdot\hat{n}) = \frac{1}{|\vec{n}|}\left[\left(\frac{\vec{n}}{|\vec{n}|}\cdot\nabla\right)|\vec{n}| - (\nabla\cdot\vec{n})\right]$$

Using this approach the model does not need to use the concept of marginal stability and stability parameter to uniquely define the steady state velocity and radius of the dendrite tip. The model indeed contains an expression for the stability parameter but the process determines its value.

The simulated composition map of a single dendrite growing from one wall at a given orientation is presented in Figure 14.18. The secondary and tertiary arms still grow perpendicular to the primary arms, which are not aligned with the x-y directions of the mesh. The effect of the upper wall prevents the side facing it to develop branching, while the lower side is fully branched. The dendrite tip approaches the parabolic shape. Another example of model output is shown in Figure 14.19 for the case of multiple equiaxed dendrites.

The classic theories for dendritic growth predict that at low Péclet numbers the radius of the dendrite tip increases with lower growth velocity (Kurz et al., 1986) as shown in Figure 14.20. The BSS model was used to simulate the influence of growth velocity on dendrite morphology. The growth velocity was changed by imposing different cooling rates (Figure 14.21). As the cooling rate decreases, the

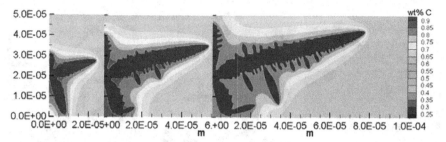

Figure 14.18. Simulation of a Fe-0.6wt%C alloy dendrite with $\theta = 15°$. $\Delta s = 0.1$ μm. From left to right: after 0.03, 0.05 and 0.07s (Beltran-Sanchez and Stefanescu, 2004). with kind permission of Springer Science and Business Media.

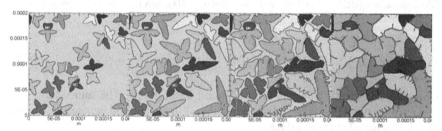

Figure 14.19. Simulation of equiaxed solidification of Al-4wt%Cu alloy showing grain boundary formation. From left to right: after 0.04, 0.08, 0.16, and 0.2s (Beltran-Sanchez and Stefanescu, 2004). With kind permission of Springer Science and Business Media.

tip radius of the dendrite becomes larger. When the radius of curvature of the dendrite tip is of the same order of magnitude as that of the whole grain the dendrite becomes a cell in directional solidification or globular in equiaxed solidification.

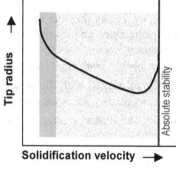

Figure 14.20. Schematic correlation between solidification velocity and dendrite tip radius.

More recently, Zhu and Stefanescu (2007) developed a computationally efficient quantitative virtual front tracking model for the two-dimensional simulation of dendritic growth in the low Péclet number regime. The kinetics of dendritic growth is driven by the difference between the local equilibrium composition, calculated from the local temperature and curvature, and the local actual liquid composition, obtained by solving the solutal transport equation:

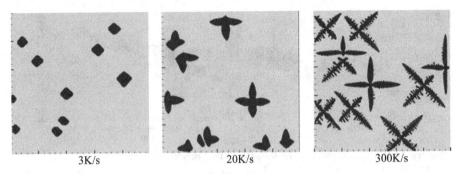

3K/s   20K/s   300K/s

Figure 14.21. Composition maps of the transition from cellular to dendritic microstructure as function of the cooling rate (different heat transfer coefficients) for equiaxed grains of Fe-0.6wt%C alloy (Beltran-Sanchez and Stefanescu, 2003). With kind permission of Springer Science and Business Media.

$$\partial C_i / \partial t = \nabla \cdot \left(D_i \nabla C_i \right) + C_i \left(1 - k \right) \partial f_S / \partial t \qquad (14.51)$$

where the subscript i indicates solid or liquid, respectively. The source term (second term on the RHS) is included only at the SL interface and denotes the amount of solute rejected at the interface.

For each interface cell with non-zero solid fraction, the compositions of liquid and solid are calculated and stored simultaneously. The solid composition is calculated with:

$$C_S = \sum_{n=1}^{N} k \, \Delta f_S (n) C_L (n) \Big/ \sum_{n=1}^{N} \Delta f_S (n)$$

where N indicates the number of iterations (time step intervals) after the cell became an interface cell. The interface equilibrium temperature and composition were calculated with equations (14.45) and (14.46), respectively.

The calculated local interface equilibrium composition C_L^* from Eq. (14.46) is compared with the local actual liquid composition C_L, calculated with the solute transport Eq. (14.51). If $C_L^* - C_L > 0$, the solid fraction of the cell is increased. According to the interface equilibrium condition, during one time step interval, the increase in solid fraction of an interface cell can be evaluated through:

$$\Delta f_S = \left(C_L^* - C_L \right) \Big/ \left(C_L^* \left(1 - k \right) \right)$$

If at time t_N, the sum of the solid fraction in an interface cell equals one, i.e., $\sum_{n=1}^{N} \Delta f_S (n) = 1$, this interface cell has fully solidified. Then, the interface cell changes its state to solid. From the calculated increase in solid fraction at each time step, the normal growth velocity of the interface can be obtained with:

$$V_n = \Delta f_S \, \Delta x / \Delta t$$

where Δx is the cell size, and Δt is the time step.

The model adopts previously proposed solutions for the evaluation of local curvature and interface capturing rules with a virtual interface tracking scheme, which make the model virtually mesh-independent. The dynamics of dendrite growth from the initial unstable stage to the steady-state stage is accurately described. Side branching develops without the need to introduce local noise. An example of the output of a 3D calculation with this model is presented in Figure 14.22.

Figure 14.22. Equiaxed dendrite generated through a CA code (M.F. Zhu, 2007 private communication).

Zhu *et al.* (2007) have summarized the applications of front tracking/CA models for the simulation of dendritic and non-dendritic microstructure evolution in semi-solid processing solidification, the influence of fluid flow on dendrite morphology, and the dendritic-eutectic solidification.

14.3 Phase field models

Phase field (PF) models represent the SL interface as a continuous transition layer with finite thickness between the two phases using an additional PF variable, which avoids explicit tracking of the phase boundaries. The form of the phase field and conservation equations is derived from thermodynamics, assuming an expression for the entropy distribution in the system and assuring positive generation of entropy. The technique can simulate the kinetics of dendritic growth. In addition, several factors affecting dendritic growth can be incorporated, such as the crystallographic orientations, motion of the boundaries during impingement (Warren *et al.*, 2000), and others.

However, the technique has significant limitations. The need to spread the change of the phase field variable over several mesh points during the numerical solution of the equations demands massive computer resources. This limits the use of this technique on very small domains or for a few dendrites. Furthermore, the technique cannot readily incorporate the solution of both heat and solute conservation equations, implying that only simplified cases of limited practical interest are addressed, such as constant cooling rate or isothermal conditions. Moreover, some of the parameters do not have direct correlation with physical data, such as diffusion coefficients and mobility parameter, therefore functional forms are assumed. They have been applied successfully to the solidification of pure metals (Karma and Rappel 1996) and alloys (Warren and Boettinger 1995, Loginova *et al.* 2001, Lan *et al.* 2003) producing very realistic dendritic growth patterns (Granasy *et al.* 2004), and validating quantitatively that the growth kinetics of the dendrite tip agrees well with the results of microscopic solvability theory (Karma and Rappel, 1998). However, the simulation results of PF models are significantly dependent on the prescribed interface thickness, which is required to be sufficiently small for accurate simulations. Because the solute-diffusion length is several orders of magnitude smaller than the heat-diffusion length, resulting in extremely thin solute boundary layers, sharp treatment of the interface is required for the simulations of dendritic growth in alloys. This imposes a very fine grid, leading to high computational times.

Karma (2001) proposed an anti-trapping current concept that allows elimination of non-equilibrium effects at the interface. Thus a relatively thick interface width can be used. Together with the efficient adaptive grid techniques, the computational speed of PF simulations has increased by several orders of magnitude. Recently, quantitative PF simulations of alloys for free dendrite growth (Ramirez *et al.* 2004, Ramirez and Beckermann 2005). Greenwood *et al.* (2004) reported simulation of cellular and dendritic growth in directional solidification of dilute binary alloys using a PF model solved on an adaptive grid. The simulated spacing of primary branches as a function of pulling velocity for various thermal gradients and alloy compositions was found to have a maximum and agreed with experimental observations.

An example of dendrite growth simulation through the PF method is given in Figure 14.23 (Granasy *et al.*, 2003). At low nucleation rate there is enough space to develop a full dendritic morphology.

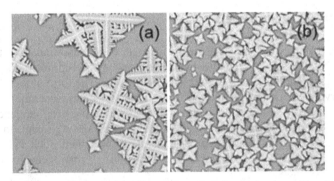

Figure 14.23. Phase field simulation of two-dimensional anisotropic multigrain solidification as a function of composition and nucleation rate in the Cu–Ni system at 1574 K. (a)–low nucleation rate; (b)-high nucleation rate (Granasy *et al.*, 2003). With kind permission of Springer Science and Business Media.

Tiaden (1999) has simulated the microstructure evolution during peritectic solidification of Fe-C alloys using a multiphase field approach. It was assumed that the process is non-equilibrium, diffusion controlled. Three phases, liquid, ferrite and austenite, were considered, and phase fields were defined for each phase. The phase field model was coupled with a solute diffusion model. An example of the calculation of the growth of four ferrite particles during constant cooling of a Fe-C alloy is shown in Figure 14.24. Anisotropy was not considered. Below the peritectic temperature single nuclei were placed on the four interfaces. It is seen that the peritectic γ grows around the primary δ by simultaneous consumption of both ferrite and liquid. The peritectic reaction is the fastest growth mechanism because at the peritectic temperature the carbon concentration in γ is higher than that in δ but smaller than that of the liquid. Thus the fastest growth will be where liquid and ferrite can directly react. Thus, the austenite grows around the ferrite.

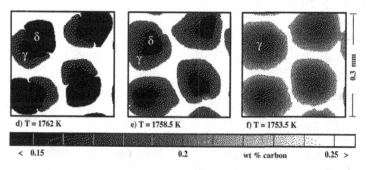

Figure 14.24. Simulation of peritectic solidification at constant cooling. The carbon concentration is illustrated by the gray scale (Tiaden, 1999). Reprinted with permission from Elsevier.

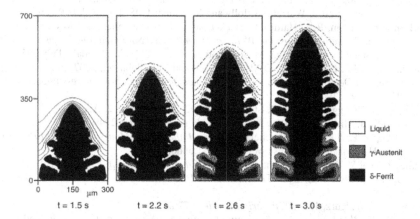

Figure 14.25. Simulation of directional peritectic solidification. The morphology of the primary ferrite becomes unstable and a dendrite evolves. Under the peritectic temperature austenite is formed and coats the dendrite (Tiaden, 1999). Reprinted with permission from Elsevier.

Results of simulation of directional solidification under constant thermal gradient are shown in Figure 14.25. The δ-dendrite produced by morphological instability growth in the liquid until the peritectic temperature is reached. At this temperature, some random nucleation of γ was imposed on the system, and then austenite continued to grow on the undercooled dendrite consuming both the ferrite and the liquid. This model was later expanded to ternary systems (see Figure 14.26).

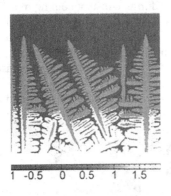

Figure 14.26. Simulation of peritectic solidification in a Fe-C-Mn system - L, δ, γ phases (MICRESS, 2004). www.micress.de

References

Aizawa T., 1978, *Imono* **50**:33
Anderson M.P., D.J. Srolovitz, G.S. Grest and P.S. Sahni, 1984, *Acta Metall.* **32**:783
Atwood R.C., S. Sridhar, W. Zhang and P.D. Lee, 2000, *Acta Mater.* **48**:405
Atwood R.C. and P.D. Lee, 2003, *Acta Materialia* **51**:5447
Avrami M., 1939, *J. Chem. Phys.* **7**:1103
Beltran-Sanchez L. and D.M. Stefanescu, 2003, *Metall. Mater. Trans.* **34A**:367
Beltran-Sanchez L. and D.M. Stefanescu, 2004, *Metall Mater Trans* **35**:2471
Campbell J., 2003, in *Modeling of Casting, Welding and Adv. Solidif. Processes X*, eds. D.M. Stefanescu, J. Warren, M. Jolly and M. Krane, TMS, Warrendale PA, p.209
Carlson K.D., Z. Lin, R.A. Hardin, C. Beckermann, G. Mazurkevich, and M.C. Schneider, 2003, in *Modeling of Casting, Welding and Advanced Solidification Processes X*, eds. D.M. Stefanescu, J. Warren, M. Jolly and M. Krane, TMS, Warrendale, Pa p.295
Catalina A.V., J.F. Leon-Torres, D.M. Stefanescu and M.L. Johnson, 2007, in *Proceedings of the 5th Decennial International Conference on Solidification Processing*, Sheffield, UK, pp.699-703
Chang S., D. Shangguan and D.M. Stefanescu, 1992, *Metall Trans.* **23A**:1333
Das A. and E.J. Mittemeijer, 2000, *Metall. and Mater. Trans.* **31A**:2049
Dilthey U., V. Pavlik, and T. Reichel, 1997, in: *Mathematical Modelling of Weld Phenomena*, H. Cerjak ed., The Inst. of Materials, London p.85
Dilthey U. and V. Pavlik, 1998, in: *Modeling of Casting, Welding and Advanced Solidification Processes-VIII*, B.G. Thomas and C. Beckermann, eds., TMS, Warrendale Pa. p.589
Dustin I. and W. Kurz, 1986, *Zeitschrift Metallkde* **77**:265
Fras E., 1975, *PhD Dissertation*, Academy of Mining and Metallurgy, Cracow, Poland
Gandin C.A., T. Jalanti and M. Rappaz, 1998, in: *Modeling of Casting, Welding and Advanced Solidification Processes-VIII*, B.G. Thomas and C. Beckermann, eds., TMS, Warrendale Pa. p.363
Ganesan S. and D.R. Poirier, 1990, *Met. Trans.* **21B**:173
Granasy L., T. Borzsonyi and T. Pusztai, 2003, in *Lecture Notes in Computational Science and Engineering*, ed H Emmerich et al., Berlin, Springer, **32**:190–5
Granasy L., Pusztai T., Warren J.A., 2004, *J. Phys.: Condens Matter* **16**:R1205
Greenwood M., Haataja M., Provatas N., 2004, *Phys Rev Lett* **93**:246101

Hesselbarth H.W. and I.R. Göbel, 1991, *Acta Metall.* **39**:2135
Hillert M., 1955, *Acta Metall.* **3**:653
Jacot A., Rappaz M., 2002, *Acta Mater.* **50**:1909
Johnson W.A. and R.F. Mehl, 1939, *Trans. AIME* **135**:416
Juric D, Tryggvason G., 1996, *J Comput Phys.* **123**:127
Karma A., 2001, *Phys Rev Lett* **87**:115701
Karma A., Rappel W.J.,1996, *Phys Rev Lett* **77**:4050
Karma A., Rappel W.J., 1998, *Phys Rev E* **57**:4323
Kanetkar C.S. and D.M. Stefanescu, 1988, in: *Modeling of Casting and Welding Processes IV,* A. Giamei, G.J. Abbaschian editors, TMS Warrendale. Pa. p.697
Kanetkar C.S., I.G. Chen, D.M. Stefanescu, N. El-Kaddah, 1988, *Trans. Iron and Steel Inst. of Japan* **28**:860
Kirkaldy J.S., 1995, in: *Mathematics of Microstructure Evolution*, L.Q. Chen et al. eds., TMS, Warrendale Pa. p.173
Kolmogorov A.N., 1937, *Bulletin de L'Académie des Sciences de L'URSS* (3):355-359
Kothe D.B., R.C. Mjolsness, and M.D. Torrey, 1991, RIPPLE: A computer Program for Incompressible Flows with Free Surface, Los Alamos National Lab., LA-10612-MS, Los Alamos, NM
Kubo K. and R.D. Pehlke, 1985, *Met. Trans.* **16B**:359
Kurz W., B. Giovanola and R. Trivedi, 1986, *Acta Metallurgica* **34**(5):823-830
Lan C.W., Chang Y.C., Shih C.J., 2003, *Acta Mater* **51**:1857
Lee P.D. and J.D. Hunt, 1997, *Acta Mater.* **45**:4155
Lee P.D., A. Chirazi, R.C. Atwood, W. Wang, 2004, *Mat. Sci. and Eng.* **A365**:57
Lee S.Y., S.M. Lee and C.P. Hong, 1998, in *Modeling of Casting, Welding and Adv. Solidif. Processes X*, eds. B.G. Thomas and C. Beckermann, TMS, Warrendale PA,p.383-390
Loginova I., Amberg G., Agren J., 2001, *Acta Mater* **49**:573
Ludwig A. and M. Wu, 2002, *Metall. and Mater. Trans.* **33A**:3673-3683
MICRESS-ACCESS, 2004, Aachen, Germany
Nastac L., 1999, *Acta mater.* **47**:4253
Nastac L., 2004, *Modeling and Simulation of Microstructure Evolution in Solidifying Alloys*, Kluwer Academic Publishers, Boston, 285p
Nastac L. and D.M. Stefanescu, 1992, *Micro/Macro Scale Phenomena in Solidification*, C. Beckermann et al. eds. HTD-Vol. 218, AMD-Vol. 139, ASME, New York p.27
Nastac L. and D. M. Stefanescu, 1996, *Metall. Trans.* **27A**:4061 and 4075
Nastac L. and D.M. Stefanescu, 1997, *Modelling and Simulation in Mat. Sci. and Eng.*, Inst. of Physics Publishing **5**(4):391
Ni J. and C. Beckermann, 1991, *Metall. Trans.* **22B**:349
Ohnaka I., J. Iwane, Y. Sako, H. Yasuda and H. Zhao, 2002, in *Proceedings of the 65th World Foundry Congress*, Gyeogju, Korea, p.639
Ohnaka I., 2003, in *Modeling of Casting, Welding and Advanced Solidification Processes X*, eds. D.M. Stefanescu, J. Warren, M. Jolly and M. Krane, TMS, Warrendale Pa. p.403
Oldfield W., 1966, *ASM Trans.* **59**:945
Pang H. and D.M. Stefanescu, 1996, in: *Solidification Science and Processing,* I. Ohnaka and D.M. Stefanescu eds., TMS, Warrendale, Pa. p.149
Péquet Ch., M. Gremaud, and M. Rappaz, 2002, *Metall. Mater. Trans.* **33A**:2095
Poirier.R., K. Yeum and A.L. Maples, 1987, *Met. Trans.* **18A**:1979
Ramirez J.C., Beckermann C., Karma A., Diepers H-J., 2004, *Phys Rev A* **69**:051607
Ramirez J.C. and Beckermann C., 2005, *Acta Mater* **53**:1721
Rappaz M. and P. Thévoz, 1987, *Acta. Metall.* **35**:1487 and 2929
Rappaz M. and C.A. Gandin, 1993, *Acta Metall. Mater.* **41**:345

Rappaz M., C.A. Gandin, and C. Charbon, 1995, in: *Solidification and Properties of Cast Alloys*, Proceedings of the Technical Forum, 61st World Foundry Congress. Beijing: Giesserei-Verlag p.49

Rappaz M., C.A. Gandin, J.L. Desbiolles and P. Thevoz, 1996, *Metall. and Mat. Trans.* **27A**:695

Sabau A.S. and S. Viswanathan, 2002, *Metall. Mater. Trans.* **33B**:243

Sahni P.S., G.S. Grest, M.P. Anderson and D.J. Srolovitz, 1983, *Phys. Rev. Lett.* **50**:263

Sako Y., I. Ohnaka, J.D. Zhu and H. Yasuda, 2001, in *Proc. of the 7th Asian Foundry Congress*, The Chinese Foundrymen's Association, Taipei, p.363

Sasikumar R. and R. Sreenivisan, 1994, *Acta Metall.* **42**(7):2381

Spittle J.A. and S.G.R. Brown, 1989, *Acta Metall.* **37**:1803

Stefanescu D.M., 2001, in *Modeling for Casting and Solidification Processing*, K. O. Yu ed., Marcel Dekker Inc., New York p.123

Stefanescu D.M. and S. Trufinescu, 1974, *Z. Metallkde.* **65**:610

Stefanescu D.M. and C. Kanetkar, 1986, in: *State of the Art of Computer Simulation of Casting and Solidification Processes*, H. Fredriksson ed., Les Edition de Physique, Les Ulis, France p. 255

Stefanescu D.M. and H. Pang, 1998, *Canadian Metallurgical Quarterly* **37**(3-4):229-239

Su K.C., I. Ohnaka, I. Yanauchi and T. Fukusako, 1984, in: *The Physical Metallurgy of Cast Iron*, H. Fredriksson and M. Hillert eds., North Holland, New York p.181

Sung P.K., D.R. Poirer, and S.D. Felicelli, 2002, *Mod. Sim. Mat. Sci. Eng.* **10**, 551

Thévoz P., J. L. Desbioles, and M. Rappaz, 1989, *Metall. Trans.* **20A**:311

Tiaden J., 1999, *J. Crystal Growth*, **198/199**:1275-1280

Tynelius K., J.F. Major and D. Apelian, 1993, *Trans. AFS* **101**:401

Udaykumar H.S., Mittal R., Shyy Wei, 1999, *J Comput Phys.* **153**:535.

Viswanathan S., A.J. Duncan, A.S. Sabau and Q. Han, 1998, in: *Modeling of Casting, Welding and Advanced Solidification Processes VIII*, B.G. Thomas and C. Beckermann eds., TMS, Warrendale, PA p. 49

Wang C.Y. and C. Beckermann, 1993a, *Mater. Sci. and Eng.* **A171**:2787

Wang C.Y. and C. Beckermann, 1993, *Metall. Trans.* **24A**:2787

Wang C.Y. and C. Beckermann, 1994, *Metall. Trans.* **25A**:1081

Wang C.Y. and C. Beckermann, 1996, *Metall. Mater. Trans.* **27A**:2754

Warren J.A., Boettinger W.J., 1995, *Acta Metall Mater* **43**:689

Warren J.A., R. Kobayashi, and W.C. Carter, 2000, in *Modeling of Casting, Welding and Advanced Solidification Processes IX*, P.R. Sahm, P.N. Hansen, and J.G. Conley Eds., Shaker Verlag., Aachen, Germany, p.CII-CIX

Wu M. and A. Ludwig, 2006, *Metall. Mater. Trans.* **37A**:1613-1631

Zhao P, Venere M., Heinrich J.C., Poirier D.R., 2003, *J Comput Phys* **188**:434

Zhu J.D. and I. Ohnaka, 1991, in: *Modeling of Casting, Welding and Advanced Solidification Processes V*, M. Rappaz, M.R. Ozgu and K. Mahin, eds, TMS, Warrendale, PA p.435

Zhu M.F. and Hong C.P., 2001, *ISIJ Int.* **41**:436

Zhu M.F., Kim J.M. and Hong C.P., 2001, *ISIJ Int.* **41**:992

Zhu M.F. and Hong C.P., 2004, *Metall Mater Trans* **35A**:1555

Zhu M.F. and D.M. Stefanescu, 2007, *Acta Materialia* **55**:1741-1755

Zhu M.F., C.P. Hong, D.M. Stefanescu and Y.A. Chang, 2007, *Metall. Mater. Trans.* **38B**:517

Zhu P. and R. W. Smith, 1992, *Acta metall. mater.* **40**:683 and 3369

Zou J. and R. Doherty, 1993, in: *Modeling of Casting, Welding and Advanced Solidification Processes VI*, T.S. Piwonka, V. Voller and L. Katgerman, eds, TMS, Warrendale, PA p.193

Xiao R., J.I.D. Alexander and F. Rosenberg, 1992, *Phys. Rev. A*, **45**:R571-R574

15

ATOMIC SCALE PHENOMENA - NUCELATION AND GROWTH

Solidification is the result of the formation of stable clusters of solid in the liquid (nucleation), followed by their growth. These are phenomena occurring at the atomic scale (nano meter) level. The present understanding of the beginning of formation of solid crystals from their liquid is based on the classical theory of homogeneous nucleation. This theory uses macroscopic concepts and classic thermodynamics to describe the appearance of the first microscopic crystals in the melt.

Nucleation and growth control the morphology and the fineness of the as-cast microstructure, and even the phases that it contains. In turn, the microstructure controls the mechanical properties of the casting. Thus, in our effort to quantify the evolution of microstructure during solidification, the ultimate goal is to be able to describe the structure as the result of movement of individual atoms. This task is yet out of reach but significant progress has been made in this direction.

15.1 Nucleation

The classic nucleation theory is a phenomenological theory that assumes that clusters of atoms or molecules form spontaneously in the matter undergoing transformation. In other words, steady state, or time independent nucleation is assumed, resulting in a constant nucleation rate. While the classical theory is applicable to the study of a wide range of nucleation phenomena, in many cases the assumption of a constant nucleation rate is wrong. Time-dependent nucleation is relevant to many first-order phase transformations including condensation from vapor, crystallization of undercooled liquids, crystallization of glass (devitrification), and others (Kelton, 1991).

15.1.1   Steady-state nucleation - homogeneous nucleation

Above or below the equilibrium transformation temperature, fluctuations in density, atomic configurations, heat content, etc., occur in the liquid. They make possible the formation of minute particles of crystalline solid (long range order), called *embryos*. Consequently, a liquid-solid interface is created, and associated with it is an interface energy. As a result, the free energy of the system increases, and, unless sufficient undercooling is available, the embryo will remelt. If the undercooling of the melt is sufficient, the embryo will survive, and will grow to form a *nucleus*. Because the nucleus has the same composition as the liquid and solid, this is called homogeneous nucleation. It can be demonstrated that the embryo must grow to a certain *critical size* in order to become stable and form a nucleus.

Consider that an embryo of radius r is formed in the liquid. This will result in a change in free energy, firstly, because of the decrease in the free energy resulting from the change of the volume of radius r from liquid state to solid, and secondly, because of the increase in the free energy due to the newly created liquid-solid interface. The change in free energy is:

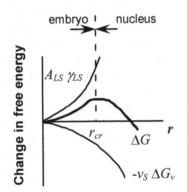

Figure 15.1. Variation of the free energy of the liquid - solid system with the radius of the embryo.

$$\Delta G = -v_S \Delta G_v + A_{LS}\gamma_{LS} = -\frac{4}{3}\pi r^3 \Delta G_v + 4\pi r^2 \gamma_{LS} \qquad (15.1)$$

where v_S is the volume of solid formed, A_{LS} is the newly created liquid-solid interface, γ_{LS} is the surface energy associated with the newly created surface of the grain. The first RHT of this equation is the driving force for nucleation. The second term represents an energy barrier. The whole equation is represented by the ΔG curve in Figure 15.1. The maximum of this curve corresponds to a radius r_{cr}, which is called the *critical radius*. If the embryo has reached size r_{cr}, then it can be seen from the figure that further increase of the embryo will result in a decrease of the free energy. This means that spontaneous growth of the embryo, which is now a nucleus, is possible. If on the contrary, $r_{embryo} < r_{cr}$, the embryo will melt, unless additional energy is removed from the system. To find the value of r_{cr}, one has to find the maximum of the curve described by Eq. (15.1), that is to equate the first derivative of this equation with respect to r, to zero, $\partial \Delta G/\partial r = 0$. Differentiating and using Eq. (2.8), the size of the critical radius is:

$$r_{cr} = \frac{2\gamma_{LS}}{\Delta G_v} = \frac{2\gamma_{LS}T_e}{\Delta H_f \Delta T} = \frac{2\gamma_{LS}T_e}{\Delta H_f(T_e - T)} = \frac{2\gamma_{LS}}{\Delta S_f(T_f - T)} \qquad (15.2)$$

Note that r_{cr} depends inversely on ΔT. For negative values of both, a sphere of liquid in a solid crystal is implied (melting).

Experimental work summarized by Turnbull (1956) has demonstrated that the undercooling required for homogeneous nucleation, ΔT_N, is, for most liquids, larger than $0.15 \cdot T_f$. For *fcc* and *bcc* metals, $\Delta T_N/T_f \approx 0.18$. However, using the droplet emulsion technique Perepezko *et al.* (1979) obtained undercooling almost twice as large as that obtained by previous investigators. According to Turnbull (1981), this may be because continuous coatings on small droplets that have different thermal contraction may generate sufficient stress to displace the thermodynamic equilibrium temperature. Using Turnbull's value for ΔT_{max} in Eq. (15.2), and remembering that the size of the nucleus is related to the number of atoms in the nucleus, n_{cr}, by:

$$n_{cr}v_a = \frac{4}{3}\pi r_{cr}^3 \qquad (15.3)$$

where v_a is the atomic volume [m^3/atom] and n_{cr} has no units. It can be calculated that a nucleus of critical radius includes several hundred atoms (see Application 15.1). Note that if the undercooling is zero, r_{cr} must be infinite, which means that solidification will not occur. In other words, the system is at equilibrium.

Eq. (15.2) can be manipulated to obtain the undercooling due to the radius of curvature, ΔT_r, as follows:

$$\Delta T_r = T_f - T_e^* = \frac{\gamma_{LS}}{\Delta S_f}\frac{2}{r} = \Gamma\frac{2}{r} \qquad (15.4)$$

where T_e^* is the temperature of local equilibrium, and Γ is the Gibbs-Thompson coefficient. It can be calculated that for most metals Γ is approximately 10^{-7} °C·m. Then, for example, for $r = 10\mu m$ the curvature undercooling is $\Delta T_r = 0.05$ °C. Thus, the liquid-solid interface energy is important only for morphologies with $r <$ $10\mu m$, *i.e.*, nuclei, interface perturbations, dendrites and eutectic phases.

The excess free energy of the critical nucleus can be calculated by substituting Eq. (15.2) into Eq. (15.1) to obtain:

$$\Delta G_{cr} = \frac{16\pi\gamma^3}{3\Delta G_v^2} = \frac{16\pi\gamma^3 T_f^2}{3\Delta H_f^2 \Delta T^2} \qquad (15.5)$$

Let us calculate the nucleation velocity. The thermodynamics of embryo formation can be related to the rate of appearance of nuclei, I (intensity or velocity of homogeneous nucleation), through the description of the population distribution of embryos. Consider an undercooled liquid. The molecules of the liquid are colliding and continually forming clusters that redissolve because they are too small to be stable. Denoting α' as a molecule of liquid and β_i' as an embryo of solid containing i molecules, the sequence of bimolecular reactions leading to nucleus formation is:

$$\alpha' + \alpha' \leftrightarrow \beta_2'$$
$$\beta_2' + \alpha' \leftrightarrow \beta_3'$$
$$\beta_3' + \alpha' \leftrightarrow \beta_4'$$
....
....
$$\beta_{i^*-1}' + \alpha' \leftrightarrow \beta_{i^*}'$$
$$\beta_{i^*}' + \alpha' \leftrightarrow \beta_{i^*+1}'$$

where a cluster containing i^* molecules is considered as a critical size nucleus which will continue to grow. The equilibrium number of embryos containing i atoms can be obtained by minimizing the free energy of the system with respect to the number n_i of such embryos of size i. The free energy change of the system upon introducing n_i embryos in a liquid containing n_L atoms per unit volume is:

$$\Delta G = enthalpy\ of\ system\ with\ embryos - enthalpy\ of\ system\ without\ embryos \qquad or$$

$$\Delta G = n_i \Delta G_i - T \Delta S_i$$

The entropy of mixing is $\Delta S_i = -k_B\, n[C \ln C + (1 - C) \ln(1 - C)]$ with $n = n_i + n_L$ $C = n_i/(n_i + n_L)$ and $1 - C = n_L/(n_i + n_L)$. Then:

$$\Delta S_i = -(n_i + n_L)k_B\left(\frac{n_i}{n_i + n_L}\ln\frac{n_i}{n_i + n_L} + \frac{n_L}{n_i + n_L}\ln\frac{n_L}{n_i + n_L}\right) \qquad or$$
$$\Delta S_i = -k_B\left[-(n_i + n_L)\ln(n_i + n_L) + n_i \ln n_i + n_L \ln n_L\right]$$

Minimizing the free energy:

$$\frac{\partial \Delta G}{\partial n_i} = \Delta G_i - T\frac{\partial \Delta S_i}{\partial n_i} = 0$$

$$\frac{\partial S_i}{\partial n_i} = -k_B\left[(n_i + n_L)\frac{1}{n_i + n_L} + \ln(n_i + n_L) - n_i\frac{1}{n_i} - \ln n_i\right]$$

$$\Delta G_i + kT\ln\frac{n_i}{n_i + n_L} = 0$$

Since $n_L \gg n_i$ the number of embryos of critical size (nuclei) is:

$$n_i^{cr} = n_L \exp\left(-\frac{\Delta G_i}{k_B T}\right)$$ (15.6)

Note that $\Delta G_i = \Delta G_{cr}$ is the excess free energy of the critical nucleus given by Eq. (15.5). Assuming that each critical nucleus grows into a crystal and is thereby removed from the distribution of cluster sizes, the velocity of homogeneous nucleation, I, is:

I = (no. of critical nuclei) x (rate of incorporation of new atoms in nuclei)   or

$I = n_i^{cr} \cdot dn / dt$, where n_i^{cr} is expressed in m^{-3}, and dn/dt in s^{-1}. Then:

dn/dt = (atomic vibration frequency) x (probability of a successful jump)   or

$$\frac{dn}{dt} = v \exp\left(-\frac{\Delta G_A}{k_B T}\right)$$

where ΔG_A is the free energy of activation for the transfer of atoms from liquid to crystal. The exponential in this equation can also be understood in terms of the fraction of atoms in the liquid, which are sufficiently activated to surmount the interface addition activation energy. Thus, the nucleation rate is proportional with the product between the probability of formation and the probability of growth of the critical nuclei:

$$I = n_L v \exp\left(-\frac{\Delta G_{cr}}{k_B T}\right) \exp\left(-\frac{\Delta G_A}{k_B T}\right) = n_L v \exp\left(-\frac{\Delta G_{cr} + \Delta G_A}{k_B T}\right)$$ (15.7)

For most metals the pre-exponential constant has the value $I_o = 10^{42}$ m^{-3}·s^{-1}. Substituting the value of ΔG_{cr} from Eq. (15.5) one can further write:

$$I = I_o \exp\left(-\frac{K_N^{hom}}{T \Delta T^2}\right) \exp\left(-\frac{\Delta G_A}{k_B T}\right)$$

and, since at low ΔT the second exponential is very small, $\exp(-\Delta G_A/k_B T) \approx 0.01$, Eq. (15.7) simplifies to:

$$I = I_o' \exp\left(-\frac{K_N^{hom}}{T \Delta T^2}\right)$$ (15.8)

with $I_o' = 10^{40}$ m^{-3} s^{-1}. An alternative derivation of this equation is presented in the inset.

Alternative derivation of Eq. (15.8) (after Flemings, 1974). The nucleation intensity can be calculated as $I = (n_i^{cr}) \cdot (v_{LS} n_s)$, where v_{LS} is the jump frequency of atoms from the liquid to the solid embryo, and n_S is the number of atoms of liquid in contact with the surface of the embryo. The jump of atoms in the bulk liquid is:

$$v_L = 6 D_L / \lambda^2 \tag{15.9}$$

where D_L is the liquid diffusivity, and λ is the jump distance. Assuming a cubic lattice, the jump from liquid to solid is $v_L/6$, that is atoms can jump only at one of the 6 faces. Further assuming that the jump distance is equal to the atomic distance, a, we obtain:

$$v_{LS} = D_L / a^2 \tag{15.10}$$

The number of atoms of liquid in contact with the surface of the embryo can be evaluated as $4\pi r_{cr}^2 / a^2$. The expression for the steady state nucleation velocity is then:

$$I = \left(\frac{D_L}{a^2}\right)\left(\frac{4\pi r_{cr}^2}{a^2}\right) n_L \exp\left(-\frac{\Delta G_{cr}}{k_B T}\right) \tag{15.11}$$

For typical metals $n_L \sim 10^{28}$ m^{-3}, $D_L \sim 10^{-9}$ m^2/s, and $a \sim 3\cdot10^{-8}$ m (Perepezko, 1988), so that we obtain the sought off Eq. (15.8):

$$I \approx 10^{40} \exp\left(-\frac{\Delta G_{cr}}{k_B T}\right) = 10^{40} \exp\left(-\frac{K_N^{hom}}{T \Delta T^2}\right)$$

Eq. (15.8) describes the nucleation rate (intensity of nucleation). I is expressed in m$^{-3}\cdot$s^{-1}. The constant K_N^{hom} can be calculated using Eq. (15.5). The nucleation rate equation shows a steep dependency of the nucleation velocity on temperature (see Figure 15.2). ΔT_{cr} is the critical undercooling, at which nucleation occurs almost instantaneously.

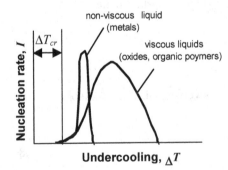

Figure 15.2. Variation of nucleation velocity with undercooling.

Experimental data on the critical undercooling at which homogeneous nucleation occurs and Eq. (15.8) have been used by Turnbull (1950) to evaluate the S/L surface energy of various metals (see Table 15.1 and Application 15.2). Note that the typical undercooling required for homogeneous nucleation derived by this method is $\Delta T_N = (0.13$ to $0.25)$ T_f. Turnbull used the substrate technique.

More recently, new experimental techniques such as emulsion of droplets, fluxing, and containerless solidification, produced undercoolings that are significantly higher than predicted by Turnbull (see data summarized by Kelton in Table 15.1). Thus a broader range for the $\Delta T_N/T_f$ seems to exist.

Table 15.1. Relationship between maximum undercooling and liquid /solid interface energy for selected metals.

Metal	T_f, K	Turnbull (1950)			Kelton (1991)		
		ΔT_N, K	$\Delta T_N/T_f$	γ_{SL}, J/m^2	ΔT_N, K	$\Delta T_N/T_f$	γ_{SL}, J/m^2
Al	934	195	0.21	0.121	175	0.19	0.108
Bi	544	90	0.17	0.054	227	0.42	0.088
Co	1767	330	0.19	0.234	330	0.19	0.238
Cu	1357	236	0.17	0.177	236	0.17	0.178
Fe	1811	295	0.16	0.204	420	0.23	0.277
Ga	303	76	0.25	0.056	174	0.57	0.077
Hg	234				88	0.38	0.031
Mn	1519	308	0.20	0.206	308	0.20	0.216
Ni	1728	319	0.18	0.255	480	0.28	0.300
Sn	505	118	0.23	0.054	191	0.38	0.075
Pb	600	80	0.13	0.033	240	0.40	0.06
Ti	1940				350	0.18	0.202

Experiments with alloys have demonstrated that they behave like pure metals, in the sense that homogeneous nucleation starts at similar undercoolings. However, experimental data on the maximum undercooling are less reliable than for pure metals because the entrained droplet ("mush quenching") technique used is susceptible of allowing some weak heterogeneous nucleation on the solid solution. The undercooling must be calculated from the liquidus temperature. Some data on alloys undercooling are provided in Table 15.2.

Table 15.2. Maximum undercooling in alloys obtained by the entrained droplet technique.

Alloy	Primary phase	ΔT_{max}, K	Reference
Cu-Pb	Cu	3	Wang and Smith, 1950
Al-10wt%Sn	Al	99	Wang and Smith, 1950
Ni-Graphite	Ni	775	Hunter and Chadwick, 1972
Ni-Graphite (+Mg)	Ni	800	Hunter and Chadwick, 1972
Fe-Graphite	Fe	745	Hunter and Chadwick, 1972
Fe-Graphite (+Mg)	Fe	280	Hunter and Chadwick, 1972
Al-Si	Al	176	Southin and Chadwick, 1978

In commercial casting alloys, homogeneous nucleation is virtually unknown. It is, however, relevant to the new processes involving rapid solidification.

15.1.2   Steady-state nucleation - Heterogeneous Nucleation

From Table 15.1 it is seen that the undercooling required for homogeneous nucleation is rather high. For example, for pure iron it can be as high as 420K. This undercooling is required because of the relatively large activation barrier required for nucleus formation (ΔG_{cr}). In fact, homogeneous nucleation is the most difficult kinetic path for crystal formation. Undercoolings of this magnitude are never observed in commercial casting alloys. The order of magnitude of the undercooling is only a few tens of degrees, at maximum. Solidification can be triggered by preexisting fragments of crystals or solid films of oxides, by the mold walls, or by additions of chemicals made on purpose (called grain refiners or inoculants). This nucleation, occurring on solid substrates foreign to the metal, is called *heterogeneous nucleation*.

Theories for homogeneous nucleation have been extended to cover heterogeneous nucleation, but the discrepancy between experimental data and theory is rather significant. This is because the mechanisms of homogeneous and heterogeneous nucleation are different (Mondolfo, 1983). Homogeneous nucleation results from the stabilization of a transient grouping of atoms, so that a nucleus consisting of many atoms is formed all at once. In heterogeneous nucleation, the atoms of the metal to be nucleated attach themselves to the best locations on the nucleant, and the nucleus grows atom by atom.

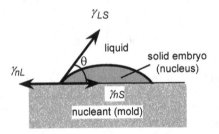

Figure 15.3. Contact angle between embryo and substrate. Note interfacial energy relationships between substrate (n), liquid (L) and solid (S).

The classic theory of nucleation shows that the nucleation velocity for the case of heterogeneous nucleation can be calculated with an equation similar to Eq. (15.8). The concept of contact angle between the growing embryo in the shape of a spherical cap and the substrate, is used. The angle θ is determined by the equilibrium condition between the interface energies of the three phases, liquid, growing solid embryo (nucleus) and substrate nucleant (mold), as shown in Figure 15.3. With this definition, the free energy of formation of a heterogeneous nucleus of critical radius is given by:

$$\Delta G^{het} = -v_S \Delta G_v + A_{LS}\gamma_{LS} + A_{nS}\left(\gamma_{nS} - \gamma_{nL}\right)$$

The surface area of spherical cap is: $A_{LS} = 2\pi r^2 (1 - \cos\theta)$

The area between the spherical cap and the substrate is: $A_{nS} = \pi r^2 \sin^2\theta$

The volume of the spherical cap is: $v_S = (\pi r^3/3)(2 - 3\cos\theta + \cos^3\theta)$

Young's equation gives: $\gamma_{nL} = \gamma_{nS} + \gamma_{LS}\cos\theta$

Substituting in the free energy of formation equation yields:

$$\Delta G^{het} = \left(-v_{sphere}\Delta G_v + A_{sphere}\gamma_{LS}\right)\frac{2 - 3\cos\theta + \cos^3\theta}{4}$$ or simply:

$$\Delta G^{het} = \Delta G^{hom} \cdot f(\theta) \quad \text{where} \quad f(\theta) = \frac{1}{4}(2 + \cos\theta)(1 - \cos\theta)^2 \tag{15.12}$$

Then, the same procedure as for the derivation of the nucleation velocity for homogeneous nucleation can be used. In the calculation of the number of embryos of critical size, n_i^{cr}, the number of surface atoms of nucleation sites per unit volume, n_a, must be substituted for the number of atoms per unit volume of liquid, n_L. An equation similar to Eq. (15.7) is derived:

$$I_{het} = n_a v\exp\left(-\frac{\Delta G_{cr}^{het}}{k_B T}\right)\exp\left(-\frac{\Delta G_A}{k_B T}\right) = n_a\frac{k_B T}{h}\exp\left(-\frac{\Delta G_{cr}^{het} + \Delta G_A}{k_B T}\right)$$

where h is Planck's constant. Calculating the free enthalpy of formation of nuclei of critical radius from Eq. (15.12) yields:

$$I_{het} = \frac{n_a}{n_L}I_o'\exp\left[-\frac{K_N^{hom}}{T\Delta T^2}f(\theta)\right] \tag{15.13}$$

An alternative derivation was proposed by Perepezko (1992) (see inset).

Alternative derivation of the nucleation rate for heterogeneous nucleation. The number of atoms in contact with the surface of the embryo can be calculated as:

$$n_s = 2\pi r_{cr}^2(1 - \cos\theta)/a^2$$

The numerator is the surface area of the spherical cap. With these changes, Eq. (15.11) becomes:

$$I_{het} = \left(\frac{D_L}{a^2}\right)\left[\frac{2\pi r_{cr}^2(1 - \cos\theta)}{a^2}\right]n_a\exp\left[\left(-\frac{\Delta G_{cr}^{hom}}{k_B T}\right)f(\theta)\right] \tag{15.14}$$

Since $n_a \sim 10^{20}$ m^{-3}, and using other numerical values as before, I_{het} can also be written as:

$$I_{het} \approx 10^{30} \exp\left[\left(-\frac{\Delta G_{cr}^{hom}}{k_B T}\right) f(\theta)\right] = 10^{30} \exp\left[-\frac{K_N^{hom} f(\theta)}{T \Delta T^2}\right] \tag{15.15}$$

Note that if the angle θ is small, the nucleation velocity can be significant, even for rather small undercoolings. For the case $\theta = 0°$, $f(\theta) = 0$ and $\Delta G^{het} = 0$. This means that when complete wetting occurs, there is no nucleation barrier. If $\theta = 180°$, that is there is no wetting, $f(\theta) = 1$ and $\Delta G^{het} = \Delta G^{hom}$. If $\theta = 30°$, $f(\theta) = 0.02$, and $\Delta G^{het}/\Delta G^{hom} = 0.02$ which shows a strong influence of the substrate. Even for $\theta = 90°$, $f(\theta) = 0.5$. A schematic comparison between ΔG^{het} and ΔG^{hom} is shown Figure 15.4. The heterogeneous nucleation rate also increases with higher number of available nucleation sites (or higher area of the sites), since I_{het} is proportional to n_a.

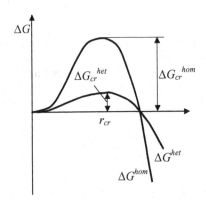

Figure 15.4. Variation of excess free energy required for homogeneous and heterogeneous nucleation with embryo radius. Note that r_{cr} is independent of the type of nucleation.

Another approach was proposed by Hunt (1984). Eq. (15.7) may be rewritten for the case of heterogeneous nucleation as follows:

$$I_{het} = (N_s - N) K_1 \exp\left(-\frac{\Delta G_{cr}^{het}}{k_B T}\right) \tag{15.16}$$

where N_s is the number of heterogeneous substrates originally available per unit volume, N is the number of particles that have already nucleated, and K_1 is a constant. Approximating liquid diffusivity as $D_L \approx a^2 v \exp(-\Delta G_A / k_B T)$, Eq. (15.7) becomes:

$$I_{het} = \frac{n_a D_L}{a^2} \exp\left(-\frac{\Delta G_{cr}^{het}}{k_B T}\right) \tag{15.17}$$

Now, the constant K_1 can be evaluated by combining Eqs. (15.16) and (15.17):

$$K_1 = \frac{n_a D_L}{a^2 (N_s - N)}$$ From Eq. (15.5): $$\frac{\Delta G_{cr}^{het}}{k_B T} = \frac{16}{3} \frac{\pi \gamma_{LS}^3 T_f^2}{\Delta H_f^2 \Delta T^2 k_B T} f(\theta)$$

For metallic systems, T is very large compared with ΔT and can be considered constant. For a given nucleant, $f(\theta)$ is also a constant; thus, one can write $\Delta G_{cr}^{het} / (k_B T) = K_2 / \Delta T^2$. Eq. (15.16) becomes:

$$I_{het} = (N_s - N) K_1 \exp\left(-\frac{K_2}{\Delta T^2}\right) \tag{15.18}$$

Again, following Hunt's suggestion, K_2 can be calculated if ΔT is defined as the nucleation temperature, ΔT_N, where the initial nucleation rate is $I_{het} = 1$ m^{-3} s^{-1}. Since for this case $N = 0$, $I_{het} = (N_s - N) K_1 \exp(-K_2/\Delta T^2)$ and Eq. (15.18) becomes:

$$K_2 = \Delta T_N^2 \ln(N_s K_1)$$

An example of the calculation of K_1 is given in Application 15.3. With I_{het} known, the final number of nuclei is calculated with:

$$N = \int_0^t I_{het} \, dt \tag{15.19}$$

where t is the time required for the number of nuclei to increase from 0 to N_s. Examples of calculation of I_{het} and N as a function of undercooling are given in Figure 15.5 (see Application 15.4 for details).

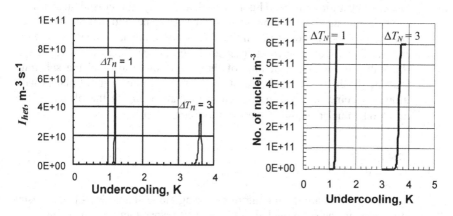

Figure 15.5. Variation of nucleation rate, I_{het}, and N_s as a function of undercooling.

The variation of the number of nuclei (nuclei volumetric density) as a function of undercooling is so steep for both nucleation undercoolings considered that, for all practical purposes, the complex nucleation equation, Eq.(15.19), can be substituted with a Dirac delta function (Stefanescu *et al.*, 1990), $\delta(T - T_N)$, having the following definition:

for $T \neq T_N$ $\delta(T - T_N) = 0$
for $T = T_N$ $\delta(T - T_N) = 1$

Thus, Eq. (15.18) can be written as:

$$\frac{\partial N}{\partial t} = N_s \cdot \delta(T - T_N) \qquad (15.20)$$

This nucleation law predicts *instantaneous nucleation*, that is, that all nuclei are generated at the nucleation temperature.

Based on the theory summarized by Eq. (15.13), an efficient heterogeneous nucleant (inoculant) should satisfy the following requirements:

- low contact angle between metal and nucleant particles or high surface energy between the liquid and the nucleant, γ_{nL}; indeed if $\gamma_{nL} > \gamma_{nS} + \gamma_{LS} \cos\theta$, the nucleus can spread on the substrate and grow; on the other hand, if $\gamma_{nL} < \gamma_{nS} + \gamma_{LS} \cos\theta$, the nucleus must shrink and disappear
- the nucleant must expose a large area to the liquid; this will increase n_a, and thus I ; this can be achieved by producing a fine dispersion of nucleant, or by using a nucleant with a rough surface geometry
- the substrate must be solid in the melt; its melting point must be higher than the melt temperature, and it must not dissolve in the melt
- because the atoms are attaching themselves to the solid lattice of the substrate, the closer the substrate lattice resembles that of the solid phase, the easier nucleation will be. This means that, ideally, the crystal structure of the substrate and the solid phase should be the same, and that the lattice parameters of both should be similar. This condition is known as *isomorphism*. They should have at least analogous crystalline planes (*epitaxy*). Actually, the crystal structures of the solidifying alloy and the substrate may be different. Then, the substrate must have one or more planes with atomic spacing and distribution close to that of one of the planes of the solid to be nucleated, *i.e.*, have a low disregistry, δ:

$$\delta = \frac{a^n - a^S}{a^S} \qquad (15.21)$$

where a_S and a_n are the interatomic spacing along shared low-index crystal directions in the solid nucleus and nucleant respectively. If this is the case, γ_{nS} is very small, and some understanding of the process may be obtained by comparing γ_{LS} and γ_{nL}. Unfortunately, these data are mostly unavailable.

- low symmetry lattice (complex lattice): while it is impossible to assign numbers to lattice symmetry, entropy of fusion can to some extent be used as a measure of lattice symmetry. In general, less symmetrical lattices have higher entropies of fusion

- ability to nucleate at very low undercooling.

The nucleation concepts introduced in the preceding paragraphs, while rather helpful from the qualitative point of view, fail to accurately predict phenomena occurring in real alloys, frequently because of lack of adequate data. Nevertheless, they are helpful in the understanding of the widely used *inoculation* or *grain refinement* processes. The terms inoculation and grain refinement refer to the same process. Inoculation is widely used in metal casting processing in order to control the grain size, and, to a lesser extent, grain morphology. Typical inoculants (grain refiners) for different casting alloys are listed in Table 15.3. Inoculation must not be confused with modification. Modification is a process related mostly to growth. The main purpose of inoculation is to promote grain refinement, while modification is used to change the morphology of the eutectic grains.

Table 15.3. Examples of typical inoculants used for casting alloys

Metal or alloy	Inoculant
Cast iron	FeSi, SiCa, graphite
Mg alloys	Zr, carbon
Cu-base alloys	Fe, Co, Zr
Al-Si alloys	P, Ti, B
Pb alloys	As, Te
Zn-base alloys	Ti

15.1.3   Time-dependent (transient) nucleation

To understand the concept of time-dependent nucleation let us consider a silicate glass during annealing (Kelton, 1991). The number of nuclei can be experimentally obtained by recording the number of small crystallites that appear as a function of the annealing time, as shown by the black dots on Figure 15.6. Assuming a steady-state nucleation rate I^{st}, the number of nuclei per unit volume, N_v, will be:

$$N_v = \int_0^t I(t)\,dt = I^{st}\,t \qquad (15.22)$$

Thus, N_v is a linear function of time with slope I^{st}. Yet, Figure 15.6 shows a non-linear experimental correlation between these two quantities between the time for the onset of nucleation, t_o, and the time to steady state, t_{st}. The time-dependent nucleation rate, $I(t)$, given by the local slope of N_v versus t, approaches I^{st} at long annealing times.

Several analytical models for transient nucleation have been proposed. Most of them can be described by an equation of the form:

$$I_{n_{cr},t} = I^{st}\left(1 - \exp(-t/\tau)\right) \tag{15.23}$$

where τ is a function of the rate of monomer addition to a cluster of size n_{cr}, $k_{n_{cr}}^+$, and of the Zeldovich factor, $Z = \left(|\Delta G'|/6\pi k_B T n_{cr}\right)^{1/2}$. Here $\Delta G'$ is the Gibbs free energy per molecule of the new phase, less that of the initial phase. Numerical calculations (Kelton, 1991) seem to indicate that $\tau = 1/2\pi k_{n_{cr}}^+ Z^2$, as suggested by Chakraverty (1966) gives the closest approximation.

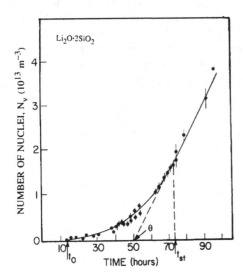

Figure 15.6. The number of nuclei produced as a function of time at 703K for lithium disilicate glass (Fokin *et al.*, 1981).

15.2   Growth Kinetics

15.2.1   Types of interfaces

Once the embryo grows to reach critical radius it will spontaneously grow. The interface between the nucleus and the liquid will move (grow) toward the liquid through a mechanism that will largely depend on the nature of the solid/liquid interface. As shown in Figure 15.7a, two types of interfaces may be considered:

- Diffuse interface (atomically rough): the contour of the liquid/solid interface is not smooth. Each step in Figure 15.7a corresponds to an atomic distance. This interface will advance in the liquid through *continuous growth*. Random incorporation of atoms onto the surface will occur, because incoming atoms have many nearest neighbors. This type of growth is typical for metals, and results in a *non-faceted* liquid/solid interface
- Flat interface (atomically smooth): liquid - solid transition occurs across a single atomic layer. The interface will advance through *lateral growth*.

This growth is typical for nonmetals, such as graphite growing in cast iron, or silicon in aluminum alloys, and results in a *faceted* interface

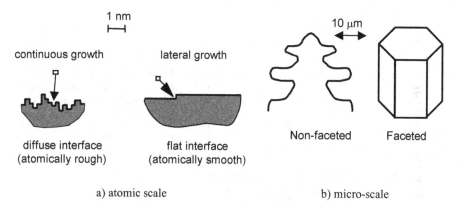

a) atomic scale   b) micro-scale

Figure 15.7. Types of liquid/solid interfaces.

An example of nonfaceted and faceted interface at the micro-scale level is given in Figure 15.7b. Whether an interface will be faceted or nonfaceted depends on the nondimensional ratio $\alpha = \Delta S_f/R$, where ΔS_f is the entropy of fusion, and R is the gas constant. According to Jackson (1958), the change in relative surface free energy when atoms are added to a smooth surface can be calculated with:

$$\frac{\Delta G}{N k_B T_e} = \alpha N_A \frac{N - N_A}{N^2} - \ln \frac{N}{N - N_A} - \frac{N_A}{N} \ln \frac{N - N_A}{N} \qquad (15.24)$$

where N is the number of atoms in a complete monolayer of surface, and N_A is the number of atoms on the surface (see derivation in inset). The relative surface free energy is plotted against the fraction of occupied sites in Figure 15.8. It can be seen that for $\alpha \leq 2$, the lowest free energy occurs when half of the available surface sites are filled; this means that the surface is rough. For $\alpha > 2$, the lowest free energy occurs either at low or at high fraction of occupied sites, that is when the interface is smooth. The correlation between the entropy of fusion and interface morphology is given in Table 15.4.

Table 15.4. Correlation between interface morphology and the entropy of fusion.

Material	$\Delta S_f/R$	Morphology
Regular metals and some organics	<2	non faceted
Semi-metals and semiconductors, Bi, Sb, Ga, Si	2.2 - 3.2	faceted
Most inorganics (*e.g.* carbides, nitrides)	>3.5	faceted

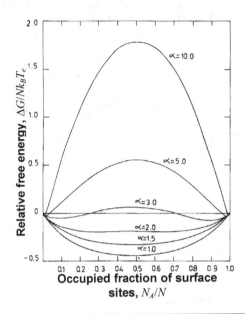

Figure 15.8. Relative surface free energy versus the fraction of occupied surface sites (Jackson, 1958). Reprinted with permission of ASM International. All rights reserved. www.asminternational.org

Derivation of Eq. (15.24). The basic assumptions are: 1) local equilibrium between solid and liquid at the interface and 2) the final structure is determined by minimization of the free energy. The approach used by Jackson (1958) was based on a nearest-neighbor bond model. The change in free energy of the interface is given by:

$$\Delta G_s = - \Delta E_o - \Delta E_1 + T \, \Delta S_{LS} - T \, \Delta S_1 - P \, \Delta v \tag{15.25}$$

where ΔE_o is the energy gained by transferring N_A atoms to the surface, ΔE_1 is the average energy gained by the N_A atoms due to the presence of other atoms on the surface, ΔS_{LS} is the entropy difference between liquid and solid, ΔS_1 is the entropy difference due to possible randomness of atoms on the surface. Since the change in volume is very small, the last RHT is zero.

If a single atom positioned on the interface has n_S nearest neighbors in the solid before attachment and a maximum number of nearest neighbors n_1, the number of nearest neighbors to an atom in the bulk solid is: $n = 2n_S + n_1$. The energy of an atom in the solid is $2E_{LS}(n_S/n)$, where E_{LS} is the change of internal energy for the atom to transfer from liquid to solid. Then: $\Delta E_o = 2E_{LS}(n_S/n) N_A$. Each of the N_A atoms has a fraction N_A/N of its nearest neighbors sites filled, where N is the number of atoms in a complete monolayer interface. If there are n_1 nearest neighbors:

$$\Delta E_1 = \Delta E_{LS}\left(\frac{n_1}{n}\right)\left(\frac{N_A}{N}\right)N_A$$

Also, we have $\Delta S_{LS} = (\Delta H_f/T_e) N_A$. The probability of arranging N_A atoms on the surface is:

$$W = \frac{N!}{N_A!(N - N_A)!}$$

Using Boltzman's relationship ($S = k_B \ln W$) and Stirling's approximation we write:

$$\Delta S_1 = k_B N \ln \frac{N}{N - N_A} + k_B N_A \ln \frac{N - N_A}{N_A}$$

Introducing all these equations in Eq. (15.25) and dividing each side by $N k_B T_e$, we obtain:

$$\frac{\Delta G_s}{N k_B T_e} = -\frac{E_{LS}}{k_B T_e} \frac{N_A}{N} \left(\frac{N_A}{N} \frac{n_1}{n} + \frac{2 n_s}{n} \right) + \frac{T \Delta H_f}{N k_B T_e^2} \frac{N_A}{N} - \frac{T}{T_e} \ln \left(\frac{N}{N - N_A} \right) - \frac{T}{T_e} \frac{N_A}{N} \ln \left(\frac{N - N_A}{N_A} \right)$$

Considering the equilibrium situation, $T = T_e$. Further, since E_{LS} is the latent heat of fusion, and using the notation:

$$\frac{E_{LS}}{k_B T_e} \frac{n_1}{n} = \Delta S_f \frac{n_1}{n} = \alpha$$

Jackson obtained Eq. (15.24). The ratio n_1/n depends on the crystallography of the interface.

As will be demonstrated in the following paragraphs, continuous growth is considerably faster than lateral growth. Indeed, it was calculated that the kinetic undercooling required for continuous growth is typically of the order of 0.01 - 0.05K, while the undercooling required for lateral growth is of the order of 1 - 2K.

15.2.2   Continuous growth

The kinetics of continuous growth, that is the growth velocity, V, can be described by the equation (Turnbull, 1949):

$$V = \mu_o \Delta T_k \quad \text{with} \quad \mu_o = \frac{\beta D_L \Delta H_f}{a k_B T_f^2} \tag{15.26}$$

where μ_o is the growth constant, ΔT_k is the kinetic undercooling required for atom attachment, ΔH_f is the latent heat of fusion, a is the amount the interface advances when an atom is added, and β is a correction factor that can be written as:

$$\beta = (a/\lambda)^2 \, 6 v_{LS} / v_L$$

where λ is the jump distance for an atom in the liquid, v_{LS} is the frequency with which atoms jump across the liquid/solid interface, and v_L is the frequency with which atoms jump in the bulk liquid. The derivation of this equation is given in the inset.

Derivation of Eq. (15.26). Turnbull (1949) described kinetics of continuous growth using classic rate theory. The continuous growth velocity is $V = a \, v_{net}$, where v_{net} is the net jump frequency across the interface. In turn, the net jump frequency with which atoms can jump from liquid to solid at the melting temperature is:

$$v_{net} = v_{LS} - v_{SL} = v_{LS}\left[1 - \exp\left(-\Delta G_v / \left(k_B T_e\right)\right)\right]$$

where ΔG_v is the activation energy for transport of atoms from liquid to solid (the difference in free energy between the liquid and the solid). Since $1 - \exp(-x) \approx x$, and using Eq (2.8):

$$v_{net} = v_{LS} \frac{\Delta G_v}{k_B T_e} = v_{LS} \frac{\Delta H_f \Delta T_k}{k_B T_f^2}$$

Substituting v_{LS} from Eq. (15.10), the continuous solidification velocity becomes:

$$V = \frac{D_L \Delta H_f}{a k_B T_f^2} \Delta T_k = \mu_o \Delta T_k$$

However, since it is believed that λ may be as much as an order of magnitude smaller than a, a correction factor, β must be attached to μ_o.

Eq. (15.26) shows a linear dependency of V on ΔT. μ_o has values between 10^{-2} and 1 m·s^{-1}·°C^{-1}. An example of calculation of μ_o and of the kinetic undercooling is given in Application 15.5.

15.2.3   Lateral growth

The kinetics of lateral growth can be described by different equations, depending on the growth mechanism. Two growth mechanisms will be discussed, *i.e.*, growth by screw dislocation and growth by two-dimensional nucleation.

Growth by screw dislocations
When growth occurs through a screw dislocation mechanism atoms are continuously added at the step of a screw dislocation, which rotates about the point where the dislocation emerges. For one dislocation, it was demonstrated that growth kinetics could be described by the parabolic equation:

$$V = \mu_1\left(\Delta T_k\right)^2 \quad \text{with} \quad \mu_1 = \frac{\left(1 + 2g^{1/2}\right)}{g} \frac{\beta D_L \Delta H_f^2}{4\pi \gamma T_f^3 k v_m} \tag{15.27}$$

where g is the diffusness parameter and v_m is the molar volume. Difuseness is defined by the number of atomic layers, n, comprising the transition from solid to liquid at T_f for a diffuse interface. For flat interfaces $g = 1$, while for diffuse interfaces $g \ll 1$. The diffusness parameter can be calculated as $g = \left(\pi^4 / 8\right) n^3 \exp\left(-\pi^2 n / 2\right)$.

In metals, the typical number of dislocations is 10^8 cm^{-2}. The growth equation presented above is however valid for this case also. Experimental evidence demonstrates that when dislocations were introduced into a crystal the growth rate increased rapidly.

Growth by two-dimensional nucleation

A two-dimensional nucleus grows by addition of atoms on the lateral sides of the nucleus (Figure 15.9). The excess free energy required for the formation of a 2D nucleus is $\Delta G = -\pi r^2 a \Delta G_v + 2\pi r a \gamma$. Then:

$$\frac{\partial \Delta G}{\partial r} = -2\pi r_{cr} a \Delta G_v + 2\pi a \gamma = 0 \quad \text{and} \quad r_{cr} = \frac{\gamma}{\Delta G_v} = \frac{\gamma T_f}{\Delta H_f \Delta T} \quad \text{with} \quad \Delta G_{cr} = \frac{\pi a \gamma^2 T_f}{\Delta H_f \Delta T}$$

Following steps as in 3D nucleation it is calculated that:

$$V = \mu_4 \exp\left(\frac{\mu_3}{3\Delta T_k}\right) \tag{15.28}$$

with $\quad \mu_3 = \mu_o \dfrac{\pi g B^2 a T_f^2}{\beta D_L} \quad$ and $\quad \mu_4 = \mu_o \left(\dfrac{\Delta H_f}{k_B T_f^2}\right)^{1/6} (\Delta T_k)^{7/6} \left(2 + g^{-1/2}\right)$

where B is Turnbull's empirical relationship between γ and ΔH_f written as $B = v_m \gamma / (a \Delta H_f)$. B is equal to 0.5 for metals and to 0.35 for nonmetals. Note that in this case the dependency of V on ΔT is exponential.

A comparison between the various growth kinetics is shown schematically in Figure 15.10. It can be seen that, for a given kinetic undercooling, continuous growth is the fastest.

15.3 Applications

Application 15.1

Calculate the number of copper atoms included into a homogeneous nucleus of critical size.

Answer:
Using Eq. (15.2) and data from Appendix B and Table 15.1 it is calculated that the critical radius of the copper nucleus is $1.035 \cdot 10^{-9}$ m. The volume of a copper atom is: $v_a = v_m / N_{Av} = 1.18 \cdot 10^{-29}$ m^3, where v_m is the molar volume and N_{Av} is Avogadro's number. Then, from Eq. (15.3) it is obtained that $n_{cr} = 394$.

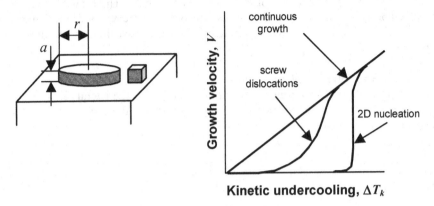

Kinetic undercooling, ΔT_k

Figure 15.9. Growth of a two-dimensional nucleus on a solid substrate.

Figure 15.10. Comparison between growth velocities by different mechanisms.

Application 15.2

The maximum undercooling observed in liquid nickel is 319 °C. Assuming homogeneous nucleation occurs at this temperature, calculate the liquid/solid interface energy for nickel.

Answer:

From Eqs. (15.8) and (15.5)) we have:
$$I = 10^{40} \exp\left(-\frac{16\pi\gamma^3 T_f^2}{3\Delta H_f^2 k_B T \Delta T^2}\right)$$

Using the data in Appendix B the correlation presented in Figure 15.11 is obtained. Note that for units consistency the volumetric latent heat expressed in J/m³ must be used. It is seen that γ_{LS} decreases rapidly at first and then much slower. Thus, it is considered that a good number for nucleation velocity is in the range of 10^4 to 10^5, which gives an interface energy of o.255 to 0.256 J/m³.

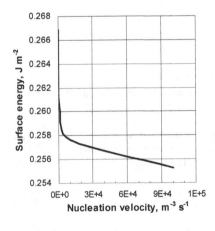

Figure 15.11. Correlation between nucleation velocity and the LS interface energy.

Application 15.3

Evaluate the constant K_1 for a Fe-C alloy nucleating on SiO_2 particles.

Answer:
Assuming that the substrates are spherical in shape with a radius r_s, the number of surface atoms of substrate per unit volume of liquid can be calculated as $n_a = 4\pi r_s^2 N_s n_s^{2/3}$. In turn, the number of atoms per unit volume of silica is:

$$n_s = \frac{\rho_{SiO_2} \cdot N_{Av}}{molarweight} = \frac{(2.65 \cdot 10^3 \, kg \cdot m^{-3}) \cdot (6 \cdot 10^{23} \, atmole^{-1})}{(28.1 + 32) \cdot 10^3 \, kg \cdot mole^{-1}} = 2.645 \cdot 10^{28} \, atm^{-3}$$

Now n_a can be calculated based on some assumptions for r_s, as follows:

for $r_s = 10 \, \mu m$ $n_a = 1.11 \cdot 10^{10} \, N_s$
for $r_s = 100 \, \mu m$ $n_a = 1.11 \cdot 10^{12} \, N_s$

At the beginning of nucleation $N = 0$ and thus $K_1 = n_a D_L / (a^2 N_s)$. Substituting $D_L = 2 \cdot 10^{-8}$ m^2/s (see Appendix B) and $a = 0.25$ nm, gives:

for $r_s = 10 \, \mu m$ $K_1 = 3.55 \cdot 10^{21} \, s^{-1}$
for $r_s = 100 \, \mu m$ $K_1 = 3.55 \cdot 10^{23} \, s^{-1}$

Similar calculations can be performed for Al-Si nucleating on Si crystals with results of the same order of magnitude (Stefanescu et al., 1990).

Application 15.4

Calculate the nucleation rate and the evolution of the number of nuclei for an Fe-C alloy assuming two different nucleation undercoolings of 1 and 3 K, and a final number of nuclei of $N_s = 6 \cdot 10^{11} \, m^{-3}$. This is equivalent to the assumption of two alloys solidifying on the same number of substrates but with different chemistry. Therefore, the two types of nuclei become active at two different nucleation temperatures.

Answer:
The calculation is straightforward. An average value of $10^{22} \, s^{-1}$ is taken for K_1. Eq. (15.18) is used to calculate I_{het} and the number of nuclei as a time summation. The results are shown in Figure 15.5.

Application 15.5

Consider a pure nickel single crystal growing with a planar interface at a velocity of 10^{-5} m/s. Assuming growth by the continuous growth mechanism, calculate the growth constant and the kinetic undercooling.

Answer:
To conserve units consistency, the growth constant in Eq. (15.26) is written as:

$$\mu_o = \beta D_L \Delta H_f / (a k_B N_A T_f^2)$$

N_A is Avogadro's number. Assuming $\beta = 1$ (which is true if $a = \lambda$ and $v_{LS} = v_L/6$) and taking $D_L = 3 \cdot 10^{-9} m^2/s$ it is calculated that $\mu_o = 7.84 \cdot 10^{-3} m \, s^{-1} \, K^{-1}$. Note that, again for units consistency, the latent heat must be expressed in J/mole. Then: $\Delta T_k = V/\mu_o = 1.28 \cdot 10^{-3} K$.

It is thus demonstrated that the kinetic undercooling is very small as compared with the other undercoolings discussed in this text, and therefore the solidification velocity of metals is not limited by interface kinetics.

References

Chakravery B.K., 1966, *Surf. Sci.* **4**:205

Flemings M.C., 1974, *Solidification Processing*, Butterworths, London

Fokin V.M., A.M. Kalinina and V.N. Filipovich, 1981, *J. Crystal Growth* **52**:115

Hunt J.D., 1984, *Mater. Sci. Eng.* **65**:75

Hunter M.J. and G.A Chadwick, 1972, *J. Iron Steel Inst.* **210**:707

Jackson K.A., 1958, in: *Liquid Metals and Solidification*, ASM, Metals Park, Ohio p.174

Kelton K.F., 1991, in: *Solid State Physics*, eds. H. Ehrenreich and D. Turnbull, Academic Press, Boston **45**:75-178

Mondolfo L.F., 1983, in: *Grain Refinement in Casting and Welds*, G.J. Abbaschian and S.A. David, eds., The Metallurgical Society of AIME, Warrendale PA p.3

Perepezko J.H., D.H. Rasmussen, I.E. Anderson and C.R. Loper, 1979, in: *Solidification of Castings and Alloys*, The Metals Society, London p.169

Perepezko J.H., 1992, in: *ASM Handbook vol. 15 Casting*, D.M. Stefanescu ed., ASM International, Metals Park, Ohio p.101

Southin R.T. and G.A. Chadwick, 1978, *Acta Metall.* **26**:223

Stefanescu D.M., G. Upadhya and D. Bandyopadhyay, 1990, *Metall. Trans.* **21A**:997

Turnbull D., 1949, *Thermodynamics in Metallurgy*, ASM, Metals Park, Ohio

Turnbull D., 1950, *J. Appl. Phys.* **21**:1022

Turnbull D., 1956, *Solid State Physics*, Academic Press, New York **3**:225

Turnbull D., 1981, in: *Progress in Materials Sciences*, Chalmers Anniversary Volume, J. W. Christian, P. Haasen and T.B. Massalski eds., Pergamon Press, Oxford p.269

Wang C.C. and C.S. Smith, 1950, *Trans. AIME* **188**:136

APPENDIX A: SOME SOLUTIONS OF THE DIFFUSION EQUATIONS

PDE	IC and BC	Solution
$\dfrac{\partial^2 \phi}{\partial x^2} = 0$	$\phi(0) = \phi_1$ and $\phi(L) = \phi_2$	becomes ordinary differential equation $\phi(x) = C_1 x + C_2$ with $C_1 = (\phi_2 - \phi_1)/L$ and $C_2 = \phi_1$
$\dfrac{\partial^2 \phi}{\partial x^2} = 0$	$\phi(x,0) = f(x)$ $\dfrac{d\phi}{dx}(0) = 0$ and $\dfrac{d\phi}{dx}(L) = 0$	becomes ordinary differential equation, initial boundary problem: $\phi(x) = C_1 x + C_2$ with $C_1 = 0$ and $C_2 = \phi(x) = \dfrac{1}{L}\int_0^L f(x)dx$ from $\rho c \int_0^L C_2\, dx = \rho c \int_0^L f(x)dx$
$\dfrac{\partial \phi}{\partial t} = \Gamma \dfrac{\partial^2 \phi}{\partial x^2}$	$\phi(x,0) = f(x)$ $\phi(0,t) = 0$ and $\phi(L,t) = 0$	separation of variables, Fourier series $\phi(x,t) = \sum_{n=1}^{\infty} B_n \sin\dfrac{n\pi x}{L}\exp\left[-\left(\dfrac{n\pi}{L}\right)^2 Dt\right]$ with $B_n = \dfrac{2}{L}\int_0^L f(x)\sin\dfrac{n\pi x}{L}dx$
$\dfrac{\partial \phi}{\partial t} = \Gamma \dfrac{\partial^2 \phi}{\partial x^2}$	$\phi(x,0) = f(x)$ $\phi(0,t) = \phi_1$ and $\phi(L,t) = \phi_2$	1) nonhomogeneous problem $\phi(x,t) = \phi_E(x) + \sum_{n=1}^{\infty} a_n \sin\dfrac{n\pi x}{L}\exp\left[-\left(\dfrac{n\pi}{L}\right)^2 \alpha t\right]$ with $\phi_E(x) = \phi_1 + \dfrac{\phi_2 - \phi_1}{L}x$ and $a_n = \dfrac{2}{L}\int_0^L [f(x) - \phi_E(x)]\sin\dfrac{n\pi x}{L}dx$ 2) $\phi(x,t) = A + B\,\mathrm{erf}\,\dfrac{x}{2\sqrt{\alpha t}}$

$\dfrac{\partial \phi}{\partial t} = \Gamma \dfrac{\partial^2 \phi}{\partial x^2}$	$\phi(x,0) = f(x)$ $\dfrac{d\phi}{dx}(0,t) = 0 \; and$ $\dfrac{d\phi}{dx}(L,t) = 0$	separation of variables, Fourier series $$\varphi(x,t) = \sum_{n=0}^{\infty} A_n \cos\frac{n\pi x}{L} \exp\left[-\left(\frac{n\pi}{L}\right)^2 \alpha t\right]$$ with $A_0 = \dfrac{1}{L}\displaystyle\int_0^L f(x)dx$ and $A_n = \dfrac{2}{L}\displaystyle\int_0^L f(x)\cos\frac{n\pi x}{L}dx$
Laplace's equation $\dfrac{\partial^2 \phi}{\partial x^2} + \dfrac{\partial^2 \phi}{\partial y^2} = 0$	$\phi(x,0) = f_1(x)$ $\phi(x,H) = f_2(x)$ $\phi(0,y) = g_1(y)$ $\phi(L,y) = g_2(y)$	for $g_2 = f_1 = f_2 = 0$ $$\phi(x,y) = \sum_{n=1}^{\infty} A_n \sin\frac{n\pi y}{H}\sinh\frac{n\pi(x-L)}{H}$$ with $$A_n = \frac{2}{H\sin(-n\pi L/H)}\int_0^H g_1(y)\sin\frac{n\pi y}{H}dy$$ The original solution is obtained by adding together four such solutions
Laplace's equation for a circular disk $\dfrac{1}{r}\dfrac{\partial}{\partial r}\left(r\dfrac{\partial \phi}{\partial r}\right) + \dfrac{1}{r^2}\dfrac{\partial^2 \phi}{\partial \theta^2} = 0$	$\phi(a,\theta) = f(\theta)$ where a is the radius	$$\phi(r,\theta) = \sum_{n=0}^{\infty} A_n r^n \cos n\theta + \sum_{n=1}^{\infty} B_n r^n \sin n\theta$$ with $A_0 = \dfrac{1}{2\pi}\displaystyle\int_{-\pi}^{\pi} f(\theta)d\theta$ $A_n a^n = \dfrac{1}{\pi}\displaystyle\int_{-\pi}^{\pi} f(\theta)\cos n\theta d\theta$ $B_n a^n = \dfrac{1}{\pi}\displaystyle\int_{-\pi}^{\pi} f(\theta)\sin n\theta d\theta$
$\rho c\dfrac{\partial T}{\partial t} = \dfrac{\partial}{\partial x}\left(k\dfrac{\partial T}{\partial x}\right) + \dot{Q}_{gen}$ for $\dot{Q}_{gen} = \alpha_1 T$		Sturm-Liouville Eigenvalue problem
Laplace's eq. in a circular cylinder		separation of variables, Fourier-Bessel functions

APPENDIX B: PROPERTIES OF SELECTED MATERIALS

Selected constants

Boltzman constant	$k_B = 1.38 \cdot 10^{-23}$ J·K^{-1}
Avogadro constant	$N_A = 6.023 \cdot 10^{23}$ mole^{-1}
Gas constant	$R = 8.31$ J·mole^{-1}·K^{-1}
Stefan-Boltzman constant	$\sigma = 5.67051 \cdot 10^{-8}$ W·m^{-2}·K^{-4}
Planck constant	$h = 6.63 \cdot 10^{-34}$ J·s

Table B.1. Equilibrium partition coefficient between solid (δ) and liquid iron, $k_{\delta/L}$, for some elements (Lesoult, 1986)

Dissolved element	Partition coefficient, k	Slope of liquidus line - m_L [K/%]
H	0.297	70700
N	0.281	5170
S	0.052	3350
P	0.16	2760
C	0.11	8000
Mn	0.73	500
Si	0.77	1300
Cr	0.94	900
Al	0.70	300

Table B1. Properties of selected metals and alloys near their melting point

Property	Units	Ni	IN-718 (18.5Fe, 52.5Ni, 19Cr, 3Mo, 5Nb)	Cu	Pb	Sn-Pb
C_o	wt%		5Nb			-
C_E	wt%		γ + Laves: 19.1			38.1
$C_{\alpha m}$	wt%					2.5
$C_{\beta m}$	wt%					81
$T_f\text{-}T_L$	°C	1455	1336	1085	328	-
ΔT_o	°C					-
T_E	°C					183
m_α	K·wt%$^{-1}$		$-0.74 \cdot 10^{(0.09Nb)}$			-0.83
m_β	K·wt%$^{-1}$					2.43
f_α	-					0.63
f_β	-					0.37
k	-		0.99-0.12Nb+0.0046Nb^2			
ΔH_f	J·kg^{-1}	$2.91 \cdot 10^5$	$2.95 \cdot 10^5$	$2.05 \cdot 10^5$		
ΔS_f	J·m^{-3}·K^{-1}			1.2×10^6		
k	W m^{-1}·K^{-1}		30.5	166		
k_S	W m^{-1}·K^{-1}	80		342	35	
c_L	J·kg^{-1}·K^{-1}			495		
c_S	J·kg^{-1}·K^{-1}	670		473		
ρ_L	kg·m^{-3}			8000		
$\rho_{S\alpha}$	kg·m^{-3}	7850	7620	7670	10300	7300
$\rho_{S\beta}$	kg·m^{-3}					10300
D_L	m^2·s^{-1}	$3 \cdot 10^{-9}$	$1 \cdot 10^{-9}$			$1.1 \cdot 10^{-9}$
D_S	m^2·s^{-1}		$56 \cdot 10^{-5} \cdot \exp(-2.8 \cdot 10^5/RT)$			
α_L	m^2·s^{-1}			$42 \cdot 10^{-6}$		
α_S	m^2·s^{-1}			$67 \cdot 10^{-6}$		
v_m^S	m^3·mol^{-1}	$7.1 \cdot 10^{-6}$		$8.3 \cdot 10^{-6}$	$18.7 \cdot 10^{-6}$	
γ_{LV}	J·m^{-2}	1.782		1.286	0.460	
γ_{LS}	J·m^{-2}	0.322		0.177	0.040	
Γ_α	m·K		$3.65 \cdot 10^{-7}$	$1.5 \cdot 10^{-7}$	$0.48 \cdot 10^{-7}$	$0.79 \cdot 10^{-7}$
Γ_β	m·K					$0.48 \cdot 10^{-7}$
θ_α	°					65
θ_β	°					35
μ_c	K·s·m^{-2}					$5.93 \cdot 10^{-9}$
μ_r	m·K					$0.207 \cdot 10^{-6}$
μ_V	K^{-2}·s^{-1}·m					$2.04 \cdot 10^{-4}$
φ	-					1
$\Delta v/v$	-			4.2×10^{-2}		

Table B2. Properties of aluminum and of selected aluminum-base alloys near their melting point

Property	Units	Al	Al-Cu		Al-Si
C_o	wt%	-	4.5	32.7	-
C_E	wt%	-	32.7	32.7	12.6
$C_{\alpha m}$	wt%	-	5.65	5.65	1.64
$C_{\beta m}$	wt%	-	-	52.5	99.98
$T_f\text{-}T_L$	°C	661	650	-	-
ΔT_o	°C	-	102	-	-
T_E	°C	-	548.2	548.2	577.2
m_α	$K \cdot wt\%^{-1}$	-	-3.6	-4.6	-7.5
m_β	$K \cdot wt\%^{-1}$	-	-	3.8	15.7
f_α	-	-	-	0.54	0.873
f_β	-	-	-	0.46	0.127
k	-	-	0.14		
ΔH_f	$J \cdot kg^{-1}$	$3.91 \cdot 10^5$			
ΔS_f	$J \cdot m^{-3} \cdot K^{-1}$	$1.02 \cdot 10^6$			
k	$W \cdot m^{-1} \cdot K^{-1}$	93			
k_S	$W \cdot m^{-1} \cdot K^{-1}$	213			
c_L	$J \cdot kg^{-1} \cdot K^{-1}$	1070			
c_S	$J \cdot kg^{-1} \cdot K^{-1}$	1170			
ρ_L	$kg \cdot m^{-3}$	2390			
ρ_E	$kg \cdot m^{-3}$				$2602\text{-}0.29 \cdot T$
$\rho_{S\alpha}$	$kg \cdot m^{-3}$		$2564\text{-}0.29 \cdot T$		
$\rho_{S\beta}$	$kg \cdot m^{-3}$	-		4000	2150
D_L	m^2/s	-	$2.8 \cdot 10^{-9}$		$4.3 \cdot 10^{-9}$
D_S	m^2/s	-	$5.54 \cdot 10^{-13}$		
α_L	$m^2 \cdot s^{-1}$	$37 \cdot 10^{-6}$			
α_S	$m^2 \cdot s^{-1}$	$70 \cdot 10^{-6}$			
$v_m^{\,S}$	$m^3 \cdot mol^{-1}$	$11 \cdot 10^{-6}$			
γ_{LV}	$J \cdot m^{-2}$	0.865			
γ_{LS}	$J \cdot m^{-2}$	0.093			
Γ_α	$m \cdot K$	$0.9 \cdot 10^{-7}$	$2.4 \cdot 10^{-7}$		$1.96 \cdot 10^{-7}$
Γ_β	$m \cdot K$	-	-	$0.55 \cdot 10^{-7}$	$1.7 \cdot 10^{-7}$
θ_α	°	-	-	65	30
θ_β	°	-	-	55	65
μ_c	$K \cdot s \cdot m^{-2}$	-	-	$4.62 \cdot 10^9$	$8.3 \cdot 10^9$
μ_r	$m \cdot K$	-	-	$0.47 \cdot 10^{-6}$	$0.94 \cdot 10^{-6}$
μ_V	$K^{-2} \cdot s^{-1} \cdot m$	-	-	$1.89 \cdot 10^{-3}$	$1.07 \cdot 10^{-5}$
φ	-	-	-	1	3.2
$\Delta v/v$	-	6.5×10^{-2}			

Table B3. Properties of iron and of selected iron-base alloys in the vicinity of their melting point

Property	Units	Fe	δFe-C		γFe-Gr	γFe-Fe₃C
C_o	wt%		0.09	0.6	4.26	4.3
C_E	wt%		perit. 0.53	4.26	4.26	4.30
$C_{\alpha m}$	wt%		-		2.08	2.11
$C_{\beta m}$	wt%		-		99.9	6.67
T,T	°C	1538	1531	1490	-	-
ΔT_o	°C		36	72	-	-
T	°C		perit. 1493	1154.5	1154.5	1147.1
m_α	K·wt%⁻¹		-81	-80	-90	-91
m_β	K·wt%⁻¹		-		470	60
f_α	-		-		0.926	0.515
f_β	-		-		0.074	0.485
k_α	-		0.17	0.35	0.49	
k_β	-				0.001	
ΔH	J·kg⁻¹		2.72·10⁵		2.6·10⁵	1.5·10⁵
ΔS	J·m⁻³·K⁻¹		1.07·10⁶			
k_L	W m⁻¹·K⁻¹	35			27.5	
k_S	W m⁻¹·K⁻¹	40				
c_L	J·kg⁻¹·K⁻¹	820				
c_S	J·kg⁻¹·K⁻¹	794			880	
ρ_L	kg·m⁻³	7000				
$\rho_{S\alpha}$	kg·m⁻³	7210			7000	7400
$\rho_{S\beta}$	kg·m⁻³		-		2110	7200
D_L	m²/s		2·10⁻⁸		1.25·10⁻⁹	4.7·10⁻⁹
D_S	m²/s		6·10⁻⁹	1·10⁻⁹	5.54·10⁻¹³	
α_L	m²·s⁻¹		6.1·10⁻⁶			
α_S	m²·s⁻¹		5.8·10⁻⁶			
v_s	m³·mol⁻¹	7.7·10⁻⁶				
γ_{LV}	J·m⁻²	1.806				
γ_{LS}	J·m⁻²	0.204				
Γ_α	m·K	1.9·10⁻⁷			1.9·10⁻⁷	1.9·10⁻⁷
Γ_γ	m·K		-	-	3.7·10⁻⁷	2.4·10⁻⁷
θ_α	°		-	-	25	50
θ_β	°		-	-	85	55
μ_c	K·s·m⁻²		-	-	151·10⁹	6.03·10⁹
μ_r	m·K		-	-	2.36·10⁻⁶	0.752·10⁻⁶
μ_V	K⁻²·s⁻¹·m		-	-	9.18·10⁻⁸	3.96·10⁻⁵
φ	-		-		5.4	1.8
$\Delta v/v$	-	3.6·10⁻²				

Table B4. Data for solidification kinetics modeling of selected alloys

Quantity	Units		Cast iron	Al-Si	IN-718 (18.5Fe, 52.5Ni, 19Cr, 3Mo, 5Nb)
T_L	°C		$1521-44 \cdot C-9.9 \cdot C^2$		
T_E	°C	st:	$1154+4 \cdot Si-2 \cdot Mn$		
	°C	met:	$1148-15 \cdot Si+3 \cdot Mn$		
μ_V	$K^{-2} \cdot s^{-1} \cdot m$	st:	$(3 \text{ to } 9.5) \cdot 10^{-8}$	10^{-6} to 10^{-7}	
	$K^{-2} \cdot s^{-1} \cdot m$	met:	$2.4 \cdot 10^{-6}$		
N_{eut}	m^{-2}	st:	$1 \cdot 10^5 + 3.36 \cdot 10^4 \, (dT/dt)$		
	m^{-2}	met:	$5 \cdot 10^5 + 1 \cdot 10^5 \, (dT/dt)$		
$N_{primary}$	m^{-2}		$4.81 \cdot 10^7 + 5.33 \cdot 10^6 (dT/dt)$ $+8.7 \cdot 10^4 \, (dT/dt)^2$		$2.39 \cdot 10^9 \, (2.31-$ $0.17(dT/dt))^2$
N_{Gr} DI	m^{-3}		$1.12 \cdot 10^{11} + 9.55 \cdot 10^{11} (dT/dt)$		

Note that the number of grains per unit area can be transformed in grains per unit volume with the following relationship:

$$N_V = 0.87 \, (N_A)^{1.5} \qquad N_V \, (m^{-3}) \text{ and } N_A \, (m^{-2})$$

Table B5. Properties of some molding materials

Material	k_m, W m^{-1}·K^{-1}	ρ_m, kg·m^{-3}	c_m, J·kg^{-1}·K^{-1}
silica sand	0.52	1600	1170
magnesite	4.15		
mullite	0.38	1600	750
zircon sand	1.04	2720	840
plaster	0.35	1120	840

Table B6. Heat-transfer coefficients in the gap at the metal/mold/air interfaces

Process	h, W·m^{-2}·K^{-1}
splat cooling	$10^5 - 10^6$
powder atomization	42 to 420
steel in cast iron or steel mold	
before gap formation	400 to 1020
after gap formation	400
ductile iron in cast iron mold (carbon coated)	1700
cast iron in sand mold	315
aluminum in copper mold	1700 to 2550
aluminum die casting	
before gap formation	2500 to 5000
after gap formation	400
sand-mold - air	6.9

Note: variations in the heat transfer coefficient before gap formation result from imperfect mold/casting thermal contact because of oxide skin, surface tension, mold coatings, etc.

References

Chen C., R.G. Thompson, and D.W. Davis, 1991, *Superalloys 718, 625 and Various Derivatives*, E.A. Loria ed., TMS, Warrendale, Pa. p.81

Flemings M.C., 1974, *Solidification Processing*, McGraw Hill

Holman, J.P., 1986, *Heat Transfer*, 6th edition, McGraw Hill, New York

Kanetkar C.S., I.G. Chen, D.M. Stefanescu and N. El-Kaddah, 1988, *J. Iron Steel Inst Jap.* **28**:860

Kurz W. and D.J. Fisher, 1989, *Fundamentals of Solidification*, 3rd ed., Trans Tech Publ., Switzerland

Magnin P. and R. Trivedi, 1991, *Acta metall. mater.* **39**(4):453

Magnusson T. and L. Arnberg, 2001, *Metall. Mater. Trans.* **32A**:2605

Nastac L. and D.M. Stefanescu, 1995, *AFS Trans.* **103**:329

Poirier D.R. and E.J. Poirier, 1993, *Heat Transfer Fundamentals for Metal Casting*, 2nd ed., TMS, Warrendale Pa.

C

APPENDIX C - PHASE DIAGRAMS

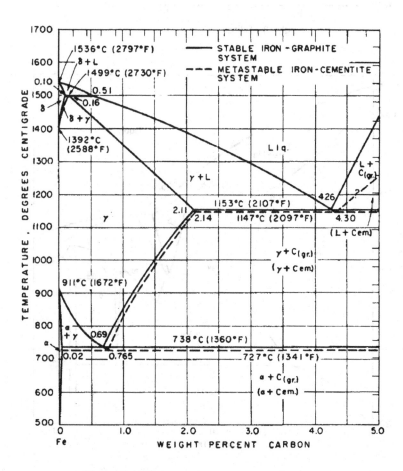

Figure C.1 The iron-carbon equilibrium diagram.

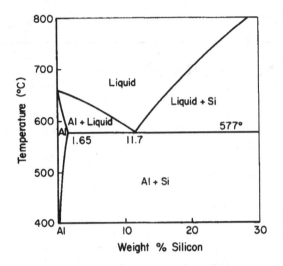

Figure C.2. The aluminum-silicon phase diagram.

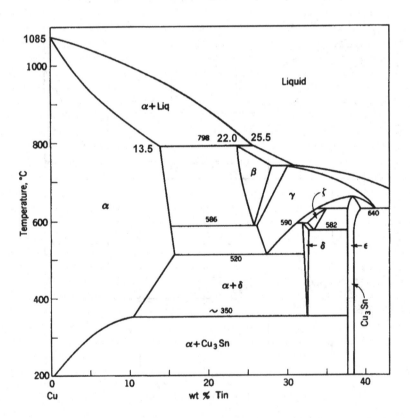

Figure C.3 The copper-tin equilibrium diagram.

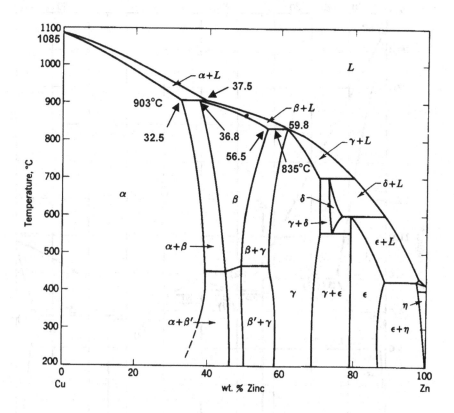

Figure C.4 The copper-zinc equilibrium diagram.

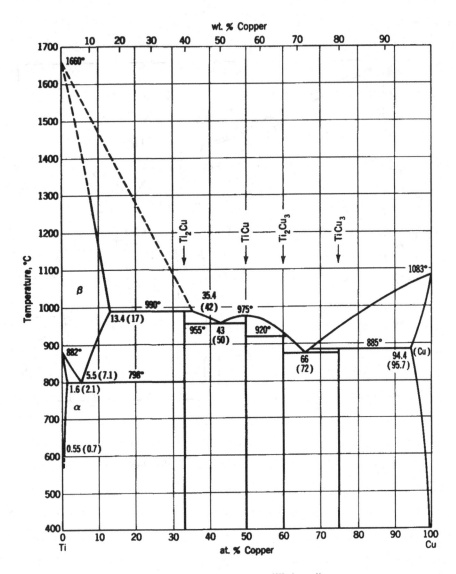

Figure C.5 The titanium-copper equilibrium diagram.

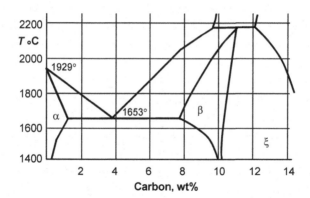

Figure C.6. Section through the vanadium-carbon phase diagram.

INDEX

A

activation barrier *368*
advection *25, 34, 335*
anisotropy *349, 352*
   interface anisotropy *171*
atomic scale phenomena *361*
atomic vibration frequency *365*
alloys
   aluminum-silicon alloys *233*
   long freezing range alloys *118*
   short freezing range alloys *118*

B

banding *41*
big bang mechanism *129*
Biot number *90*
Blake-Kozeny model *59*
boat method *81*
boundary
   boundary layer *11, 13, 33-45*
   boundary layer equation *34*
   boundary conditions *30-31, 79*
   diffusion boundary layer *33*
Boussinesq approximation *56, 57*
Bridgman directional solidification *47*
Brody-Flemings model *42, 137*
buoyancy
   buoyancy-driven convection *172*
   solutal buoyancy *25, 60*

C

Carman-Kozeny equation *111*
capillary
   capillary controlled growth *164*
   capillary flow *49*
   capillary length *164, 171*
casting
   casting modulus *87, 121*

chill-casting *83*
   die-casting *88*
   investment mold/cast *84*
   sand mold *84*
   strip castings *178*
cast iron *1, 18*
   eutectic solidification *226*
   gray-to-white transition *231*
   solidification *219*
cellular
   cellular automaton models *346*
   cellular growth *157*
   hexagonal cellular substructure *159*
Chvorinov criterion/rule *113, 121*
Clapeyron equation *15*
coalescence *134,182*
coarsening *179*
   arm coarsening *141*
   coarsening constant *181*
   dynamic coarsening *179*
collision limited growth model *17*
columnar *46*
   columnar grains *129,186*
   columnar growth *144, 146, 186*
   columnar mushy zone *135*
   columnar solidification *63*
   columnar structure *67, 149, 184, 185*
   columnar-to-equiaxed transition *183*
composites
   cast metal matrix composites *283*
   irregular fibrous composite *269*
   regular fibrous composite *269*
conservation
   conservation of energy *26, 30, 74*
   conservation of mass *26, 51*
   conservation of momentum *27, 109*
   conservation of species *108*
   conservation of solute atoms *37, 40*
consolute temperature *266*
contact angle *50, 368*
control volume *106*
convection *285, 299, 302*